Water Quality Modeling That Works

Wu-Seng Lung

Water Quality Modeling That Works

 Springer

Wu-Seng Lung
University of Virginia
Charlottesville, VA, USA

ISBN 978-3-030-90482-1   ISBN 978-3-030-90483-8 (eBook)
https://doi.org/10.1007/978-3-030-90483-8

This Springer imprint is published by the registered company Springer Nature Switzerland AG
The registered company address is: Gewerbestrasse 11, 6330 Cham, Switzerland

To Kathy

Preface

My motivation for writing this book is grounded in memories of my first professional work experiences, at Hydroscience, where I began working in 1975. It was at Hydroscience where I had the privilege of meeting and working with Dr. Donald J. O'Connor on estuarine water quality modeling, and where I quickly learned from our weekly meetings that I was simply yet another just out of school individual running a model, and not actually doing any modeling. Dr. O'Connor's approach involved gaining important physical insights into the receiving water system by examining the field data gradually developed a model for the water system in his thoughts. Such a practice really intrigued me and fundamentally changed the way I approach water quality modeling. While working under Dr. Robert V. Thomann on Lake Ontario eutrophication modeling, he always stressed the importance of configuring the model with the correct values of coefficients supported by data. How to interpret the modeling results is equally or perhaps even more important. Dr. Dominic M. Di Toro has reminded me that model performance can only be judged from how well they reproduce observations and how parameters are fit to the field data to estimate the necessary coefficients. When all is said and done, modeling without data is a waste of time. This enriching experience at Hydroscience and the Manhattan (all three were faculty at Manhattan College in New York City) doctrine has stayed with me throughout my professional career.

In 1987, Dr. Thomann predicted that water quality modeling would benefit from a significant increase in spatial and temporal resolutions in the years to come. He cautioned, however, that obtaining accurate results would continue to hinge on the mastering of underlying modeling skills by the modeler. Thus, almost four decades later—with technological advances such as machine learning—when we ask whether the role of skilful modelers can be replaced by, in essence, computing on "autopilot," the answer is unequivocally negative. Today, many want to configure and run MIKE 21, for example, without data support and do not care about the results correct or not, and even worse, they often times manipulate the computations to get the results they want. In addition, they just learn to use the interface to run the model and do not care what is behind it. I have seen this practice occur in many places and hope to deliver a strong message to reverse this trend.

Water quality modeling and photography are, in certain ways, a similar exercise. The frameworks used—water quality models and camera systems—are parallel devices. The rapid advancement of digital technology and many accessories have produced more and more sophisticated camera systems these days, just like many highly sophisticated water quality modeling frameworks do. However, these modern marvels cannot replace the human thoughts behind the camera that set the lighting and composition, an art based on the skills of the photographer. Along these same lines, water quality modelers have field data, a form of science, something much more reliable than art. The use of field data to enhance the skills for the modeling analysis is the key of this book.

A data analysis plays several significant roles in a water quality modeling study. It enables the modeler to better understand the existing water quality conditions with available data, independently quantifies key kinetics rates to configure the model, provides checkpoints in the model calibration, and identifies any potential data gaps. However, the step of performing data analysis is often ignored, leaving the modeler a significant amount of guesswork on model configuration in assigning key parameters and coefficients. This book presents techniques to independently quantify model coefficients with strong data support such as mass transport coefficients, kinetic coefficients in BOD/DO, and eutrophication modeling of various water systems.

A new set of technical issues on BOD/DO modeling also challenges the modelers these days. A refined concept of the BOD based on long-term BOD data to quantify the BOD/DO kinetics of CBOD and nitrification for the modeling analysis is presented. Laboratory tests to obtain spatially variable deoxygenation rates in the receiving water are demonstrated. The persistent eutrophication problem is still with us today. To assist modelers, a large number of eutrophication modeling studies of rivers, lakes, reservoirs, and estuaries are presented in this book. Each displays different features of complex interactions in the system: switching from DO endpoint to chlorophyll a endpoint, from nutrient limitation to flow limitation for algal growth, and seasonal limiting nutrient between nitrogen and phosphorus, turbidity maximum mediated eutrophication, and algal-related diurnal DO variations. Understanding these physical insights is much more meaningful than simply running models without data support.

Over the past three decades, the mixing zone determination has become a popular subject in implementing water quality standards and has created a greater push for a modeling analysis to support water quality management. This book presents the mixing zone modeling analysis supported by field data with applications to whole effluent toxicity (WET), metals, temperature, color, and estrogens. Once again, field data are crucial to support this modeling work. Acidification and eutrophication are the two extremes in a wide spectrum of water quality condition in ambient systems. Acidification has been closely associated with energy production, resulting in the water–energy nexus (e.g., pH and sulfate modeling of acid mine drainage). Interestingly, the pH-mediated sediment phosphorus release that led to the 1983 blue-green algal blooms in the Potomac Estuary was the key to that successful investigation. This book also presents a succinct version of carbonate equilibrium leading into pH modeling using an engineering approach with a number of case studies.

Modeling the fate and transport of endocrine disrupting chemicals (EDCs) and pharmaceutical and personal care products (PPCPs) in water systems is becoming an urgent matter given the sharp increase in the number and widespread use of these so-called emerging chemicals. First, key processes and their kinetics of modeling the fate and transport of metals in watersheds and receiving waters are presented. The focus is also on estrogens, an EDC most commonly found in domestic wastewaters but generally ignored by regulatory agencies to date. Modeling estrogens in the South River Watershed in Virginia is presented on this topic. Instantaneous sorption equilibrium is visited for pharmaceuticals. Significant concentration differences predicted for the dissolved triclosan due to the slow sorption kinetics are predicted for the Patuxent Estuary.

Finally, the model is well calibrated and verified with field data. The modeler proceeds to produce model predictions under future conditions for water quality management. Can the modeler claim a victory and go home? No, not by a long shot! A key question the modeler can expect to face from decision makers is: How do you know your model prediction results are correct? No model verified under existing conditions can fully operate as a framework free of uncertainty. Open boundary conditions are difficult to assign for tidal systems (i.e., an estuary) under future conditions when interior load reductions influence the boundary conditions. Is the sediment system in equilibrium with the external loads? Reality check applications such as model post-audit are presented to illustrate the difficulties a modeler can potentially face.

Instead of discussing the running of models without data support, this book emphasizes the physical insights and field data support required to successfully perform water quality modeling. The goal is to reduce the degree of art in the exercise and significantly increase science and engineering in the modeling analysis. Environmental engineers and scientists engaged in quantifying the water quality impacts of pollutants to specific water systems will find this book valuable in their day-to-day practices. This book reinforces the critical importance of properly understanding the physical attributes of water systems. This is also what sets this book apart from the volumes currently available in the water quality modeling field—nearly all other books in the field are categorized as textbooks and, unlike this book, offer few practical examples or exercises to follow. Therefore, this book targets the advanced modelers with the urge and desire to master their skills.

In closing, I would like to thank my Ph.D. Advisor, Prof. Raymond P. Canale at the University of Michigan, who introduced me to this new (at that time) and exciting (still at this time) field filled with unforgettable experiences. My training at Michigan was invaluable. I would also like to thank him for introducing me to Hydroscience upon the completion of my Ph.D. I have also been fortunate to participate in several international studies sponsored by NATO, Universitas 21, and University of Virginia's Global Study Program. The Metro Council of Minneapolis and St. Paul provided funding and volumes of data to work on the upper Mississippi River and Lake Pepin. Funding for the Patuxent Estuary modeling work was provided by the State of Maryland and the Smithsonian Environmental Research Center. Many other sponsors have contributed to the work presented in this book. I am also indebted to my long-time colleague and friend, Dr. Jan-Tai Kuo of National Taiwan University (NTU) on many river and reservoir modeling studies in Taiwan. Preparation of this manuscript could not be completed without the assistance of my former Ph.D. students. Dr. Sen Bai at Tetra Tech provided many excellent suggestions as to the content of the book and assisted with fecal coliform modeling. Dr. Alex Nice of AECOM provided immeasurable assistance on the Patuxent Estuary eutrophication and metals modeling work. Dr. Dong Liu foresaw the importance of non-instantaneous equilibrium of pharmaceuticals in water systems. Dr. Xiaomin Zhao's (now at Paradigm Environmental) work on tracking estrogens in watersheds laid the groundwork for future studies. I was also fortunate to work with Dr. Chien-Hung Chen (now at Stantec in Taiwan) on the Danshui River BOD/DO study as his Ph.D. thesis Advisor at NTU. The assistance of Dr. Shih-Kai Ciou (with Sinotech Engineering Consultants in Taiwan) on providing the data and model results of the Feitsui Reservoir in Taiwan is really appreciated. Dr. Wai Thoe of Hong Kong Environmental Protection Department provided material on *E. coli* modeling of Hong Kong beaches. I am also grateful to my Technical Editor, Monica Wedo, for improving this manuscript; her work is really appreciated.

Charlottesville, USA   Wu-Seng Lung

Contents

Abbreviations

μg/L	Micrograms per Liter
μmho/cm	Micromhos per Centimeter
1Q10	1-day 10-year Low Flow
7Q10	7-day 10-year Low Flow
ACSA	Augusta County Service Authority
ADMI	American Dye Manufacturing Manufacturers Institute
AFO	Animal Feeding Operations
AHOD	Areal Hypolimnetic Oxygen Deficit
AIZ	Allocated Impact Zone
AOC	Assimilable Organic Carbon
ATE	Acute Toxicity Endpoint
BOD	Biochemical Oxygen Demand
BOD_5	5-day Biochemical Oxygen Demand
BPA	Biphenyl A
CAFO	Concentrated Animal Feeding Operations
$CBOD_5$	5-day Carbonaceous Biochemical Demand
$CBOD_u$	Ultimate Carbonaceous Biochemical Demand
CCC	Criterion Continuous Concentration
CCMS	Committee on the Challenge of Modern Society
cfs	Cubic Feet per Second
CFU	Colony Forming Unit
CMC	Criterion Maximum Concentration
cms	Cubic Meters per Second
COD	Chemical Oxygen Demand
COVID-19	Coronavirus
CPCB	Central Pollution Control Board
CRE	Caloosahatchee River Estuary
CSO	Combined Sewer Overflow
CTE	Chronic Toxicity Endpoint
CWA	Clean Water Act
DIN	Dissolved Inorganic Nitrogen

DIP	Dissolved Inorganic Phosphorus
DJB	Delhi Jal Board (Delhi Water Board)
DMA	Dimethylarsenate
DO	Dissolved Oxygen
DOC	Dissolved Organic Carbon
DOM	Dissolved Organic Matter
E. coli	*Escherichia coli*
E1	Estrone
E2	Estradiol
E2α	17α-estradiol
E2β	17β-estradiol
EDC	Endocrine Disrupting Chemical
EE2	17α-ethinylestradiol
ENR	Enrofloxacin
EPA	Environmental Protection Agency
HEPP	Hydroelectric Power Plant
ICPRB	Interstate Council of the Potomac River Basin
lbs/day	Pounds per Day
MCL	Maximum Contaminant Level
MDNF	Momentum-Dominated Near Field
MFL	Minimum Flow Limit
mg/L	Milligrams per Liter
mgd	Million Gallons per Day
MMA	Methylarsonate
MOU	Memorandum of Understanding
MPCA	Minnesota Pollution Control Agency
MPN	Most Probable Number
MRLC	Multi-Resolution Land Characteristics Consortium
MWCOG	Metropolitan Washington Council of Governments
NATO	North Atlantic Treaty Organization
NBOD	Nitrogenous Biochemical Oxygen Demand
ng/L	Nanograms per Liter
NM	New Mexico
NOAA	National Oceanic and Atmospheric Administration
NOEC	No Observable Effect Concentration
NP	Neptunium
NPDES	National Pollutant Discharge Elimination System
Ortho-P	Orthophosphate
PEM	Potomac Estuary Model
POC	Particulate Organic Carbon
POM	Particulate Organic Matter
PON	Particulate Organic Nitrogen
POP	Particulate Organic Phosphorus
PPCP	Pharmaceutical and Personal Care Product
PPO$_4$	Particulate Phosphate

RMZ	Regulatory Mixing Zone
SFWMD	South Florida Water Management District
SOD	Sediment Oxygen Demand
TCMP	2-chloro-6-(trichloromethyl) pyridine
TCS	Triclosan
TIC	Total Inorganic Carbon
TKN	Total Kjeldahl Nitrogen
TMDL	Total Maximum Daily Load
TP	Total Phosphorus
TRC	Total Residue Chlorine
TSS	Total Suspended Solids
TU_a	Acute Toxicity Unit
TU_c	Chronic Toxicity Unit
TWBC	Tidal Basin and Washington Ship Channel
UPRC	Upper Potomac River Commission
USGS	United States Geological Survey
UV	Ultraviolet
UVa	University of Virginia
VDEQ	Virginia Department of Environmental Quality
VEC	Valued Ecosystem Component
VIMS	Virginia Institute of Marine Science
VPDES	Virginia Pollutant Discharge Elimination System
WET	Whole Effluent Toxicity
WLA	Wasteload Allocation
WQBEL	Water Quality-Based Effluent Limit
WWTP	Wastewater Treatment Plant
ZID	Zone of Initial Dilution

Chapter 1
Challenges in Modeling for Water Quality Management

In his paper entitled: System Analysis in Water Quality Management—a 25 Year Retrospect, Thomann (1987) predicted a significant increase in spatial and temporal resolutions of water quality models in the years to come. A decade later, he confirmed his prediction in another landmark paper entitled: The Future Golden Age of Predictive Models for Surface Water Quality and Ecosystem Management (Thomann 1998). He indicated that the remaining challenges lie in how to obtain the correct model results supported by data for use in water quality management. Given field data, what is the number one challenge in water quality modeling? The answer: what if the model results do not reproduce the field data.

Water quality modelers must rely on science and data, the two fundamental pillars to support the modeling analysis, by increasing the objectiveness of our modeling work. Accordingly, water quality modeling would let the data speak while letting the science to guide. The design and focus of this book is on performing water quality modeling with these two goals in mind. Our emphasis is not on models, regardless of whether they are simple or complex. Rather, we focus on modeling analyses with strong scientific and data support. It is noted that modeling cannot generate data; rather, modeling is used to interpret data and to generate predictions with such data that have been used to properly calibrate and verify the model. Without strong data support, these modeling frameworks will not otherwise provide a meaningful analysis. Can sophisticated models (or modeling frameworks) replace the role of a skillful modeler by, in essence, putting them on "auto-pilot"? The answer is unequivocally negative. Spatial resolution is not the key for a successful modeling analysis for water quality management. Instead, modeling skills with strong data support are crucial to obtaining defensible results for water quality management. As such, readers of this book should not expect to see merely a parade of modeling frameworks. Instead, readers should expect, and enjoy, an active discussion about the salient features of the modeling analysis, albeit a simple task of deriving a single, key model coefficient value.

W.-S. Lung, *Water Quality Modeling That Works*,
https://doi.org/10.1007/978-3-030-90483-8_1

1.1   The Potomac Estuary Algal Bloom

This chapter begins with an interesting, classic example of using a water quality model to solve a mystery of a eutrophication issue, which demonstrates the essence of modeling, not just running models, for water quality management. The progression of water quality problems in the Potomac Estuary (Fig. 1.1) started in the late 1960s when extensive algal blooms developed in addition to a depressed dissolved oxygen (DO) condition in the Washington, DC vicinity (Thomann 1987). Following a series of significant reductions in biochemical oxygen demand (BOD) and ammonia loads in the 1970s, the low DO problem was eliminated.

When it comes to nutrient reductions to mitigate algal blooms in the Potomac Estuary, the debate has always been between nitrogen and phosphorus (i.e., which one is the limiting nutrient). It was first argued that nitrogen was the limiting nutrient

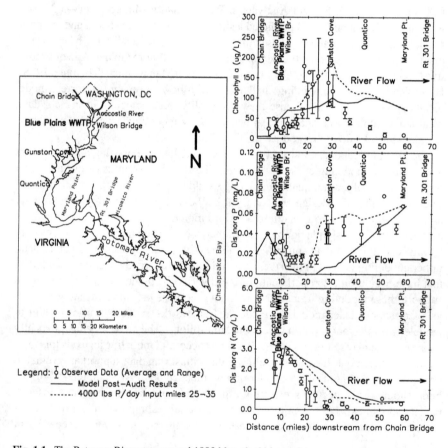

Fig. 1.1   The Potomac River estuary and 1983 blue algal bloom data

and that nitrogen should be controlled. However, algal blooms in the form of the blue-green algae, *Microcystis aeruginosa*, can fix nitrogen from the atmosphere, defeating the purpose of controlling nitrogen. Conversely, concern was also expressed over the release of phosphorus from the sediment. Eventually, the phosphorus removal strategy was founded on the notion that with sufficient reduction of phosphorus, it could be made the limiting nutrient (Thomann 1987). Since it was considered cheaper to remove phosphorus than nitrogen, the phosphorus removal program was instituted. Such action led to significant reductions in point source phosphorus input to the Potomac Estuary from 24,000 lbs/day in the late 1970s to less than 2000 lbs/day by early 1980s at a cost of US$1B for phosphorus removal at municipal wastewater treatment plants (WWTPs).

By early 1980s, the total wastewater flows from nine major WWTPs to the Potomac Estuary amounted to 446 million gallons per day (mgd) with the Blue Plains WWTP (see the location of this facility in Fig. 1.1) as the single largest discharger at 310 mgd. Following a major algal bloom in the Potomac in 1977, an intensive effort was then undertaken to update the modeling framework for eutrophication, resulting in the Potomac Eutrophication Model (PEM) (Thomann and Fitzpatrick 1982). This model, including sediment–water interactions, was calibrated and verified using 7 years of data and was then used to analyze nutrient control alternatives (Thomann 1987). In the meantime, the phosphorus concentrations in the Potomac Estuary continued to decrease and algal blooms disappeared.

With the phosphorus reductions almost fully in place, another major bloom occurred in the summer of 1983 in the form of *M. aeruginosa* with chlorophyll *a* measured at about 300 μg/L in the main channel and 800 μg/L in the embayments. A collective effort between government agencies, the Metropolitan Washington Council of Governments (MWCOG), the state environmental agencies (Maryland, Virginia, and Washington DC), and Interstate Council of the Potomac River Basin (ICPRB) was initiated to investigate. Understandably, a number of questions were asked:

1. What was the cause of the bloom?
2. The Blue Plains WWTP effluent total phosphorus (TP) concentration was around 0.4 mg/L in 1983. Would the situation be relieved when the Blue Plains WWTP effluent TP levels reached the target of 0.2 mg/L?
3. What are the most likely mechanisms responsible for the bloom?
4. Would the PEM be able to reproduce the bloom?

The investigation panel put these questions to the PEM. When the PEM was applied with the summer 1983 meteorological and hydrological data, it tracked the onset of the bloom up to about 100 μg/L by the end of July, but then failed to reproduce the further intensification of the bloom (see the comparison of model results and data in the top right plot of Fig. 1.1).

The model results did not match the field data—a challenge to the modelers. An immediate responsive action is to increase the algal growth rate with the hope to grow more algae, right? However, the second plot of Fig. 1.1 gives us the answer. The model could not grow more algae because it ran out of food, i.e. dissolved inorganic phosphorus (DIP) around the area approximately 20 miles downstream from the Chain Bridge (see the middle right plot of Fig. 1.1). Lacking DIP was the principal reason for the failure of the PEM to capture the full bloom. Interestingly, PEM results over-predicted the dissolved inorganic nitrogen (DIN) in the Potomac Estuary (see the bottom right plot of Fig. 1.1).

The under-predicting DIP and over-predicting DIN suggested that if the PEM had extra phosphorus, the bloom would go higher, which would take up more nitrogen, resulting in lower calculated DIN and a better match with the DIN data. Next, using the PEM as a tool, the investigation panel asked the PEM for an additional amount of phosphorus to sustain the bloom. Model sensitivity analyses responded that there were additional phosphorus sources of about 4000–8000 lbs/day needed to overcome the shortage of phosphorus. The dashed lines in Fig. 1.1 shows the effect of including this source: matching the chlorophyll a, DIP, and DIN data much better. The PEM partially explains the observed data but had not provided all the answers to the above questions. What this modeling analysis shows is that changing model coefficient values must have justifications. The interplay between model results and data led the investigation panel to seek additional phosphorus sources for the summer 1983 algal bloom.

1.2   Searching for Additional Phosphorus Sources

A possible phosphorus source could be diffusive release of DIP from the sediment to the water column. A phosphorus release at levels of about 40–80 mg P/m^2/day was necessary to produce 4000 lbs/day of the phosphorus to sustain the bloom (Thomann et al. 1985). It is known that such releases can occur in substantial amounts under anaerobic conditions. However, as noted, the DO in the Potomac was generally above 3–4 mg/L. Excessive oxygen was being produced by the bloom via photosynthesis, therefore there was no shortage of DO in the water column, nor in the sediment. In response, Di Toro and Fitzpatrick (1984) proposed a hypothesis for an aerobic sediment release of this order for the Potomac Estuary.

The mechanism proposed for the Potomac Estuary event during the summer of 1983 is related to the high pH that occurred during August and September. The sorption of phosphorus is highly pH dependent. If the pH rises in the overlying water, this effect will diffuse through the sediment–water interface and reduce the sorption of DIP to the iron oxides/hydroxides. Although the sorbents are still present, their ability to sorb is reduced due to the high pH and therefore, the phosphorus flux should increase. If the sorption potential of the oxides/hydroxides is completely eliminated, then high pH aerobic DIP flux should be essentially equal to the anaerobic DIP flux. In both of these cases, the sorption barrier has been removed by dissolution in the anaerobic case, or the sorption capacity has been eliminated by the high pH in the aerobic case. The result of either of these mechanisms is a large increase in the DIP flux to the overlying water. It should be noted that this mechanism should not affect the flux of ammonia to the overlying water since its sorption to the oxides/hydroxides is orders of magnitude less than for phosphorus. Thus, the sorption hypothesis predicts that only phosphorus flux will be increased by high pH events.

1.3 pH Rise During the Bloom

Examination of the DIP and TP data in the Potomac Estuary shows that it was not until the August and September surveys that the increase in DIP and TP between milepoints 20 and 40 was readily evident. Coincidently, that corresponded to the time in which pH first rose above 9.0, as shown in Fig. 1.2. Note that during July

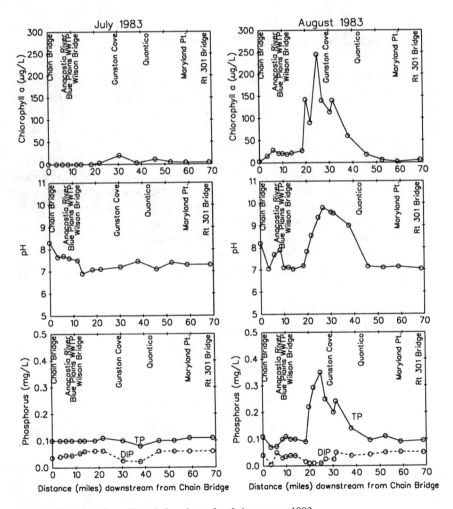

Fig. 1.2 Chlorophyll *a*, pH, and phosphorus levels in summer 1983

the pH was generally below 7.5 between milepoints 25 and 40, while during August and September the pH rose to above 9.0 and was as high as 10.0 in this region. The TP shows a relatively flat profile for July followed by a sharp rise in TP between milepoints 20 and 40.

The preceding hypothesis by Di Toro and Fitzpatrick (1984) supported by the high pH in the Potomac Estuary in August 1983, may have resulted in a subsequent high aerobic phosphorus release from the sediment. By reproducing the environmental conditions (i.e. high pH in the aerobic overlying water) for the sediment from the Potomac Estuary, Seitzinger (1983, 1984, 1985) was able to measure an aerobic release flux of phosphorus matching the range of 4000–8000 lbs/day, thereby confirming the aerobic release from the sediment under high pH conditions.

1.4   Impact of Nitrification in Wastewater

Since the pH levels recorded in 1983 were among the highest ever measured in the Potomac Estuary, the next question then to be asked is "Why was the pH unusually high in 1983?" What was different between that year and previous years? The high pH in 1983 leads one to examine the alkalinity of the estuary since alkalinity is the acid-neutralizing capacity of a system. Because the photosynthetic reaction removes CO_2 from the water with a potential increase in pH, the resulting actual pH change will depend on the initial alkalinity (Stumm and Morgan 1996). The most readily identifiable difference in the structure of the estuary system between 1983 and earlier years is that by 1983 nitrification facilities were on line at the Blue Plains WWTP, the major point source input (Thomann et al. 1985). Nitrification reduces the alkalinity of the effluent and therefore alkalinity changes may be a result of treatment practices.

Figure 1.3 shows the longitudinal profiles of alkalinity levels for several years from 1969 to 1983. These latter data for 1982–83 show that since the late 1970s there has

Fig. 1.3 Historical alkalinity trend in the Potomac River estuary

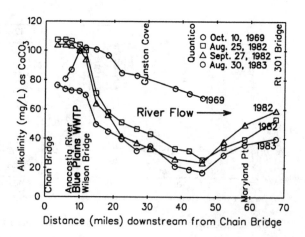

been a significant change in the alkalinity of the waters of the estuary. During the summer months of 1983, the alkalinity of the water entering the estuary from the upper watershed ranged from 80 to 110 mg/L as $CaCO_3$. The alkalinity coming from wastewater treatment plant discharges might be expected to increase the alkalinity of the upper estuary usually by about 20 mg/L above the alkalinity from the upper basin. With the installation of nitrification at the Blue Plains WWTP, the alkalinity mass has been significantly reduced (see Tables 1.1 and 1.2). The nitrification process consumes alkalinity in the water column as presented in the carbonate equilibrium (see Chap. 5 acidification modeling). Such treatment generally uses about 7.14 mg/L alkalinity per mg/L ammonia nitrogen oxidized (Thomann et al. 1985). These data suggest that the buffering capacity of the estuary has been reduced and that this may be related to the operation of the nitrification process.

Table 1.1 Alkalinity mass in the upper Potomac estuary (Thomann et al. 1985)

Sampling date	River flow (cfs)	Estimated wastewater alkalinity (lb/day)	Alkalinity mass in upper 30 miles (lbs)
10/10/1969	2100	369,000	49,380,000
8/22/1977	1600	323,000	45,170,000
8/25/1982	4200	250,000	37,968,000
9/27/1982	1800	250,000	29,430,000
7/5/1983	6140	140,000	37,224,000
8/30/1983	1900	140,000	24,820,000

Table 1.2 Monthly average alkalinity (mg/L) as $CaCO_3$ in blue plains WWTP effluent (Thomann et al. 1985)

Month	1975	1978	1980	1981	1983	1984
Jan		91	105	131	33	29
Feb		101	123	103	37	33
Mar		102	115	87	26	36
Apr		116	100	84	30	33
May	93	112	105	95	27	39
Jun	128	113	117	76	26	54
Jul	107	99	112	88	35	51
Aug	124	99	110	61	47	44
Sep	101	120	116	85	51	70
Oct	111	143	120	98	60	76
Nov	123	148	120	108	43	
Dec	120	118	110	113	31	
Avg	113	114	113	94	37	47

Prior to the 1980s, most of the nitrogen discharged was in ammonia form. Ammonia was oxidized bacterially to nitrates, potentially creating about 250,000 lbs/day of acid. However, the utilization of nitrates by algal cells resulted in the production of alkalinity about equal to that of acid produced by the bacterial oxidization of ammonia, and thus no change in either alkalinity, pH, or the buffering capacity took place. Beginning in the early 1980s, most of the wastewater ammonia was nitrified at the WWTPs and thus a potential acidic load to the estuary was removed. Therefore, one would have anticipated that alkalinity in the upper estuary in 1982 and 1983 would have increased due to increased alkalinity by the production of alkalinity when the algal cells assimilated the nitrates in the estuary. This increase did not occur. In fact, there was a net further reduction in the buffering capacity in the estuary. Why?

The additional reduction in alkalinity may be related to the low pH and low alkalinity in the wastewater effluents, especially at the Blue Plains WWTP. There could have been some iron discharged from the Blue Plains WWTP, which further decreased that alkalinity when the iron in the wastewater discharge mixed with receiving waters and precipitated into the sediment.

It has been well established that bicarbonate content determines whether natural waters are well or poorly buffered (Ruttner 1963). Carbonate-carbonic acid mixtures have a remarkable and important characteristics of preventing major fluctuations in pH when reacting with other acid-salt combinations. The pH of water is determined by the reaction between CO_2 and carbonate, more specifically, by the H^+ ions arising from the dissociation of H_2CO_3 and OH^- ions arising from the hydrolysis of bicarbonate. In well-buffered water the change in pH due to the addition of either an acid or base is very small in proportion to the amount of acid or base added.

1.5   Environmental Conditions to Kick-Off the Bloom

Thomann et al. (1985) pointed out the following observations related to the summer 1983 algal bloom in the Potomac River:

1.   River flows were 30–80% of normal
2.   Percent sunshine ranged from 70 to 80%, generally higher than in previous years
3.   Wind speed averaged about 6.5 miles per hour, significantly lower than in previous years
4.   Two-dimensional estuarine mass transport.

With the low flows, adequate nutrients, and ideal environmental conditions, an initial algal bloom began in July and early August of 1983. This initial algal bloom reduced the amount of carbon dioxide and bicarbonate, thereby further increasing the pH. The lack of buffering capacity allowed the pH to increase significantly in the upper estuary from milepoint 20 to 40 in the reach where the initial bloom was observed. Once the enhanced bloom began, the pH increased to over 9.5. At this pH level, the release of both nitrogen and phosphorus from the sediments was significantly enhanced, thus self-perpetuating the blooms. In previous years when blooms occurred, the pH did not increase over 8.0. However, in 1983, the pH increased from about 7.0 near Woodrow Wilson Bridge (MP 12) to over 9.5 between MP 25–30 (Fig. 1.2). These are the areas where the highest concentrations of phosphorus were measured in late August 1983.

The investigation team also summarized the impact of the two-dimensional mass transport in the Potomac Estuary on the initiation of the algal bloom as follows (Thomann et al. 1985). Increased spring runoff into the estuary followed by a rapid decline in flow in mid-July of 1983 resulted in increased stratification in the water column and strength of the lower estuary circulation. Sediment release point at the end of the salinity intrusion acts as a source for the stimulation of phytoplankton in this area. The phosphorus source is further enhanced by anoxic conditions in the lower reaches. Both nitrogen and phosphorus sources at the end of the salt-water intrusion are also enhanced by particulate associated nutrients being recycled in a manner similar to the "turbidity maximum" phenomena. Settling rates of particulates, including phytoplankton, are impeded (or may even be reversed) by vertical velocities in the salinity intrusion area and hence may remain in the water column (i.e., not settled into the sediment). The ability to calculate these phenomena via a one-dimensional model with longitudinal dispersion in the PEM may be inadequate, as it does not capture this complex downstream hydrodynamic circulation.

1.6 Model Enhancement

While results from the PEM with the support of water quality data had led to the discovery of the additional phosphorus loads needed to sustain the algal bloom, the missing mechanism of aerobic phosphorus release under high pH conditions would need to be included in the model for future use. The following mechanisms are therefore required to refine the PEM:

1. Aerobic phosphorus release under high pH levels
2. pH-alkalinity equilibrium in the water column to model pH, alkalinity, and acidity
3. Two-layer mass transport.

While much discussion has focused on the first two items, the unique two-dimensional estuarine mass transport must be factored in as well. Particulate matter discharged to the estuary (and the nutrient associated with the particles) settles from the upper layer to the lower layer in the water column. It is then transported upstream to a convergence region where vertical transport of the solids occurs; a "turbidity maximum" occurs in the area near the limit of salinity intrusion. This phenomenon has been studied in some detail with respect to the suspended solids in estuaries (Thomann et al. 1985). Lung and O'Connor (1984) and O'Connor and Lung (1981) gave a comprehensive analysis on turbidity maximum for a number of estuaries in the U.S. (see Sect. 4.5). The Patuxent Estuary model by Lung and Nice (2007) is another example of this two-dimensional mass transport affecting nutrient remaining in suspension near the location of the turbidity maximum. It is clear that the location of turbidity maximum is where the chlorophyll *a* peak occurs in the Potomac Estuary (between MP 20–30). The original PEM was a one-dimensional segmentation with lateral (vertically mixed) side segments for embayments, and therefore incapable of mimicking the two-dimensional mass transport; the original PEM must be upgraded to account for this phenomenon.

1.7 Summary and Conclusions

The initial effort of reproducing the 1983 bloom failed with the PEM. However, a closer look at the model results helped to identify missing phosphorus source(s) in the Potomac Estuary. Further, the modeling analysis was able to quantify the amount of missing phosphorus loads needed to substantiate the bloom in the summer of 1983. Once again, the importance of water quality data could not be underestimated. The interplay of modeling results and data eventually led to solving this mystery; neither modeling nor data alone was the silver bullet to the investigation.

The 1983 Potomac Estuary algal bloom investigation turned out to be an interesting scientific probe. The investigative panel consisted of leading scientists on phytoplankton and algal bloom, with specific expertise on blue-green algae and the Potomac Estuary. Their systematic efforts led by Thomann identified the cause of the bloom. Continued probing with the aid of the PEM provided answers to additional questions on the bloom.

Note that the sophistication of the PEM in the 1980s is no match with today's water quality modeling frameworks. The spatial discretization of the PEM was limited due to computation power at that time. None of these matter in this case, however. The failure of the original PEM to capture the bloom intensification in August and September is a clear example where mechanisms not included in a model formulation become apparently significant after some level of treatment (phosphorus removal and nitrification in this case) had been installed (Thomann and Mueller 1987). Some of the highly used modern models do have these features.

The inability of PEM to capture the bloom intensification in 1983 does not suggest that eutrophication modeling is of little value. On the contrary, the PEM provided one of the more important bases to zero in on the hidden mechanism and missing source of phosphorus. The key to this successful investigation proved to be the high caliber investigative panel members, who relied as much on water quality data as the supporting modeling analysis. Water quality modeling is not just pushing buttons. Physical insights into the system uncovered by the investigation team were the most important asset. The serendipitous outcome of this example is turning the water quality model that does not work into water quality modeling that works. In addition, water quality modeling must be supported by two pillars: research (science) and monitoring (data), which are the center of discussions in this book.

1.8 Setting the Stage for This Book

It is the author's strong conviction that responding to the challenges in modeling for water quality management is not about using the most sophisticated models. Rather, the modeler's skill and data support are of the utmost importance for overcoming any challenges. Therefore, this book will not discuss water quality modeling frameworks, either proprietary or freely distributed codes. Instead, this book promotes the approach of emphasizing modeling analysis with strong support of field data. Running models without data support, with very few exceptions, is totally meaningless and a waste of time. In addition, we repeatedly demonstrate physical insights into the modeled system and its key mechanisms throughout this book. We do not run models simply because we want to. We do modeling for a very practical mission; to provide answers to decision makers for developing water quality management strategies. This book is designed to help modelers with the necessary skills to fulfill this mission and to address the challenge: What if the model results do not reproduce the data?

Data analysis is a first task that many water quality modelers often neglect to perform prior to the modeling analysis. The significance of performing a thorough data analysis is two-fold: it not only provides significant insight into the interaction of the water quality constituents, but also enables a water quality modeler to conceptualize the formulation of this proposed "model." This is significant because a water quality model of a given system is developed progressively, within the modeler's mind. The data analysis portion, contained in the second chapter of the book, sends a strong message to the reader: data analysis is critical to the success of water quality modeling.

The first two water quality endpoints addressed in this book are DO and chlorophyll *a*. The impairment of DO is caused by BOD, a conventional pollutant. Given the many new twists presented by BOD in modern day stream modeling, this topic deserves renewed attention and re-focus, particularly as it relates to polluted streams by highly intense discharges one after another along the streams. Despite the long history of controlling problems arising out of DO, which are connected intimately with primary productivity and sediment effects, problems arising from DO still tend to be considerably more complex than generally believed by Thomann (1987). Next is the water quality issue of nutrient/eutrophication, which is not only tied to the DO endpoint, but also leads to the chlorophyll *a* endpoint. The problems related to nutrient/eutrophication are the most difficult models with which we have encountered due to complexity of the plant biology, the non-linear interactions between nutrients and aquatic plants, and the interactions of the sediment (Thomann 1987). The case studies presented in the following sections represent a unique and interesting modeling analysis to resolve eutrophication problems.

References

Di Toro DM, Fitzpatrick JJ (1984) A hypothesis to account for the 1983 algae bloom. Tech Memo, HydroQual, Inc, NJ

Lung WS, Nice AJ (2007) A eutrophication model for the Patuxent Estuary: advances in predictive capabilities. J Environ Eng 133(9):917–930

Lung WS, O'Connor DJ (1984) Two-dimensional mass transport in estuaries. J Hyd Eng 110(10):1340–1357

O'Connor DJ, Lung WS (1981) Suspended solids analysis of estuarine systems. J Environ Eng 107(1):101–120

Ruttner F (1963) Fundamental of limnology. University of Toronto Press, Toronto

Seitzinger SP (1983) Sediment-water phosphorus exchanges in the Potomac River and embayments. Final report prepared for Limno-Tech, Inc

Seitzinger SP (1984) The effect of pH on sediment-water phosphorus fluxes in the Potomac River. Preliminary project report to Academy of Natural Science, Philadelphia, PA

Seitzinger SP (1985) The Effect of oxygen concentration and pH on sediment-water nutrient fluxes in the Potomac River. Academy of Natural Sciences of Philadelphia, PA

Stumm W, Morgan JJ (1996) Aquatic chemistry, chemical equilibria and rates in natural waters, 3rd edn. Wiley, New York

Thomann RV (1987) System analysis in water quality management—a 25 year retrospect. In: Beck MB (ed) System analysis in water quality management. Pergamon Press, Oxford, pp 1–14

Thomann RV, Mueller JA (1987) Principles of surface water quality modeling and control. Harper & Row, New York

Thomann RV (1998) The future "golden age" of predictive models for surface water quality and ecosystem management. J Environ Eng 124(2):94–103

Thomann RV et al (1985) The 1983 algal bloom in the Potomac Estuary. Report prepared for Potomac Strategy State/EPA Management Committee, Washington, DC

Thomann RV, Fitzpatrick JJ (1982) Calibration and verification of a model of the Potomac Estuary. HydroQual Report to Department of Environmental Services, Washington, DC

Chapter 2
Data Analysis

A data analysis serves many roles in supporting water quality modeling. At a minimum, it can be a simple task of examining available data to get a preliminary understanding of the water system. Next, a data analysis prior to selecting and configuring modeling framework(s) is a necessary step in modeling for water quality management. A complete understanding of the system to be modeled is essential to setting up the model with the available data. Furthermore, some key processes can be independently quantified (i.e. assigning model coefficient values) prior to model configuration. Third, a data analysis also compiles necessary information for model calibration and verification. At the fullest extent, the modeler could gain additional physical insights into the system via an interplay between model results and data. Many model input elements ranging from environmental conditions, hydrological conditions, mass transport coefficients, to kinetics parameters can be independently derived from a data analysis prior to model configuration. This chapter presents a wide spectrum of data analyses ranging from a simple review of existing data to derivation of model coefficients associated with physical, chemical, and biochemical processes. A properly performed data analysis could also identify data gaps needed for fine tuning the model. Through these examples, the importance of this often neglected task will be demonstrated.

2.1 Data Screening—Lake Peipsi and Vrhovo Reservoir

Under the North Atlantic Treaty Organization (NATO) Committee on the Challenges of Modern Society (CCMS), an ecological system modeling study of coastal lagoons was launched in 1998. The University of Virginia (UVa) participated to provide water quality expertise to the program. Waters from the Baltic States provided some interesting sites for application of water quality modeling to assist the management of these water bodies. Lake Peipsi, the largest trans-boundary lake in Europe, lying on the border between Estonia and Russia in northeast Estonia (see Fig. 2.1) was

© The Author(s), under exclusive license to Springer Nature Switzerland AG 2022
W.-S. Lung, *Water Quality Modeling That Works*,
https://doi.org/10.1007/978-3-030-90483-8_2

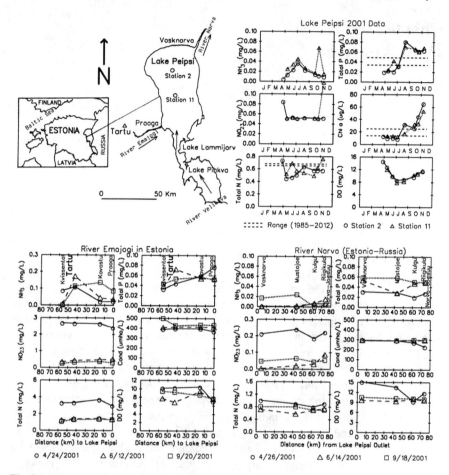

Fig. 2.1 Water quality of Lake Peipsi, River Emajogi, and River Narva

one of the water systems selected for a preliminary water quality screening. As the fifth largest lake in Europe, it has a surface area of 3555 km² with a mean depth of 8.3 m. With a volume of 25 km³ and an annual average flow of 265 m³/s (cms), the lake has a residence time about 3 years. This flow comes from watersheds of River Emajogi (30%) in Estonia and River Velikaya (70%) in Russia. River Velikava flows into Lakes Pihkva and Lammijarv, which in turn discharge into Lake Peipsi. The lake has been widely studied for eutrophication concerns (Kangur et al. 2003; Noges et al. 2004; Kangur and Mols 2008; Buhvestova et al. 2011; Blank et al. 2017). The 2001 water quality data of Lake Peipsi, River Emajogi (major inflow), and River Narva (lake outflow) were made available by Ministry of the Environment for the review. Figure 2.1 presents data of key water quality constituents for Lake Peipsi, River Emajogi (the feeding stream), and River Narva (the outflow waterway), respectively, from surveys in April, June, and September of 2001.

Nutrient inputs of ammonia, nitrite/nitrate, total nitrogen, and total phosphorus from River Emajogi enters Lake Peipsi at Praaga following a journey of 80 km. The April 24, 2001 data show higher total nitrogen and nitrite/nitrate levels from River Emajogi than June and September data. The lake is frozen from late November until the second half of April. Thus, the snowmelt from the Emajogi delivers higher nutrient loads to the lake. Nutrient levels measured at Stations 2 and 11 in Lake Peipsi are more or less comparable. Also shown are the ranges of total P, total N, and chlorophyll *a* levels from 1985 to 2012 (Blank et al. 2017). Note that nutrient concentrations at these two in-lake stations are much lower than those from River Emajogi, indicating in-lake assimilation. The Lake empties its flow into River Narva flowing northward before reaching the sea in Russia. Water quality data measured in River Narva in 2001 are also shown in Fig. 2.1. Nutrient concentrations in River Emajogi are higher than in River Narva, an indication that Lake Peipsi is serving as a sink for the loads from River Emajogi. Mean Secchi depths are 1.8 m and 0.7 m for Lake Peipsi and Lake Pihkva, respectively, another indication of poorer water quality in the upstream lake.

The above data analysis provides a preliminary understanding of the system but the extent of the existing data is not sufficient to support a eutrophication modeling effort for management use. A water quality monitoring program is needed to sample the lake for a comprehensive spatial (multiple stations in the lake) and temporal (bi-month from April to November) coverage of a lake this size. The two major inflows (from Rivers Emajogi and Velikava) must be sampled to provide concurrent coverage. Natural nutrient loads to Lake Pihkva are significantly higher than in Lake Peipsi. Therefore, Lake Pihkva needs to be included in the monitoring as well as its nutrient levels are markedly higher than in Lake Peipsi (Blank et al. 2017; Kangur et al. 2003). The long residence time of 3 years for Lake Peipsi suggests that this sampling program should be carried out for a minimum of 7 to 8 years (2.5 times the residence time) to detect any measurable changes in the water quality and to support the modeling analysis.

Another NATO/CCMS pilot study was "Modeling Nutrient Loads and Response in River and Estuary Systems" in Vrhovo Reservoir in Slovenia. The Vrhovo Reservoir (Fig. 2.2) is located on the Sava River, which is the longest river (220 km) in central Slovenia. The river continues its path through Croatia and Serbia and eventually discharges into the Danube River at Belgrade. It has a watershed area of 10,746 km^2 in Slovenia, the largest in the country. Nutrient loads from point sources such as, to a considerable extent, untreated municipal wastewater and industrial wastewater of the towns along the river, and tributaries draining distant settlements (nonpoint sources) were the most significant pollutants for the Sava River. The impact of these loads is visible in the polluted water as well as in the river sediment. Of particular concern is oxygen depletion in the waterways due to organic wastes and nutrients enrichment, i.e., eutrophication and the resulting impacts on dissolved oxygen from nutrient loads. The Vrhovo Reservoir on the Sava River was formed as part of the Vrhovo Hydroelectric Power Plant (HEPP). The construction of the Vrhovo HEPP began in 1987, representing the first HEPP in the series of six HEPP's on the portion of the Sava River between Zidani Most and the border with Croatia. The role of

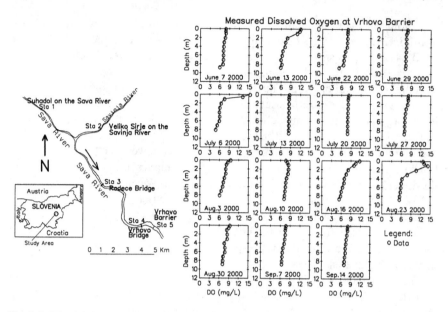

Fig. 2.2 Dissolved oxygen in Vrhovo Reservoir, Slovenia

this series of HEPP's is the production of peak energy, thus, they will operate in a daily accumulation regime. The contributing river basin of the Vrhovo Reservoir is 7189 km^2 and the average discharge is 235 m^3/s. The reservoir lies at 191 m above the sea level; the lowest water level is 182.88 m above sea level and the gross fall is 8.12 m. The surface area of the reservoir is 1.43 km^2 and its volume is 8.65 × 10^6 m^3. The hydraulic retention time of the reservoir ranges from hours up to a few days, depending on the flow rate through the reservoir. The measured conductivity levels at Vrhovo Barrier, i.e. the dam site, from June to September 2000 show very stable but somewhat high conductivity around 400 μmho/cm throughout this period.

To support a modeling study, a field-monitoring program was launched in 2000 to collect the necessary data. At that time, five water quality-monitoring stations were identified (Fig. 2.2):

- Upstream boundary at Suhadol on the Sava River (Station 1) and at the Veliko Širje Water Gauge Station on the Savinja River (Station 2).
- Two sampling stations along the reservoir; the first at Radeče Bridge (Station 3) and the second at Vrhovo Bridge (Station 4).
- Sampling station at the downstream boundary located at Vrhovo Barrier (the dam site) (Station 5).

Field measurements were performed weekly in the summer of 2000 (in June, July, August, and September). Additional measurements were performed in December 2000, and January, April, May, and June of 2001. In February and March of 2001, the discharges were too high for sampling. Measurements were then conducted every two weeks in July and August 2001. The following water quality parameters were measured: temperature, pH, conductivity, DO, BOD_5, ammonium, nitrite, nitrate, total nitrogen, orthophosphate, orthophosphate, total phosphorus, chlorophyll *a*, suspended solids, TOC, and transparency.

Conductivity and temperature in the reservoir was measured in the water column at each meter in 2000 and 2001 at three stations: Radeče Bridge, Vrhovo Bridge, and Vrhovo Barrier (at the dam). The 15 vertical profiles of conductivity at Vrhovo Barrier measured from June 7, 2000 to September 14, 2000 show no thermal stratification in the water column. Dissolved oxygen (DO) is clearly stratified in the water column at Vrhovo Barrier during the same period (Fig. 2.2) while the lowest DO levels were measured at the Vrhovo Bridge, the deepest point in the reservoir (Cvitanič et al. 2002). During the summer months with low discharges, the measured DO concentration immediately above the reservoir bottom was very low, close to zero. Similar observations were obtained from January 24, 2001 to August 23, 2001 at these three locations. (Again, the DO stratification is clearly present at Vrhovo Bridge and Vrhovo Barrier (Cvitanič et al. 2002).

The above analysis suggests that a 2-D modeling analysis (longitudinal/vertical) of the reservoir was needed to resolve the vertical gradients of DO, which is closely related to other water quality constituents such as algal biomass and nutrients in the water column. A modeling study of the Vrhovo Reservoir using the available data is presented in Chap 4, eutrophication modeling.

2.2 Algae/DO/pH Interactions

Diurnal DO fluctuation caused by significant algal growth is a well-known water quality problem in many rivers and streams, also causing periodic pH swings. Such phenomena are recently observed in the Love River in Taiwan, which originates in Renwu, Kaohsiung and flows 12 km (7.5 mi) from the Back Port Bridge through the City of Kaohsiung to Kaohsiung Harbor (Fig. 2.3), the largest sea port of Taiwan. It is of great cultural significance to the people of the city and plays an important role in its economy and tourism.

Although the water quality of the Love River has been improving since the 1979 cleanup effort, recent observations of significant algal growth are raising new concern on this effluent dominated river. A monitoring effort was launched in spring 2020 to sample three main stem stations (Haokang Bridge at upstream, Dragon Heart Bridge at mid-portion, and Kaohsiung Bridge at downstream) and five tributaries on a monthly basis. The water quality data at the three main stem stations on June 15, 2020 are summarized in Fig. 2.3, showing key constituents: conductivity, suspended solids, 5-day BOD (BOD_5), total Kjeldahl nitrogen (TKN), ammonia, nitrite/nitrate, total phosphorus (TP), and DO with the diurnal average and range of the concentrations over a period of 14 h (at hourly sampling) from sunrise to sun down. The conductivity profile shows the transition from freshwater to saline water in the downstream direction. The total suspended solids (TSS) profile shows progressive increase

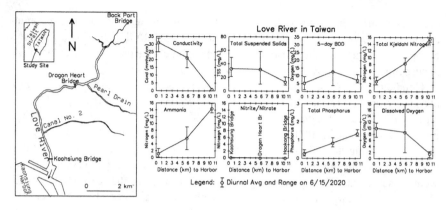

Fig. 2.3 Water quality data of love River in Taiwan

in concentrations in the downstream direction. While BOD levels are not high in this survey, its daily fluctuation is significant as noted in the following discussion. At the upstream station, nutrient levels are very high, providing significant loads for the receiving water. TKN concentrations, with the majority in ammonia form, are high in the upstream area, progressively decreasing in the downstream direction. Nitrite/nitrate are very low in the entire length of the river (a sign of nitrification inhibition noted in the following discussion). The total phosphorus trend is similar to the ammonia trend, showing a decrease in the downstream direction with sizable daily fluctuations. BOD concentrations are reasonable in the upstream portion, with the BOD levels fluctuating from 3 to 28 mg/L at the Dragon Heart Bridge. Finally, DO levels are very low in the upstream portion and are insufficient to support nitrification (resulting in high ammonia—low nitrite/nitrate) in the river. In addition, the daily range of DO between and 18 mg/L at Dragon Heart Bridge suggests a closer look into this phenomenon is required.

Figure 2.4 shows the diurnal water quality data at two main stem stations: Dragon Heart Bridge and Kaohsiung Bridge on June 15, 2020 from 0800 to 2400 h. The driving forces are temperature, BOD, ammonia, and total phosphorus, while the water quality response is shown in pH and DO. At the Dragon Heart Bridge in the mid-section of the river, nutrient concentrations (ammonia and orthophosphate, part of total phosphorus) are significantly reduced to support algal growth in mid-day (early afternoon hours), thereby causing a pH increase over one unit, quite significant! The high BOD values reflect the detection of carbon in algal biomass since the water samples were not filtered. The concurrent DO rise in the afternoon further confirms the photosynthetic activities in this portion of the river, resulting in a diurnal fluctuation of DO over 15 mg/L! The picture is clear: nutrient uptake supports the significant algal biomass production, resulting in strong diurnal pH and DO swings.

Further substantiation of significant algal growth at Dragon Heart Bridge is a sharp increase (24 mg/L) of BOD by mid-day. Assuming that the algal biomass

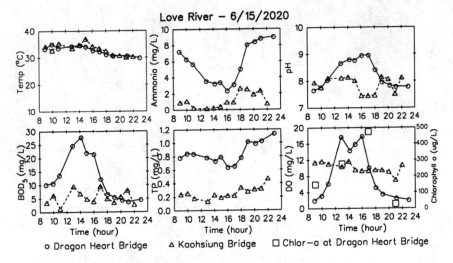

Fig. 2.4 Diurnal water quality fluctuations in Love River

carbon is highly labile (i.e., speedy assimilation for the very fresh labile carbon in algal biomass) and the carbon to chlorophyll *a* ratio is 25:1, the algal chlorophyll *a* in the BOD samples is calculated as 360 µg/L, which is reasonably close to the chlorophyll *a* rise of 347 µg/L (increased from 127 to 474 µg/L, data shown in square points in Fig. 2.4). The sharp pH rise of one unit during the day also supports the observation of very significant algal growth in that area. Finally, the data at Kaohsiung Bridge show the water quality progressively returns to average conditions: pH levels are normal and DO concentrations are high, perhaps influenced by the Kaohsiung Harbour.

Data plots in Figs. 2.3 and 2.4 offer a preliminary assessment of the eutrophication conditions in the Love River. While the temporal variations of water quality are clear, necessary data to support the diurnal modeling of pH, BOD_5, and DO in the Love River are not available, thereby requiring a new data collection effort:

a. Monitoring stations in the main stem need to be increased from three to six
b. Automatic, continuous sampling of water temperature, electrical conductivity, pH, salinity, and DO at these main stem stations
c. Hourly measurements of chlorophyll *a* during the day and bi-hourly in the night
d. Concurrent measurements of unfiltered and filtered BOD_5
e. Sediment oxygen demand measurements.

Another case of algae/DO/pH interactions is in the Santa Fe River near the City of Santa Fe in north-central New Mexico (Fig. 2.5). The river picks up the discharge from the Santa Fe WWTP and eventually enters the Rio Grande River at Cochiti Pueblo. Based on the monitoring since restoration of the La Bajada mine, it has been determined that the Santa Fe River met the numeric water quality standards for gross alpha. Later monitoring (fall 1998 through 1999) has also demonstrated that the

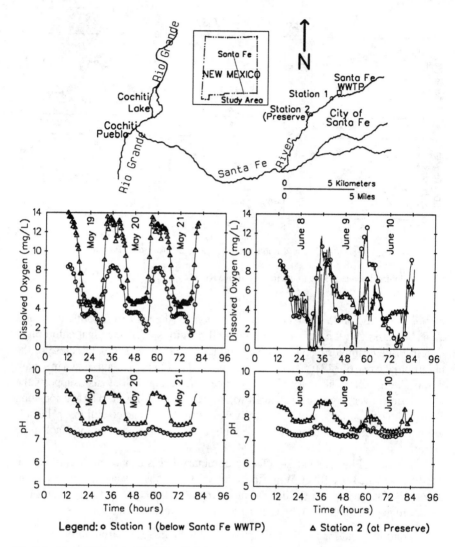

Fig. 2.5 Diurnal DO and pH in Santa Fe River, New Mexico

Santa Fe River met water quality standards for total ammonia. Therefore, a TMDL was not required for gross alpha and total ammonia. The total residual chlorine and stream bottom deposits TMDLs were completed in 1999. The remaining TMDL then focused on pH and DO in the Santa Fe River below the discharge of the Santa Fe WWTP. Monitoring for this system has shown daily variations in pH from 7.6 to 9.1 and DO from 60 to 210% saturation. Between the Santa Fe WWTP to Cochiti Pueblo, the river was identified as a water quality-limited stream due to pH and DO and was once included in New Mexico's Clean Water Act Section 303(d) list for TMDL development.

A water quality modeling study was launched to analyze the water quality problem in the Santa Fe River near Santa Fe, NM as part of the TMDL study, supported by intensive field monitoring. Figure 2.5 shows the DO and pH data from two 72-h intensive (sampling at 15-min intervals) surveys at two locations: station 1 (below the Santa Fe WWTP) and station 2 (at Preserve) during May and June 1999, respectively. The DO and pH fluctuations were observed shortly downstream of the wastewater treatment plant discharge and intensified from Station 1 to Station 2. A maximum daily swing of DO reaching 10 mg/L was observed at Station 2 during the May 1999 survey. The diurnal cycles were due to significant attached (benthic) algae, despite advanced treatment at the Santa Fe wastewater treatment plant. Field observations suggested that the significant diurnal fluctuation of dissolved oxygen and pH is caused by the excessive growth of benthic algae in the river channel bed. During the summer low flow conditions, the Santa Fe WWTP (with an average flow of 4 mgd) provides the bulk of the river flow (i.e. primary source below the plant) and sufficient nutrients to support the growth of the attached algae.

pH changes over a single unit were also detected. The maximum DO of 14 mg/L is even more impressive as the stream is located over 5000 ft above sea level, thereby under a much reduced DO saturation cap. This data review implied that traditional eutrophication models with daily averaged algal growth kinetics are not suitable for this modeling study. In addition, the benthic algae do not move with the river flow. Instead, they remain at the river bottom receiving nutrients deposited in the water column. A pH-alkalinity module needs to be linked up with the benthic algal model to address these unique features in the Santa Fe River. Results from the modeling analysis based on these data will be presented in Chap. 4 (for benthic algae and DO) and Chap. 5 (for pH).

2.3 Arsenic in the Patuxent Estuary

Toxic trace metals found in receiving waters have been a serious environmental concern with regulatory agencies. Public health issues with arsenic in drinking water has prompted the U.S. EPA to lower the maximum contaminant level (MCL) allowed in drinking water from 50 to 10 μg/L as of January 2006. Further, regulatory agencies are very likely to incorporate sediment quality and standards for trace metals like arsenic, copper, and cadmium into the management and analysis of water resources (Lung 2001). There is a need to employ models to track the fate and transport of metal in receiving waters, such as the Patuxent Estuary. As the first step of developing an arsenic model for the Patuxent Estuary, arsenic data from Reidel et al. (2000) were analyzed by Nice (2006). The dataset includes dissolved concentrations of arsenate, arsenite, methylarsonate (MMA), and dimethylarsenate (DMA) at 15 water quality monitoring stations located in the Patuxent River and Estuary.

Nine samples were collected from each monitoring station over a period ranging from May 24, 1995 to October 29, 1997. Figure 2.6 shows the concentrations of arsenic species along the Patuxent Estuary from upstream to downstream on nine

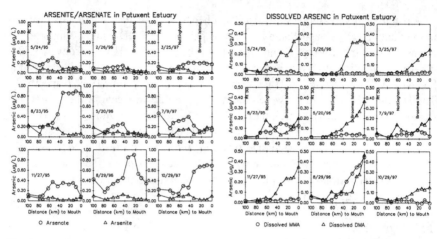

Fig. 2.6 Arsenic species in the Patuxent Estuary

dates during this period. The concentrations are plotted against station locations from upstream to downstream, which allows for visualization of both spatial and temporal trends. Two behavioral characteristics of arsenic concentrations have been observed in the estuary:

1. Substantial amounts of methylated arsenic, primarily DMA, were present in the lower estuary during winter, believed to be caused by the uptake and methylation of arsenic by algal growth (Reidel et al. 2000; Sanders and Reidel 1993),
2. A mid to lower estuary maximum of arsenic occurred in the summer, thought to be due to release of arsenic from the sediments during anoxic conditions in the bottom waters (Reidel et al. 2000).

These observations suggest that a formulation for uptake and methylation of arsenic by phytoplankton is needed for incorporation into the model. In addition, summertime flux of arsenic from the sediment requires an interactive two-layer sediment module linked with the water column (Nice 2006). The arsenic modeling analysis of the Patuxent Estuary is presented in Sect. 7.6.

2.4 Minimum Flow to Protect Water Quality

In 2000, the South Florida Water Management District (SFWMD) initiated an effort of setting a minimum flow limit (MFL) for the Caloosahatchee River Estuary (CRE) to protect the ecological system. The Caloosahatchee River is one of the two main drains of Lake Okeechobee (see Fig. 2.7). It runs from the lake outlet to the Gulf of Mexico via Ft. Myers. It is defined as the surface waters that flow through the Franklin Lock and Dam S-79 structure (see Fig. 2.7), combined with tributary contributions below the S-79 structure that collectively flow southwest to San Carlos Bay. The

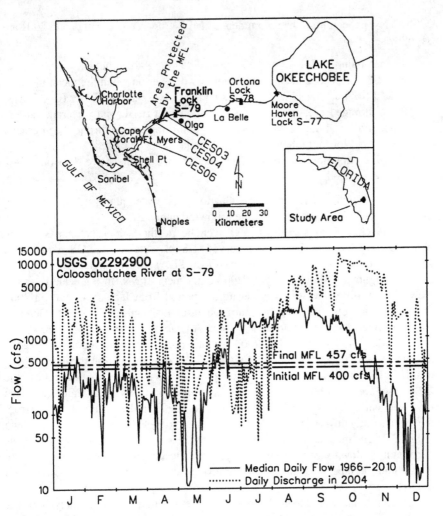

Fig. 2.7 Water quality sampling stations and flows at S-79 in the Caloosahatchee River

surface area of the CRE is 67.6 km^2 with an average depth of 2.7 m (a very shallow waterway). The Franklin Lock and Dam (S-79 structure) located near Olga, Florida serves as the upstream boundary for the CRE.

The first scientific peer review of the Caloosahatchee Minimum Flow was organized by SFWMD in September 2000, producing a review report (Edwards et al. 2000) approving the scientific approach, Valued Ecosystem Component (VEC) used in establishing the MFL. The basic assumption of the VEC is that flows or salinities appropriate for VEC are not detrimental to other important estuarine organisms. Subsequent status updates and progress reports have demonstrated a continuing effort by the District to set the MFL for the Caloosahatchee River Estuary. Based on the

2000 review panel report, the recommended MFL criteria recommended in 2001 are as follows:

1.   A mean monthly flow of 400 cfs measured at the S-79 structure.
2.   An MFL exceedance occurs during a 365-day period when the mean monthly flow at S-79 declines below 400 cfs (i) the Caloosahatchee River (C-43) West Basin Storage Reservoir is not yet operational; or (ii) the daily average salinity concentration at the Ft. Myers salinity monitoring station has not been greater than 10 ppt for more than 55 consecutive days.
3.   An MFL violation occurs when an exceedance occurs more than once in a five-year period.

An analysis of the CRE as a basic evaluation of the water quality conditions, using the data from 2004, is presented in the following paragraphs.

Daily discharge flows from Lake Okeechobee to the Caloosahatchee River are obtained from the U.S. Geological Survey (USGS) database. USGS maintains a gaging station 02,292,900 at S-79 near Olga, FL. The discharge records started in 1966. Figure 2.7 shows the year 2004 daily mean flows (in log scale to easily view the low flows). Most flows during the period from January to July in 2004 are below 3000 cfs, followed by sharp increases in August, September, October, and November during the hurricane season. A peak daily mean flow of 13,000 cfs was recorded on October 1. The 2004 Atlantic hurricanes provided a very deadly, destructive, and hyperactive season for south Florida. Then the flows reduced significantly in December 2004. Except for the period from mid-June to mid-August, the 2004 flows are much higher than the historical median flows in the Caloosahatchee River, indicating a wet year for the Caloosahatchee River basin. For a comparison, the median daily flows of this gage over the period from 1966 to 2010 are also shown in Fig. 2.7. The median flows are below the 400 cfs set by the regulatory agency as the minimum flow for over one half the year. Hurricanes sharply increase the flow in the summer months. Such high flows are expected to push salt further downstream in the estuary.

It is important to understand the water system with a thorough analysis of the water quality conditions. Such an effort called for a search of available hydrological, environmental, hydrodynamic, and water quality data. Figure 2.8 presents the nutrient concentrations at S-79, strongly influenced by the nutrient export from Lake Okeechobee and representing the nutrient loads to the Caloosahatchee Estuary for the period from 1999 to 2008. The concentration scales are the same between the nutrient components for phosphorus and nitrogen. Orthophosphorus represents the commending majority of total phosphorus in the loads. Organic nitrogen (= TKN-ammonia) is the dominating nitrogen component while ammonia contributes only a small portion of total nitrogen to the estuary. No obvious temporal trends are observed from the data plots. Nutrient concentrations are not high enough to reach alarming levels.

Figure 2.9 shows the water quality at three stations in the estuary: CES03 (upper estuary), CES04 (mid estuary), and CES06 (lower estuary). TSS levels at CES04 are the highest among the three locations, suggesting turbidity maxima, a phenomenon

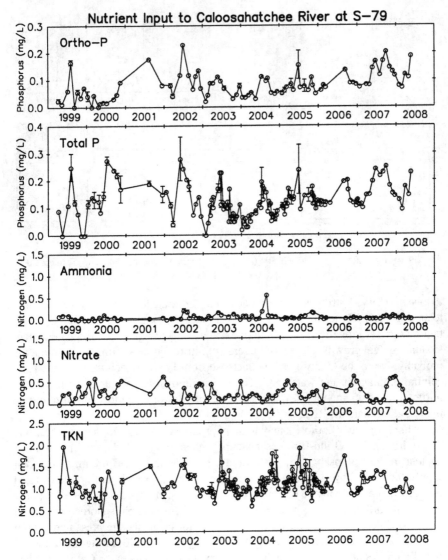

Fig. 2.8 Nutrient input at S-79 to the Caloosahatchee Estuary

showing peak TSS levels in the middle of the estuary usually observed in many partially mixed estuaries. Data at CES04 also show significant vertical TSS concentration variation (i.e., the highest concentrations are usually measured at the bottom of the water column due to settling of suspended solids). Light attenuation in the water column is closely affected by the TSS levels. Figure 2.9 also shows the nitrite/nitrate concentrations in the Caloosahatchee at the same three locations. Slight temporal attenuation of the nitrite/nitrate levels along the Caloosahatchee results in a gradual decrease of the concentrations in the downstream direction from the upper estuary

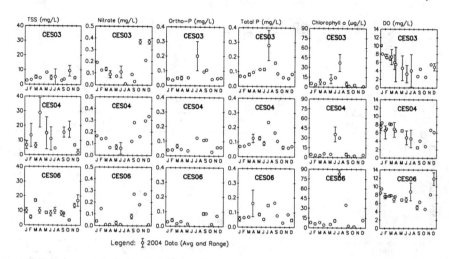

Fig. 2.9 Water quality at CES03, CES04 (mid), and CES06 in the Caloosahatchee Estuary, 2004

to lower estuary. A strong seasonal trend of nitrite/nitrate levels is displayed, leading to an extremely low level in the summer and reaching the potential of nitrogen limitation for algal growth in July. Estuarine waters typically have a nitrogen deficient to support algal growth. The sharp increase of nitrite/nitrate starting in September is probably due to the freshwater flow increase from Lake Okeechobee and the agricultural drainage in the watershed. The ammonia data are not as complete as the nitrite/nitrate data and their levels are significantly lower than those of nitrite/nitrate in the Caloosahatchee River.

Orthophosphate (Ortho-P) and total phosphorus concentration plots are also shown in Fig. 2.9. Ortho-P represents approximately 50% of the total phosphorus content in the Caloosahatchee. In general, the orthophosphate (being a key food supply for algae) levels are sufficient to support algal growth and not considered as a growth-limiting factor based on the data. The difference between total phosphorus and orthophosphate is primarily associated with the phosphorus in the algal biomass. Chlorophyll *a* concentrations in the Caloosahatchee are presented next in Fig. 2.9, showing an alarming level above 80 μg/L at CES06 during the month of July. Figure 2.9 also presents the dissolved oxygen (DO) data in the Caloosahatchee Estuary in 2004. Gradual depression of DO at CES03 and CES04 from the beginning of the year to the summer months is shown. The DO levels at CES06 are slightly higher, showing a peak DO level in early July corresponding the peak chlorophyll *a* (see Fig. 2.9) due to significant photosynthetic activities in the water column.

Following a second peer review by Buskey et al. (2017) and subsequent public hearings, the final rule on the MFL was set in September 2019 in 40E-8.221 MFLs: Surface Waters.

(2) Caloosahatchee River. The MFL for the Caloosahatchee River is the 30-day moving average flow of 457 cubic feet per second (cfs) at S-79.

(a)   A MFL exceedance occurs during a 365-day period when the 30-day moving average flow at S-79 is below 457 cfs.
(b)   A MFL violation occurs when a MFL exceedance occurs more than once in a 5-year period.

The flow, combined with tributary contributions below S-79, shall be sufficient to maintain a salinity gradient that prevents significant harm to mobile and immobile indicator species within the Caloosahatchee River. If significant harm occurs once the Caloosahatchee MFL recovery strategy is fully implemented and operational, the recovery strategy and MFL will be reviewed in accordance with Rule 40E-8.421, F.A.C. mobile and immobile species shall be monitored as described in the recovery strategy. The initial MFL of 400 cfs and the final MFL of 457 cfs are shown in Fig. 2.7.

2.5 Water Quality Improvement Following Treatment Upgrade

Data analyses presented in previous sections are designed to obtain a basic under-standing of the water systems for a variety of water quality concerns. The next level of data analysis is assessing water quality improvements following mitigated measures mandated by regulatory agencies. The effectiveness of wastewater treat-ment upgrades are judged on whether the new effluent limits can meet the water quality standards in the receiving water. Since significant investments are made to reduce pollutant loads and improve the water quality, evaluation of water quality improvements subsequent to treatment upgrade is important due to the substantial construction and operating cost. This analysis is directed determining before and after responses of the upper Mississippi River in the twin city area of Minneapolis and St. Paul, Minnesota following the installation of nitrification at the Metro Plant WWTP (see Fig. 2.10) in 1985 to improve the dissolved oxygen levels in the river. As a permit condition, a water quality assessment under a low flow condition must be conducted following this upgrade. The low flow water quality survey, including the Metro Plant effluent, was conducted in the summer of 1988 when river low flows occurred. It is wise to check that the investments are paid off by collecting the neces-sary receiving water data and comparing the before and after water quality conditions in the river, as mandated by the regulatory agency.

First, a comparison of the Metro Plant effluent quality between pre- and post-nitrification is summarized in Table 2.1. The comparison shows that the $CBOD_u$ load reduction is over 55,950 kg/day, a substantial amount resulting from the treat-ment upgrade. Further, the ammonia load reduction is about 6525 kg/day, leading to increase in nitrite and nitrate loads in the effluent. The low ammonia loads and high nitrite/nitrate loads after upgrade is very clear, transferring nitrification from the receiving water to the Metro Plant.

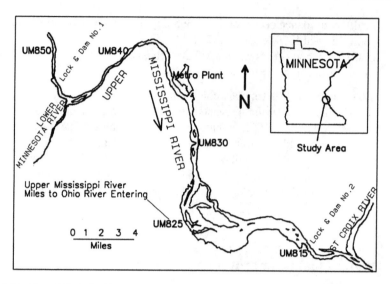

Fig. 2.10 Upper Mississippi River near Minneapolis and St. Paul

Table 2.1 Metro plant waste loads before and after upgrade

Survey	Flow (m³/s)	CBOD$_u$ (kg/day)	TKN (kg/day)	NH$_3$ (kg/day)	Nitrite/Nitrate (kg/day)
Secondary treatment	8.41	74,388	9632	7250	62.6
Secondary with nitrification	10.5	18,438	3532	727.3	11,476

The Metro Plant data before and after implementing the nitrification process are presented in Fig. 2.11 showing the ammonia, nitrate, and CBOD$_5$ concentrations from 1985 to 1989. The initial implementation of the nitrification process started as a partial year operation from June to the end of September in 1985. The data show that ammonia concentrations are significantly reduced while the nitrate levels in the effluent are sharply increased. Note the same concentration scales of nitrogen are used for ammonia and nitrate in Fig. 2.11 to give a better perspective. This first phase aimed at improving the DO levels in the summer months. The full year implementation started in 1988 as the ammonia levels in the plant effluent were significantly reduced all year around with even lower ammonia concentrations in 1989. It is interesting to see that full scale nitrification operation has contributed to additional reduction of the CBOD$_5$ concentrations in the effluent.

What is the water quality benefit, particularly the improvement of DO in the river? Fig. 2.12 presents a comparison of the upper Mississippi River water quality in terms of CBOD, ammonia, nitrite/nitrate (NO$_2$ + NO$_3$), and DO concentrations from upstream at Lock & Dam No. 1 (around UM848) to Lock & Dam No. 2 (at UM815)

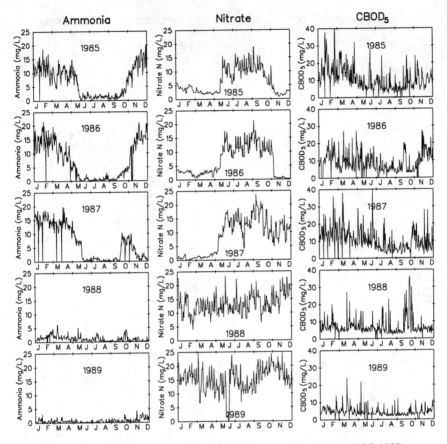

Fig. 2.11 Measured Ammonia, Nitrate, and CBOD$_5$ in metro plant effluent (1985–1989)

between the pre-nitrification and post-nitrification operations at the Metro Plant. The CBOD profiles show that the Metro Plant (at UM835) contributed minimal increase in the CBOD concentration in the river downstream from the effluent discharge. Note CBOD$_5$ was measured in 1976 and CBOD$_u$ was measured in 1988 (Fig. 2.12). The rise of CBOD concentrations (from unfiltered water samples) toward Lock & Dam No. 2 is due to the increased algal growth in the pool behind the dam. A striking comparison is shown in the ammonia and nitrite/nitrate profiles between these two surveys. In 1976, the ammonia concentration in the river increased sharply following the Metro Plant input. At the same time, the nitrite/nitrate concentrations were low in the river, showing no increase from the Metro Plant discharge. By summer 1988, the ammonia concentration in the river was significantly reduced following the treatment upgrade. The corresponding sharp increase in the nitrite/nitrate concentrations in the river following the Metro Plant input is clear—nitrification works! Both ammonia and nitrate concentrations decrease progressively in the downstream direction due to algal growth in the river, particularly approaching Lock & Dam No. 2, in both the

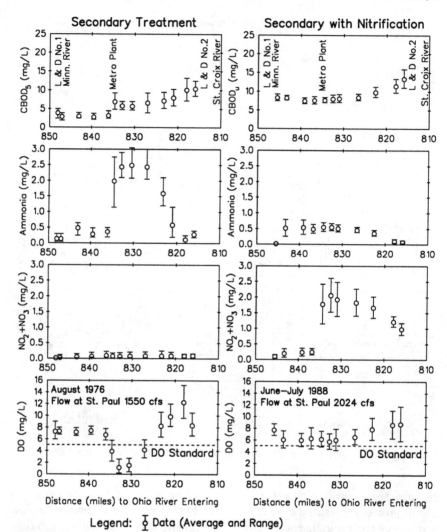

Fig. 2.12 Water quality of upper Mississippi River before/after metro plant upgrade

1976 and 1988 surveys. Finally, a comparison of the DO profiles in the river shows a significant improvement following the Metro Plant treatment upgrade. The minimum DO level is above the water quality standard of 5.0 mg/L in during the summer months of 1988. The 1976 anoxic condition, located at about 5 miles downstream from the Metro Plant, is eliminated.

Figure 2.13 shows the historical DO levels in the upper Mississippi River from 1964 to 2007. DO standard violations in the upper Mississippi River, particularly during low flow conditions, have not been observed since 1988. Data analyses presented in this section paved the way for a full model post-audit to demonstrate

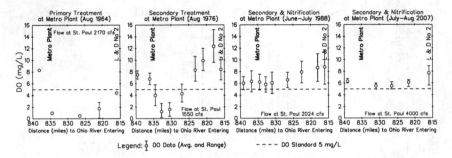

Fig. 2.13 Water quality improvement of the upper Mississippi River over four decades

to the regulatory agency the treatment upgrade benefit. Results of the post-audit modeling analysis are presented in Chap. 3.

Ammonia is not 100% ionized (i.e. dissociated) in water and the un-ionized fraction is toxic to fish and has always been a concern in receiving waters. The percentage of un-ionized ammonia is a function of pH and temperature following the chemical equilibrium. Within the temperature range of 0–50°C and a pH range of 6.0 to 10.0, the following expression (Thurston 1979) is used to evaluate the equilibrium constant:

$$pK_a = 0.0901821 + 2729.92/(273.2 + T) \tag{2.1}$$

where K_a is the equilibrium constant, $pK_a = -\log K_a$, and T is water temperature in °C. The fraction of un-ionized ammonia is given by the following expression:

$$f = \frac{1}{10^{(pK_a - pH)} + 1} \tag{2.2}$$

The spatial profiles of un-ionized ammonia and pH in the upper Mississippi River for the August 1976 period are shown in the left column of Fig. 2.14. Water temperature was 25.5°C in the river. The August 1976 un-ionized ammonia profile reveals contravention of the standard of 0.04 mg/L used by the Minnesota Pollution Control Agency (MPCA) for the upper Mississippi River. At the upstream end of the study area, the entry of the Minnesota River near UM843 raises the ammonia level in the upper Mississippi River, which is followed by a slight decrease due to nitrification in the river. The Metro Plant (without nitrification at that time) effluent discharge again raises the un-ionized ammonia level above the standard. Ammonia oxidation via microbial activity converting ammonia to nitrates produces approximately 100,000 lbs/day of acid in the river, causing a sharp decline of pH downstream of the Metro Plant. [A similar input of 250,000 lbs/day of acid to the Potomac Estuary from the Blue Plains WWTP prior to the nitrification process is reported in Chap. 1]. In the upper Mississippi River, the increase in pH downstream of UM830 is attributable to algal growth by shifting the alkalinity composition from bicarbonate to carbonate

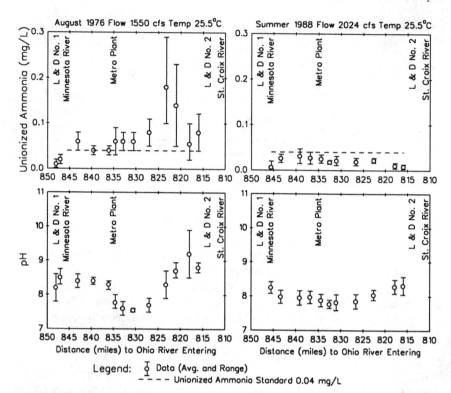

Fig. 2.14 Un-ionized Ammonia in the upper Mississippi River (before and after nitrification at the metro plant)

and from carbonate to hydroxide in the pool behind Lock & Dam No. 2. Downstream of UM820 the decrease in un-ionized ammonia is due to the overall decrease in total ammonia which results from the phytoplankton uptake of ammonia. Algal photosynthesis tends to raise the pH, thereby increasing the fraction of un-ionized ammonia (see Eqs. 2.1 and 2.2) to compensate for the loss of total ammonia to a certain extent. By 1988 (see the right-hand column of Fig. 2.14), the nitrification process was on line at the Metro Plant, resulting in a significant decrease in ammonia levels in the effluent. The slight increase in pH downstream of UM825 toward Lock & Dam No. 2 was caused by mild phytoplankton growth, consuming ammonia and thereby lowering the un-ionized levels in the river. Since 1988 the un-ionized ammonia levels in the upper Mississippi River are below the standard of 0.04 mg/L. Today, ammonia toxicity is no longer a water quality issue in U.S. rivers and streams due to water pollution control established in the 1970s. However, this problem still exists in low DO streams receiving significant loads of organic carbon and nitrogen in under-developed countries.

2.6 Trophic State Analysis

Today's eutrophication modeling frameworks are capable of configuring sophisticated water column kinetics with fine temporal and spatial resolutions. However,
they need substantial amounts of data to calibrate and verify the model results. On
the other hand, there are simple calculations that provide results to meet certain regulatory demands with minimum data support. The Vollenweider (1968) phosphorus
loading plot falls into this category. It achieved the historical status of providing key
technical input to the Great Lakes Water Quality Agreement between the United
States and Canada signed at Ottawa on April 15, 1972, specifying allowable phosphorus loading for Lakes Erie and Ontario. It was the first methodology linking
external phosphorus loadings to a measure of eutrophication. The graphical plot of
phosphorus loading (in grams/m^2/yr) as a function of mean depth of the lake with
a general division into eutrophic or oligotrophic lakes provided a basis for decision
making by environmental managers. For a given depth of the lake, the "allowable"
loading can be read directly from the plot. Subsequently, Dillon and Rigler (1974)
enhanced the analysis by introducing the lake detention time, thereby significantly
improving the methodology. The underlined assumptions of this analytical tool are:

1. The lake is completely mixed
2. The lake is at a steady-state condition (i.e. at an equilibrium between the
    phosphorus load and in-lake concentrations)
3. There is a net loss of phosphorus to sedimentation in the lake

In reality, the first two assumptions are not always met, particularly for large lakes
and impoundments. The time to response to nutrient control measures is missing as
the analysis is for steady-state conditions. Assuming that the lake phosphorus concentrations are in equilibrium with the external loads rarely holds. Further, the assumption that the total phosphorus concentrations in the water column are in equilibrium
with the sediment is usually nonexistent. This methodology is too crude of an analysis
to permit meaningful conclusions to be drawn about the effects of nutrient reduction
on phytoplankton biomass (Thomann 1977) when significant management decisions
are at stake. However, in situations with very limited data for relatively small lakes,
impoundments, or reservoirs, we can still use this methodology to assess the trophic
status.

The theoretical background of the Vollenweider-Dillon plot for a completely
mixed lake under steady-state conditions is:

$$LA = QC + LRA$$

where

L total phosphorus loading rate (gm/m^2/yr)
A lake surface area (m^2)
Q through flow rate (m^3/yr)
C concentration of total phosphorus (mg/L = gm/m^3 = 1000 μg/L)

R   fraction of total phosphorus loads loss to sediment

Solving the above equation for *C* yields

$$C = \frac{L(1 - R)/\rho}{z}$$

where
ρ = flushing rate of the lake (per yr) = lake volume/outflow rate
z = mean depth (m) = lake volume/ lake surface area
A plot of $L(1 - R)/\rho$ versus z on log–log scales yields an intersect on the vertical axis as the steady-state phosphorus concentration in the lake.

The following example demonstrates use of the plot in assessing eutrophication potential for an existing reservoir. Information is given for the South Fork Rivanna Reservoir in Albemarle County, VA: reservoir surface area = 1.58 km^2; reservoir volume = 7.11 × 10^6 m^3; and mean depth = 4.5 m. Monthly inflows (m^3/s) to the reservoir in 1980 are:

Jan	Feb	Mar	Apr	May	June	July	Aug	Sept	Oct	Nov	Dec
16.72	6.85	22.84	29.64	8.81	3.60	2.83	2.52	1.11	1.19	2.18	1.84

Outflow volumes (m^3) from the reservoir in 1980: South Fork Rivanna River = 250 × 10^6; Rivanna Water & Sewer Authority Diversion = 11 × 10^6; and evaporation loss = 1.42 × 10^6. Total phosphorus loads (kg) to the reservoir in 1980 are:

	Jan	Feb	Mar	Apr	May	June	July	Aug	Sept	Oct	Nov	Dec
Point	469.1	409.1	203.2	132.7	44.55	56.36	404.1	413.2	395.0	479.1	285.9	125.9
Nonpoint	15,812	383.6	16,124	21.71	2353	600.0	1077	909.1	131.4	233.6	551.4	220.0

Fig. 2.15 Trophic State of
South Rivanna and Walnut
Creek Reservoir

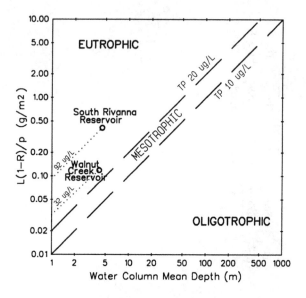

Output of phosphorus from the reservoir via the South Fork Rivanna River was 24,036 kg in 1980. From the above data, the total phosphorus load to the reservoir from both point and nonpoint sources is 63,524 kg in 1980. The total phosphorus output of 24,036 kg yields: $L(1 - R)/\rho = 0.416$ g/m^2) based on the hydraulic retention time of the reservoir at 0.0274 yr, i.e. $\rho = 36.5$ per yr. With a mean depth of 4.5 m, the South Fork Rivanna Reservoir can be located as eutrophic on the Vollenweider-Dillon plot (Fig. 2.15). Dividing the total phosphorus output of 24,036 kg/yr by the outflow of 0.26×10^9 m^3/yr yields an in-lake concentration (i.e. the annual average total phosphorus concentration in the reservoir) of 0.092 mg/L (92 μg/L), which is the intercept on the left vertical axis. A 45° line (1:1 slope) starting with 0.092 on the vertical scale of the plot passes through the center of the point for South Fork Rivanna Reservoir on the plot.

In another application, the Albemarle County Engineering Department was planning to impound the Walnut Creek for recreation use in 1985. One of the potential concerns of such a project was eutrophication, following the damming of the creek with the nutrients coming from the watershed. Facing limited water quality data for the study, the watershed nutrient export data gathered for the South Rivanna Reservoir were used for this study due to their proximity in Albemarle County. The analysis resulted in a total phosphorus load of 381.4 kg/yr to the proposed impoundment. The $L(1 - R)/\rho$ value is therefore calculated as 0.12 g/m^2 of impoundment surface area. With an average water depth of 4.11 m, the Walnut Creek Reservoir is located in Fig. 2.15, showing a trophic status of mesotrophic-eutrophic, much less eutrophic than the South Rivanna Reservoir. The calculated steady state total phosphorus concentration is 0.032 mg/L (32 μg/L), as the intercept on the vertical axis of the plot. Such results convinced Albemarle County to move forward with the impoundment in the following year. During the past three decades, the Walnut Creek

Reservoir has been one of the county's major recreation areas without eutrophication problems.

2.7 Mass Transport in Receiving Waters

The next level of data analysis is independently deriving key mass transport coefficients using available data prior to full scale water quality model configuration. This derivation not only saves time and effort but also lets the modeler feel comfortable to ease into the model kinetics, which usually are much more complicated to configure. The first set of parameters to be estimated include time of travel in a river; longitudinal, lateral, and vertical dispersion; lateral mixing; and light extinction in the water column.

2.7.1 Hydraulic Geometry

Key hydraulic geometry such as velocity and depth in streams can be derived from available data as shown in the South Fork South Branch Potomac River in West Virginia (Fig. 2.16). Flows measured at USGS gaging station 01,608,000 at Moorefield are readily available from the routine monitoring. Velocity and depth measurements, while not as frequent as the flow monitoring, are obtained from the USGS database. Figure 2.16 also shows the regression between mean velocity and flow ($V = 0.1227 Q^{0.4457}$) as well as between mean depth and flow ($D = 0.2963 Q^{0.3088}$). These regressions, represented by straight lines when plotted on a log–log scale, can be very useful in configuring one-dimensional mass transport models for streams.

2.7.2 Time of Travel

Data from a dye release study by USGS in the upper Mississippi River in the twin city area of Minneapolis and St. Paul in Minnesota are presented in Fig. 2.17. Concentration and time scales are kept the same between the plots to provide a good perspective in understanding the dispersion process. The first two plots show the results from the dye study conducted from July 12 to July 30, 1976 with Rhodamine WT dye released at the Metro Plant at 0600 h on July 12. Dye concentrations in ppb (μg/L) were measured at five downstream locations: at I-494, County Hwy 22, Shiely Dock, above Spring Lake, and finally at US Hwy 61 bridge (see these locations in Fig. 2.17). The leading edge of the dye was measured around 1800 h at I-494 and its peak concentration of 6 ppb at midnight of July 12. Flows through Spring Lake are slow, delaying the travel time. The leading edge of the dye concentration did not show up at US Hwy 61 Bridge until almost one week after its release at the Metro Plant. The spread

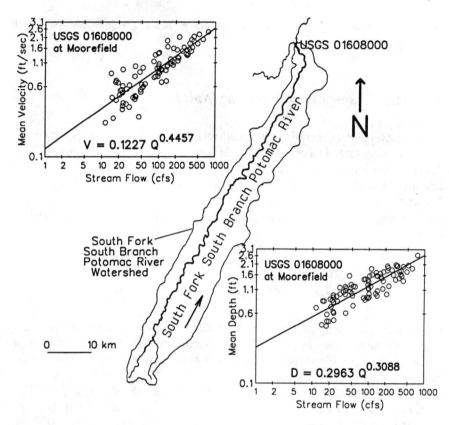

Fig. 2.16 Hydraulic geometry of South Fork South branch Potomac River

of the dye concentration profile took 11 days at this location. By the time the trailing edge passed through US Hwy 61 Bridge to complete the study, it had taken 18 days (see the second plot in Fig. 2.17) since the release on July 12. The river flow at St. Paul during the study ranged from 3260 to 3400 cfs. The third plot of Fig. 2.17 shows the dye data collected on the second study from May 11 to 21, 1977 at a range of flows from 4670 to 4800 cfs (higher than the flow in 1976). The water traveled faster under this higher flow, resulting in more compacting (i.e. less spread) of the dye concentration profiles (see the third panel of Fig. 2.17). It took 11 days to complete this study in 1977 as opposed to 18 days in the 1976 study. Data from the dye studies are useful to develop time of travel as well as longitudinal dispersion for the upper Mississippi River.

Following the steps for the dye analysis by Martin and McCutcheon (1999), Table 2.2 summarizes the derivation of the longitudinal dispersion as follows.

As expected from the dye data plots, the longitudinal dispersion coefficient at US Hwy 61 Bridge is higher than that above Spring Lake due to the much longer spread of the dye concentrations. Dye data in the third plot of Fig. 2.17 show less

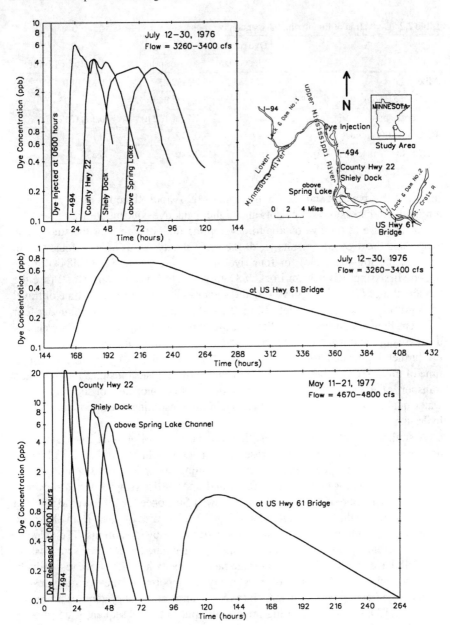

Fig. 2.17 Rhodamine WT dye data of upper Mississippi River

Table 2.2 Derivation of longitudinal dispersion coefficient in supper Mississippi River

	Dye release at		
	Metro plant	Above spring lake	US Hwy 61 bridge
Mile point	UM835.2	UM826.5	UM814.0
Peak conc. (ppb)		3.4 (at 84 h)	0.9 (at 195 h)
Non-peak (ppb)		1.5 (at 102 h)	0.3 at (312 h)
River velocity (mi/hr)		0.112	0.112
Calculated dispersion coef (mi^2/hr)		0.220	0.333

spread of the dye concentration in the upper Mississippi River, resulting in smaller longitudinal dispersion coefficients and higher peak dye concentrations.

Another use of the dye data is to develop the time of travel for the study area. Processing the dye study results yields plots of velocity versus flow (in four straight lines on the log–log scale) for four river stretches (e.g., UM847-UM844) of the upper Mississippi River from Lock & Dam No. 1 to Lock & Dam No. 2 (see the top left plot of Fig. 2.18). Converting the velocity-flow relationship to time of travel is shown in the bottom left plot of Fig. 2.18, showing the time of travel under four different flow conditions (at St. Paul) in the upper Mississippi River. The velocity decreases in the downstream direction and causes the time of travel curves to bend up, further augmented approaching the Lock & Dam No. 2, due to the increase in time of travel (see the bottom left plot in Fig. 2.18). Physical insights into the mass transport of the upper Mississippi River prepares the modeler for the subsequent water quality modeling analysis, once again demonstrating the importance of the hydraulic geometry data.

A similar plot of velocity versus flow for the lower Minnesota River using the data from USGS and MPCA are presented in the top right plot of Fig. 2.18 for two stretches from Shakopee to the river mouth joining the upper Mississippi River. [A map showing the lower Minnesota River and upper Mississippi River is shown in Fig. 2.17]. The time of travel versus river flow at Shakopee for the lower Minnesota River is shown in the bottom right plot of Fig. 2.18. The time of travel from Shakopee to the river mouth increases substantially under low flow conditions. The velocity versus river flow plots for the upper Mississippi River and lower Minnesota River in Fig. 2.18 have the same log scales for the mean velocity and river flow, respectively for easy comparison. The time of travel analysis supported by the dye study results reveals significant physical insights into mass transport in the system (as presented in Figs. 2.17 and 2.18), providing important input to the subsequent water quality modeling analysis.

Figure 2.19 shows the time of travel plots for the Kalamazoo River in Michigan and the Roanoke River in Virginia. The same temporal and spatial scales are used in these two plots for a good comparison. Also note the time of travel differences between large rivers (upper Mississippi River and the Minnesota River in Fig. 2.17) and small rivers (the Kalamazoo River and Roanoke River). These results should give us a reasonable starting point on the mass transport modeling analysis. Velocities

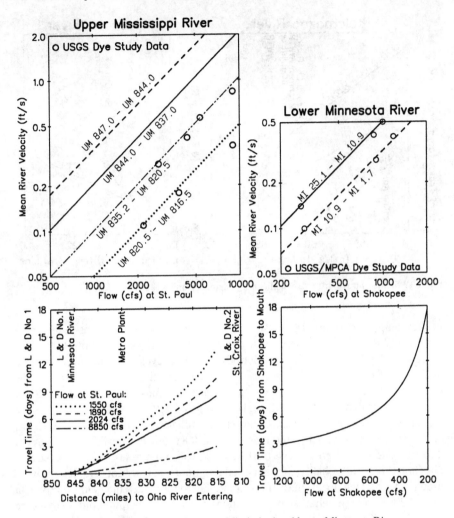

Fig. 2.18 Velocity and time of travel for upper Mississippi and lower Minnesota Rivers

and dispersion coefficients obtained from the above dye data would serve as initial assignment for the mass transport, which will be checked and/or confirmed with a mass transport model.

2.7.3 Using Dye Data to Derive Dispersion Coefficients

A dye study was conducted in the hydraulic (physical) model of the James River Estuary located at the US Army Corps of Engineers Waterways Experiment Station

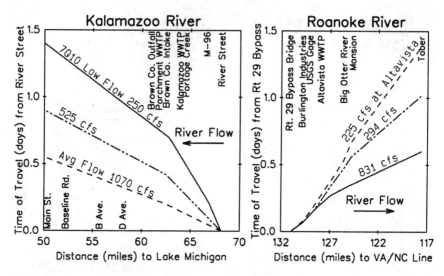

Fig. 2.19 Time of travel from dye studies for Kalamazoo and Roanoke Rivers

in Vicksburg, Mississippi. In the lab, dye was injected at two locations in the physical model: Richmond and Hopewell at a constant freshwater flow of 1000 cfs at Richmond. Dye was injected over a period of 100 tidal cycles and monitored for 120 tidal cycles from the start of the release. Dispersion coefficients were adjusted in the mass transport model to mimic the dye concentrations measured in the James River Estuary physical model to reproduce the dye spatial pattern. Figure 2.20 shows the model results matching the dye distribution very well for both release points. The calibrated tidal dispersion coefficients increase from 0.8 to 5.0 mi^2/day in the downstream direction due to tidal effect. The freshwater flow also increases in the downstream direction (Fig. 2.20). Finally, the calibrated longitudinal dispersion coefficients were used in the (mathematical) mass transport model to reproduce the field salinity data (Fig. 2.20) as a verification step. As demonstrated, both hydraulic model dye data and field salinity data played an important role in this analysis.

In the past two decades, regulatory agencies have been paying attention to mixing zones of small waste flows to set stringent permits requiring that the water quality standards be met at the point of discharge (i.e. at the end of pipe). To offer variations from this requirement, many state regulatory agencies allow mixing zones in streams, rivers, and estuaries if the dischargers can demonstrate regulatory requirements are met. Such a demonstration in general would come from a mixing modeling analysis with data support. One of the key model coefficients in river mixing zone modeling that can be independently derived from field data analyses is the lateral dispersion coefficient across the river from the discharge bank to the opposite bank. Rhodamine WT dye again plays a key role in the hydrographic survey of the discharge site to derive this coefficient for the mixing zone modeling study. Figure 2.21 shows the two-dimensional distribution of the measured dye concentrations in Moores Creek

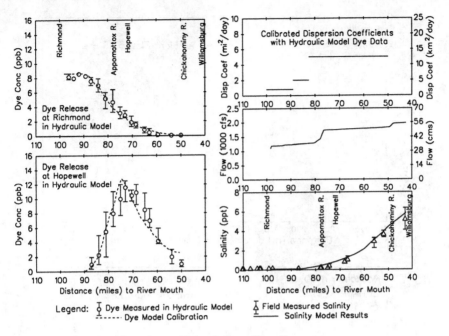

Fig. 2.20 Longitudinal dispersion coefficient in James River Estuary

in Charlottesville, VA, which receives the treated wastewater from the City of Charlottesville's WWTP. The dye concentration released in the effluent is 75 μg/L (ppb) while the background dye level is zero. The data show the dye moving along the river as well as dispersing across the river at the same time. The mixing length is about 300 m downstream from the outfall (i.e. uniform dye concentrations across the river width). While the Rhodamine WT red dye has tendency to be absorbed in the receiving water, its effect is generally minimum for small wastewater flows in a relatively small size mixing zone. An alternative to Rhodamine WT dye is electrical conductivity which has the following advantages:

1.   It is generally considered as a conservative substance in the receiving water.
2.   Conductivity levels in domestic wastewater treatment plant effluents are generally in the range of several hundred μmho/cm while the background conductivity level in receiving waters is below 100 μmho/cm. Concentration gradients of this order are ideal to track longitudinal and lateral dispersion.
3.   Minimum effort is required to measure electrical conductivity in the field via devices such as a Hyrdolab. At a minimum, measuring conductivity offers additional support for a dye study.

Figure 2.21 shows the conductivity levels concurrently measured with the dye sample taken in the field. The WWTP effluent (flow 0.31 cms) has a conductivity level of 395 μmho/cm while the background conductivity (upstream ambient flow 5.66 cms) is 100 μmho/cm, offering a good concentration gradient to see the spread

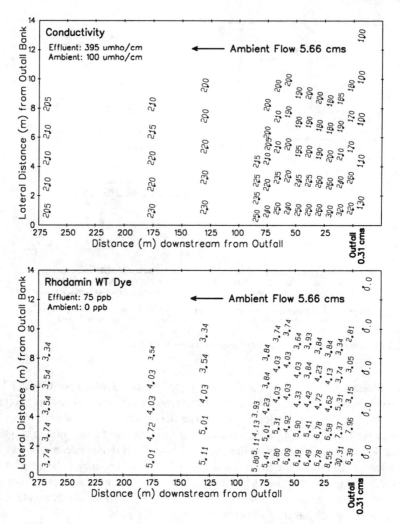

Fig. 2.21 Measured conductivity and dye concentrations in Moores Creek resulting from the WWTP Effluent in Charlottesville, VA

of a conservative substance downstream from the wastewater discharge. Case studies presented in Chap. 8 demonstrate the use of dye and conductivity data to support mixing zone modeling analyses for regulatory water quality management.

The concern of total residue chlorine (TRC) in the Lynchburg WWTP effluent raised the water quality concern in the non-tidal portion of the James River. Both Rhodamine WT dye and TRC concentrations were measured in a sampling program to support a mixing zone modeling study. Figure 2.22 shows that Rhodamine WT dye and TRC were measured at eight transects along a stretch of approximately 3800 ft downstream from the discharge point in the James River. Samples across each transect

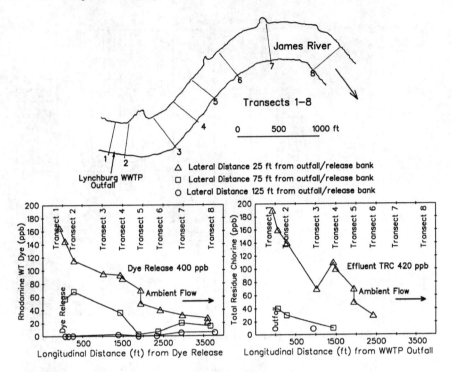

Fig. 2.22 Rhodamine WT dye and total Chlorine Residue in James River

were collected. Data from transect 1 provides the background concentrations. Both plots in Fig. 2.22 show longitudinal and lateral spread of the substance. The dispersion coefficients first derived from a 2-D mass transport model by reproducing the dye data were used to derive the die-off rate (in day^{-1}) of the TRC.

Another key model parameter that can be determined using existing data prior to running the water quality model is the vertical dispersion coefficient, K_z. A good example is demonstrated in Platte Lake in 1997. Figure 2.23 shows the hydrothermal structure of the water column, displaying the spring and fall overturns as well as thermal stratification during the summer months, a typical hydrothermal structure in a temperate climate. The lake is thermally stratified from June to September in the water column, inhibiting the vertical exchange of substances that is characterized with small vertical dispersion coefficients. During the spring and fall overturn periods, on the other hand, the water column is characterized with high vertical dispersion coefficients. During the winter months, reversed albeit slight, temperature stratification is observed in the water column (see temperature profiles in January and February of Fig. 2.23) due to the unique property of water having a maximum density at 4°C. Based on the work by Jassby and Powell (1975), Lung (2001) developed a simple procedure to calculate K_z from the following equation:

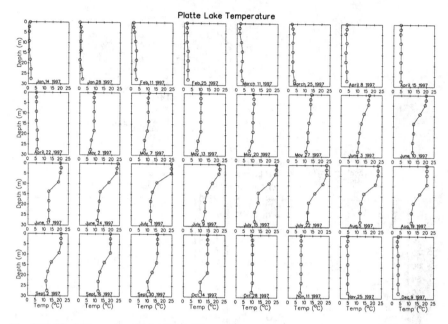

Fig. 2.23 Vertical temperature profiles in Platte Lake, Michigan–1997

$$K_z = \frac{\sum_h^0 \frac{\Delta T}{\Delta t}}{\frac{\Delta T}{\Delta z}} \tag{2.3}$$

For example, using the temperature data from June 24 and July 1,

where

$\Delta T / \Delta t$ temporal temperature gradient between June 24 and July 1 at each depth
$\Delta T / \Delta z$ vertical temperature gradient at each depth

This calculation will generate the K_z values at each depth for June 24 and July 1, respectively. That is, vertical profiles of K_z are calculated. The accuracy of this procedure is highly dependent on the spatial and temporal resolutions of temperature measurements throughout the entire year.

2.7.4 Confirmation of Mass Transport Coefficients

Results of verifying the 1-D mass transport in the Blackstone River, MA with a chloride model in the water column for a 45 mile distance under two flow conditions in 1991 are shown in the left columns in Fig. 2.24. Point source inflows and chloride loads entering the Blackstone River are from Wright et al. (1998). By reproducing the chloride concentrations, the flows and longitudinal dispersion coefficients are

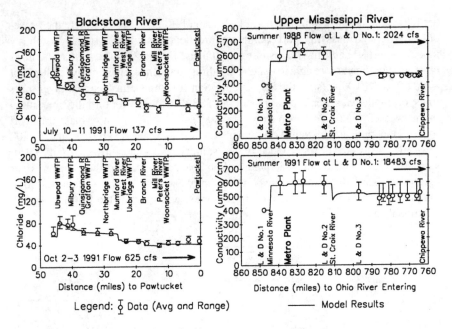

Fig. 2.24 Verification of time of travel with mass transport modeling

verified by the mass transport model, which is configured with the U.S. EPA QUAL2 modeling framework (Brown and Barnwell 1985).

Another confirmation of mass transport is demonstrated along an 85-mile stretch in the upper Mississippi River from Lock & Dam No. 1 to the confluence with the Chippewa River. In this study, conductivity (being a conservative substance) is modeled to verify the mass transport under two flow conditions (low flow in the summer of 1988 and very high flow in the summer of 1991—see plots in the right columns of Fig. 2.24). Under the low flow (2024 cfs), the high conductivity from the Metro Plant raises the conductivity in the river by 65 μmho/cm. When the river flow is very high (18,483 cfs) this impact is significantly minimized (see the bottom right plot of Fig. 2.24). Similar response to the St. Croix River input is also displayed, i.e. less impact under very high flow conditions. The longitudinal dispersion coefficients developed from this mass transport analysis are ready for the next step, water quality modeling.

Salinity is another conservative substance often utilized to calibrate mass transport models for estuaries, particularly prior to a full application of a sophisticated hydro-dynamic model. Figure 2.20 shows the 1-D salinity modeling of the James River Estuary. A common mass transport pattern in many partially stratified estuaries is the two-dimensional (longitudinal-vertical) mass transport, reported in numerous estuarine modeling studies (Prichard 1969; Dyer 1997). A unique mass transport analysis to derive mass transport coefficients with the given salinity distribution was first reported by O'Connor and Lung (1981) and Lung and O'Connor (1984) for

the Sacramento-San Joaquin Delta in California Fig. 2.25). To complete the procedure, the derived mass transport coefficients are incorporated into a two-layer salinity
model to verify these mass transport coefficients. Figure 2.25 shows that the mass
transport coefficients are able to match the salinity data under two freshwater flows of
4400 cfs and 10,000 cfs for the Sacramento-San Joaquin Delta. The two-layer mass
transport model was then used to model the TSS distribution for turbidity maximum
in the system. Figure 2.25 also shows the successful verification of mass transport
for the Hudson River in New York under two flow conditions: 3200 cfs and 18,400
cfs. Low freshwater flows produce higher salinity levels and further salinity intrusion inland as seen in both Sacramento-San Joaquin Delta and the Hudson River
Estuary. This two-layer mass transport of salinity and TSS, which explains turbidity
maximum, TSS, and eutrophication in estuaries, will be revisited in Chap. 4 with
additional cases.

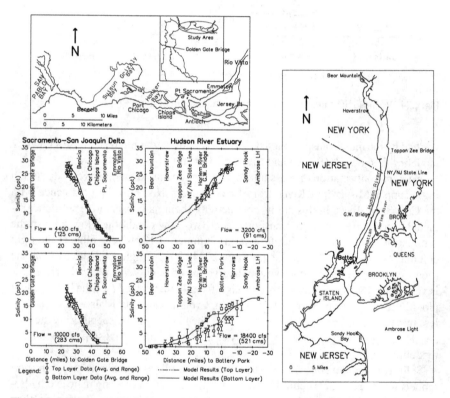

Fig. 2.25 Salinity modeling of Sacramento-San Joaquin delta and Hudson Estuary

2.8 In-Stream BOD Deoxygenation

Once the mass transport is determined, the next level of data analysis is to quantify kinetics coefficients using available data. The first one to consider is the dissolved CBOD deoxygenation rate, characterizing the decay of organic carbon in the form of a first-order consumption of dissolved oxygen in the receiving water. It should be pointed out that organic carbon sorbed on particles, which settle in the water column and reach the riverbed, is not reflected in this rate. Instead, they are accounted for in the BOD budget to show up as a combined BOD loss rate. Such a process is becoming less important as modern treatment plants would have very low levels of solids in their effluents.

A simple method of quantifying the deoxygenation coefficient is tracking the decay of dissolved CBOD in the receiving water. Using loading rates instead of concentrations accommodates the tributary CBOD input as well as river flow changes along the receiving stream. The CBOD loading rates decrease in the downstream direction because of attenuation. This simple exercise works well for rivers receiving single wastewater discharges. Figure 2.26 shows this analysis for the upper Patuxent River and upper Mississippi River, respectively. There is no other point source BOD input below the Parkway WWTP in the upper Patuxent River. There are two point source BOD loads below the Metro Plant to the upper Mississippi River, but their loads are insignificant compared to the Metro Plant input. The slope of the regression straight line of the loading rate plots (in semi-log scale) versus travel time yields the first-order in-stream deoxygenation rates for these two rivers.

A similar analysis for ammonia attenuation yields the nitrification rate in the upper Patuxent River using the data in the 1970s when treatment levels were marginal (see the top right plot of Fig. 2.26). Note that both the CBOD deoxygenation rate and nitrification rate decrease in the upper Patuxent River following the treatment upgrade. The rate reduction in the upper Patuxent River before and after treatment upgrade is most likely due to the improved solids removal. On the other hand, the dissolved $CBOD_u$ data were used in the upper Mississippi River analysis, yielding an accurate estimate of the deoxygenation rate (see the bottom plot of Fig. 2.26) because the Metro Plant had already achieved secondary treatment with nitrification.

Generating the plots in Fig. 2.26 is based on a mass balance model for the receiving water:

$$M = M_o e^{-k_d \frac{x}{U}} \qquad (2.4)$$

where M and M_o are mass rates of BOD in lbs/day or kg/day at x and x_o, respectively; k_d is the in-stream deoxygenation rate; U is the average velocity; and $\frac{x}{U}$ is time of travel (the horizontal axis of the plots in Fig. 2.26). Such an analysis of deriving the constant BOD deoxygenation rates is relatively straightforward with sufficient data support, particularly for a single point source discharge. Complicated situations of spatially variable k_d rates in rivers receiving multiple BOD loads of variable strength

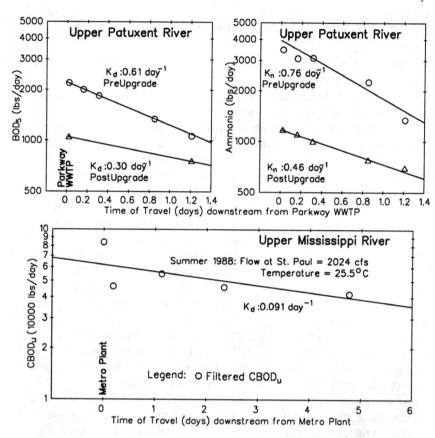

Fig. 2.26 Derivation of in-stream BOD Deoxygenation coefficients In upper Patuxent River and upper Mississippi River

would make the above simple procedure non-applicable and would thereby require more a robust approach to quantify the deoxygenation coefficient (Chap. 3).

2.9 Nonpoint Source Loads

Pollutant loading rates play a key role in the fate and transport modeling of receiving waters. The top three nonpoint loading rates to the upper Mississippi River (see the study area in Fig. 2.10) are shown in Fig. 2.27: upstream at Lock & Dam No.1, the Minnesota River, and the St. Croix River for both 1988 (a low flow year) and 1991 (a very high flow year) conditions. Seasonal variations of flow, total P, and nitrate loads are displayed. Having a sizable agricultural watershed, the Minnesota River tops the other two sources in providing major nutrient (particularly nitrate) loads to the upper Mississippi River in the spring and summer months of 1991. Under very high flows,

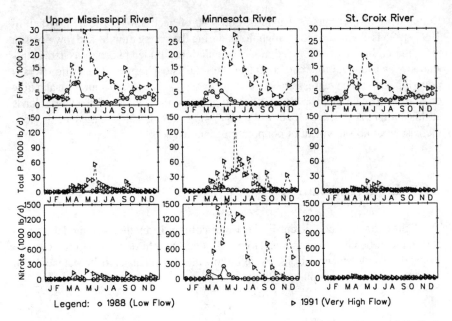

Fig. 2.27 Flows and Nutrient loads in upper Mississippi River, Minnesota River, and St. Croix River in low and very high flow years

the Minnesota River nutrient loads are dominating the system, easily offsetting the benefits from phosphorus removal at the Metro Plant. A data analysis of this nature provides a warning to the management that point source nutrient control at the Metro Plant alone will not be sufficient to curb eutrophication in Lake Pepin. Additional physical insights from the modeling analysis are available in Chap. 4.

While watershed models are used to generate nonpoint loads to modeling studies, existing flow and water quality data can be used to develop loads. Figure 2.28 shows daily stream flows at a USGS gaging station (Patuxent River at Unity) located in the Patuxent River watershed in Maryland providing flows to the Triadelphia and

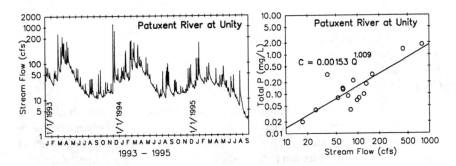

Fig. 2.28 Deriving Phosphorus loads in the Patuxent River watershed

Rocky Gorge Reservoirs (see map in Fig. 4.40 of Sect. 4.8.3). The hydrograph of 1993–1995 is characterized by high flows during the spring months and low flows in the summer. In lieu of watershed model calculations, limited phosphorus data at this station was available to create a plot of the total phosphorus concentration versus flow (Fig. 2.28). The regression between the total phosphorus concentration and flow on a log–log scale yields a straight line to generate the nonpoint phosphorus loads at this station. A positive slope of the regression line indicates that concentrations increase with flows, typical of nonpoint nutrient loads.

2.10   Light Attenuation

The light extinction coefficient in the water column is another key model parameter that can be derived from available light data, Secchi depth measurements, or suspended solids concentrations. By definition, the light intensity in the water column is related to the water depth as follows:

$$I_z = I_o e^{-K_e z} \qquad (2.4)$$

where I_z is light intensity at depth z and I_o is surface light intensity. A plot of natural log (i.e. base e) of $\frac{I_z}{I_o}$ versus z yields a slope as the light extinction coefficient, K_e. A whole year's measurements of light intensity in the water column of White Lake, Michigan from May 1973 to February 1974 are presented in Fig. 2.29, showing exponential decrease of light intensity with depth. On the same plot, $LOG(I_z/I_o)$ versus z showing a straight line, whose slope is K_e (Fig. 2.29), ranging from 0.844 m^{-1} to 1.395 m^{-1}, typical values in many natural waters.

   In lieu of light measurements, K_e can be derived from the following relationship by Di Toro (1978):

$$K_e = 0.052N + 0.174D + 0.031Chla \qquad (2.5)$$

where

K_e	light extinction coefficient (m^{-1})
N	non-living suspended solids (mg/L)
D	detritus (non-living organic suspended solids) (mg/L)
$Chla$	chlorophyll a concentration (μg/L)

   The nonvolatile suspended solids (the fixed inorganic particulate) both absorb and scatter light, whereas the organic detritus and phytoplankton biomass mainly absorb light. The last term in Eq. 2.5 is referred to as the algal self-shading effect. First, the detritus concentration is derived by subtracting the dry weight of the algal biomass from the volatile suspended solids. As an approximation, the ratio of algal biomass dry weight to algal chlorophyll a is 100:1. That is, a 100 μg/L chlorophyll

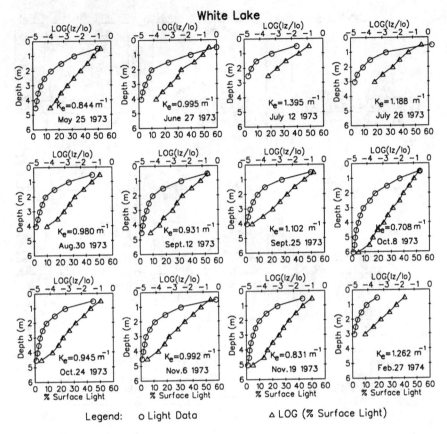

Fig. 2.29 Light data and light extinction coefficient in White Lake, Michigan

a concentration is equivalent to 10,000 μg/L, or 10 mg/L, of suspended solids. With this conversion, one can calculate the detritus concentration for use in Eq. 2.5. The calculated K_e values based on the first two terms of Eq. 2.5 for the upper Mississippi River at three locations are shown in Fig. 2.30. The light extinction coefficients were derived from suspended solids data for the upper Mississippi River for 1988, 1990, and 1991 at three locations: UM839.1, UM815.6, and UM796.1 in the upstream to downstream direction. These coefficients and the total suspended solids concentrations for these three years are also presented in Fig. 2.30. The temporal pattern of K_e follows that of total suspended solids closely. The effect of river flow on K_e is shown. Note that 1988 was a dry year with very low flows in this stretch of the Mississippi River followed by a wet year in 1990 and a very wet year in 1991. Light extinction coefficients are very high in these two high flow years, caused by high levels of suspended solids due to the high river flows. Both the scales of total suspended solids and light extinction coefficients are the same for the three locations during these three years, offering a good perspective into the flow related

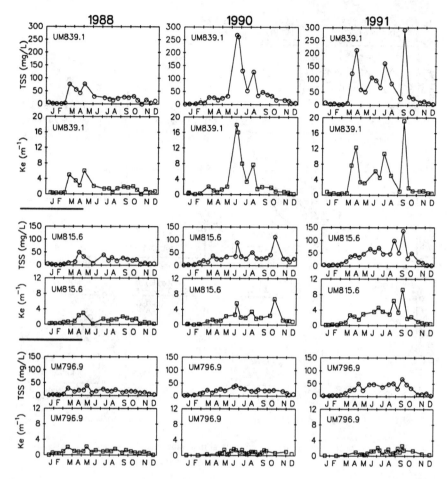

Fig. 2.30 TSS concentrations and light extinction coefficients: upper Mississippi River, 1988–1991

K_e values. The effect of the river flow on the total suspended solids concentration is very clear. High levels of total suspended solids during high flow years 1990 and 1991 significantly raised the turbidity in the water column of the upper Mississippi River when compared with the 1988 condition. [A total suspended solids concentration of 300 mg/L can push the light extinction coefficient up to 20 m^{-1}, over 14 times the K_e in White Lake!] Such a high level of light extinction is progressively reduced in the downstream direction from UM839.1 to UM796.9 toward Lake Pepin as solids continue to settle in the water column. Figure 2.31 shows light extinction coefficients in Lake Pepin at UM786 and UM764 in 1990 and 1991 (the high flow years).

Light extinction coefficients can also be quantified from Secchi depth data using the approximation: $K_e = C/H$, where C is the regression constant and H is the Secchi depth. C has a range from 1.3 to 5 (Homes 1970) with a typical value of 1.7. The Secchi depth data from Platte Lake, located on the west coast of the lower

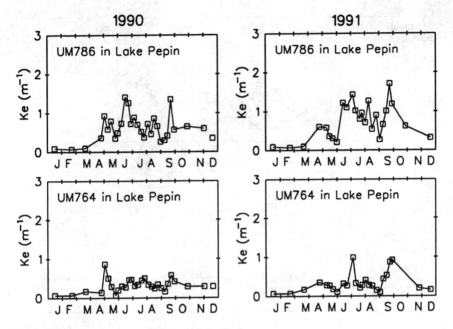

Fig. 2.31 Light extinction coefficients in Lake Pepin

peninsula of Michigan, during the period of May to September from 1980 to 1999 are plotted in Fig. 2.32. The obvious water quality problem in the lake was the reduced transparencies encountered during the summer months (May/June). Some scientists believed that the reduced transparency was the result of marl (calcium carbonate) precipitate. A seasonal low of Secchi depth occurs in June every year, resulting in high K_e values. Note that Secchi depths started to increase in 1984 with some high values reaching 7 m in the 1990s. The K_e values are calculated as $1.7/H$. The reduced transparency was the result of marl (calcium carbonate) precipitate. Canale et al. (2003) reported seasonal variation of the Secchi depth in Platte Lake in 2000 ranging from 1 to 5 m with the minimum value measured in June.

In addition to low transparencies in the months of May and June, the overall Secchi depths in 1980 to 1983 were low, with maximum K_e reaching $2 \, m^{-1}$. While the Secchi depths increase sharply during the period of May to September, low transparencies in May/June has been a persistent problem. For example, K_e exceeded $2 \, m^{-1}$ in June 1995, indicating the lowest transparency ever. Another water quality problem in Platte Lake is low dissolved oxygen in the hypolimnion during the summer months, a eutrophication problem to be addressed in Chap. 4.

Settling velocity of suspended solids is another key parameter that can be quantified with available data prior to running the water quality model. The relationship between the settling velocity and the solids particle distribution is based on Stoke's law:

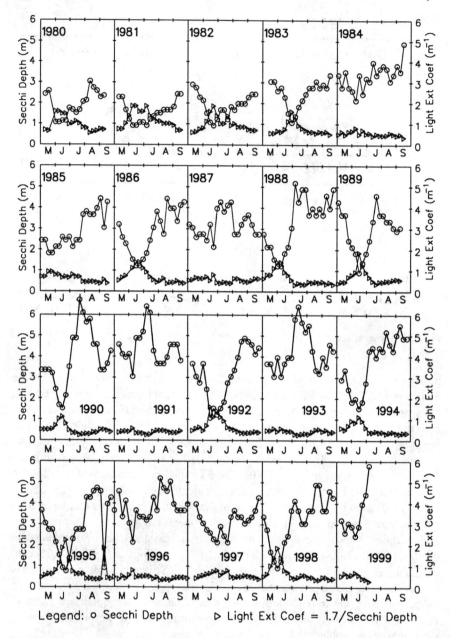

Fig. 2.32 Secchi depths and light extinction coefficients in Platte Lake (1980–1999)

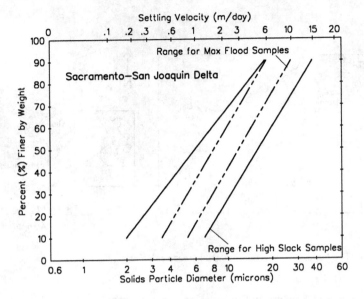

Fig. 2.33 Particle size and settling velocity distribution in Sacramento-San Joaquin Delta

$$v_s = \frac{\rho_s - \rho}{18\mu} d^2 g \tag{2.6}$$

where v_s = settling velocity; ρ_s and ρ = densities of particles and water, respectively; μ = molecular viscosity of water; d = particle diameter; and g = gravitational acceleration. For the Sacramento-San Joaquin Delta, the following empirical relationship can be derived (Lung and O'Connor 1984):

$$\rho_s = 2.06d^{-0.14428} \tag{2.7}$$

in which ρ_s is in gm/cm^3 and d in μm (microns). Combining Eqs. 2.6 and 2.7, the settling velocity for any particle size can be calculated. Figure 2.33 shows the distribution of settling velocity by weight, which is associated with an observed size distribution from the Sacramento-San Joaquin Delta in California. The mean settling velocity is about 2.45 m/day (= 8 ft/day) in the water system and this rate is constant down the length of the estuary. This settling velocity is used to model the turbidity maxima in this system as presented in Chap. 4.

2.11 High COD and BOD Loads

Delhi's Yamuna River (Fig. 2.34) in India, the largest tributary to River Ganges, is portrayed in Indian legend as a natural paradise of lilies, turtles, and fish enjoyed by

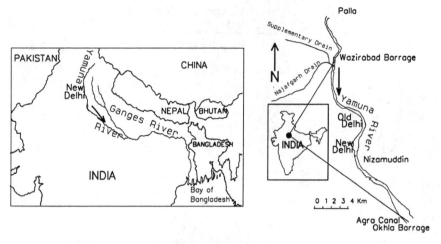

Fig. 2.34 The Yamuna River, India and study reach at Delhi

the flute-playing Krishna and his adoring gopis. However, almost all its waters are now diverted above the capital for irrigation. In the dry season before the monsoon, what flows from greater Delhi's 25 million inhabitants towards Agra and the Taj Mahal is a huge mixture of sewage and industrial waste, foaming hideously into windblown heaps of spume as it passes the Okhla barrage. The crisis afflicting their emblematic river has not escaped Indians themselves. Narendra Modi, the Hindu nationalist leader who swept to power in the 2014 election, promised a cleanup (Financial Times, February 13 2015).

Under the Global Study Program at UVa, a study team was formed in 2015 to investigate the water quality of the Yamuna. In July 2016, UVa and the Delhi Jal Board (Delhi Water Board, or DJB) signed a memorandum of understanding (MOU) to launch a study for the Yamuna. This MOU detailed the modalities and general conditions regarding the collaboration between UVa and DJB towards developing viable solutions in wastewater treatment, cleaning and development of natural water streams (i.e. water bodies and natural storm water drains), and related issues through various activities/studies identified jointly. UVa would share its knowledge, research, and field planning and design with a goal to enhance the overall environmental coordination. The water quality team of the study focused on the following questions:

1. What is the cause-and-effect relationships between the pollutant loads and the response in the river under the present conditions?
2. What modeling effort is needed to develop a water quality management plan?

The study area is the 25-km stretch of the river between Wazirabad Barrage and Agral Canal/Okhla Barrage, receiving pollutant loads from Old Delhi and New Delhi on the west bank of the river (Fig. 2.34). The first task was analyzing the existing water quality data provided by India's Central Pollution Control Board (CPCB) to understand the water system. Water quality data of fecal and total coliform bacteria, conductivity, COD, BOD$_5$, ammonia, and dissolved oxygen from Palla to Okhla Barrage in 2017 are summarized in Fig. 2.35. The river water quality is in great shape upstream of Palla prior to water diversion at the Wazirabad Barrage.

Coliform levels are extremely high in the 25-km Delhi stretch starting immediately below Wazirabad Barrage with maximum counts reaching 1.1×10^8 and 1.4×10^8 for fecal coliforms and total coliforms, respectively, easily exceeding their water quality criteria set by CPCB at 2500 MPN/100 ml and 5000 MPN/100 mL (Fig. 2.35). Domestic and industrial wastewaters from the two main drains, Najafgarh and Supplementary, enter the river below Wazirabad in addition to sizable wastewater inputs draining Old Delhi and New Delhi on the west bank of the Yamuna help to create this condition. Water diversion all but eliminates the dilution capability in this stretch of the river. Another indicator of an overloaded river system is the electrical conductivity level, over 2000 μmho/cm. For comparison, typical conductivity levels from domestic wastewater treatment plant effluents in U.S. are in the range between 500 to 700 μmho/cm. Note that the annual average conductivity at upstream Palla reaches 580 μmho/cm. Both COD and BOD$_5$ concentrations are relatively low at Palla but increase sharply by Nizamuddin with maximum levels reaching 144 mg/L and 80 mg/L, respectively. The 3 mg/L criteria for BOD$_5$ is violated throughout the 25-km stretch of the Yamuna. Ammonia levels are quite high, even above the

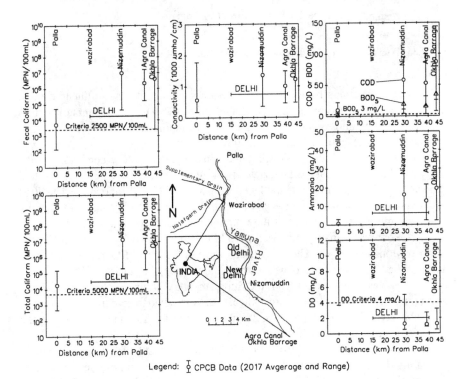

Fig. 2.35 Water quality of the Yamuna River near Delhi, India 2017

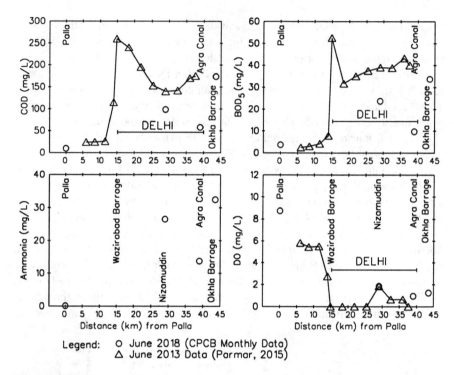

Legend: o June 2018 (CPCB Monthly Data)
          △ June 2013 Data (Parmar, 2015)

Fig. 2.36 Data to support BOD/DO modeling of the Yamuna River

effluent concentrations from a typical secondary treatment plant effluent without nitrification. The background DO concentrations at Palla are reasonable but drop sharply to minimum levels below 1 mg/L at Nizamuddin (the water quality criteria for DO is 5 mg/L), approaching an anaerobic system.

While the data in Fig. 2.35 display the general water quality conditions in the study reach, specific data was reviewed for the modeling analysis. Figure 2.36 shows COD, BOD_5, ammonia, and DO data in June 2013 and June 2018. The 2013 data have additional sampling points between Wazirabad and Nizamuddin, thereby offering a more precise picture of the water quality condition in this sub-section of the river. Poor water quality are observed in May and June, the dry months before the summer Monsoon season. The instant, sharp rise of COD and BOD_5 concentrations at Wazirabad Barrage is due to the substantial input from the Najargarh Drain, which in turn receives sizable input from the Supplementary Drain. Such powerful BOD loads (plus very low DO levels) drove the DO at Wazirabad to zero immediately. The anaerobic condition lasted the entire 25-km section of the Yamuna, representing the worst-case situation for the Yamuna.

In addition to high levels of BOD in the Yamuna, the COD levels are quite high, indicating significant amounts of carbon not biologically decomposable. Figure 2.36 presents the BOD_5 and COD data during the periods of 2013–2015 and 2017–2018 obtained from the CPCB. At Pella, which is located upstream of the Delhi section of the river, both BOD_5 and COD data are low, indicating a minimum degree of water pollution with organic and inorganic carbon in the river. The Najafgarh drain, with an approximate average flow of 25 cms entering the Yamuna River immediately downstream of Wazirabad Barrage (about 15 km below Pella), has COD and BOD values much higher than the Yamuna. Its BOD_5 levels are well above 30 mg/L while the COD levels could reach 300 mg/L. The Najafgarh drain provides the single largest BOD/COD loads into the Yamuna in the Delhi area. At Nizamuddin, which is located approximately 15 km below the entering of the Najafgarh drain, the BOD_5 and COD levels are somewhat reduced from those of the Najafgarh drain, suggesting certain degrees of attenuation in the river stretch between the Najafgarh Drain and Nizamuddin. Correlation of COD versus BOD_5 indicates a COD to BOD_5 ratio of 3.0, meaning that two thirds of the COD are inorganic carbon and ammonia. While the ammonia concentrations in the Yamuna are high (significantly over 10 mg/L), the BOD_5 measurements would not include nitrification as low DO concentrations in the water inhibits the nitrification process. [Nitrification inhibition is germane to these wastewater drains only and should not be applied to other systems.] Assuming a $CBOD_u$ deoxygenation rate of 0.22 day^{-1} for the Yamuna River, we can obtain a $CBOD_u$ to $CBOD_5$ (or BOD_5) ratio at 1.5, leaving the remaining portion (approximately 50%) of the total COD due to inorganics, primarily from non-domestic wastewater input.

In addition to the Najafgarh Drain, there are five major urban drains with average flows above one cms entering the Yamuna in the Delhi area. In the order of downstream direction along the Yamuna River the drains include: the Barapulla, Tugkakabod, Shahdara, Old Agra Canal at Okhla, and Old Agra Canal near Kalindi Kunj. Figure 2.37 shows COD versus BOD_5 plots of these drains with each of the flows indicated for a comparison with the Najafgarh. The Old Agra Canal at Okhla shows

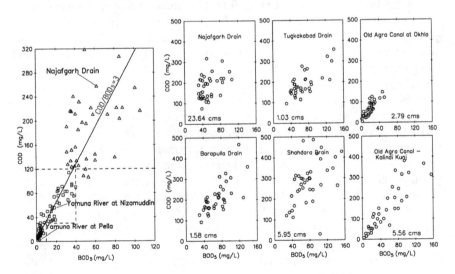

Fig. 2.37 COD and BOD$_5$ in the Yamuna River and Major Drains in Delhi Area

Fig. 2.38 BOD_u to BOD_5
ratio in Yamuna River

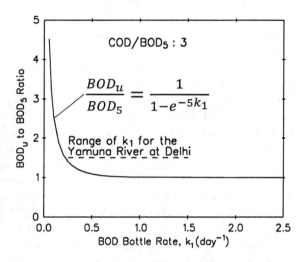

the lowest COD concentration among these major drains. Again, the $CBOD_u$ to
$CBOD_5$ ratio of 3.0 still holds for these drains. Figure 2.38 summarizes the analysis
of COD and BOD_5 of the Yamuna River. With a COD to BOD_5 ratio at 3.0, the
maximum BOD_u to BOD_5 ratio is approximately 1.5, which is associated with a
minimum BOD bottle rate, k_1 at 0.22 day^{-1} as indicated in Fig. 2.38 A higher k_1
rate closer to 2.0 day^{-1}, approaching a $CBOD_u$ to $CBOD_5$ ratio of 1.0 is possible,
implying that the organic carbon is highly labile, subject to very rapid decomposition by bacteria. Such high k_1 rates, associated with untreated wastewaters from
the Najafgarh Drain and other tributary drains, imply rapid assimilation of organic
carbon with significant deoxygenation in the Yamuna River. Thus, k_1 ranging from
0.22 to 2.0 day^{-1} is a reasonable range for the Yamuna River at the present time.

A key implication of the high loads of oxygen consuming input is their deoxygenation rate, k_d in the receiving water. Unfiltered samples for COD and BOD$_5$ tests
from the Yamuna River contain carbon sorbed onto suspended particles, thereby
ruling out using the analysis in Fig. 2.26 to derive the filtered CBOD in-stream k_d.
On the other hand, the BOD$_5$ results would not include ammonia, as nitrification
does not take place in the first five days of the test due to fact that the river receives
mostly untreated and partially treated wastewaters. In addition, the low DO levels
could not provide the oxygen needed to oxidize ammonia in the river. The increasing
ammonia levels in the 25-km study stretch (see map in Fig. 2.34) substantiate this
observation. The receiving water DO budget can therefore focus on CBOD deoxygenation only for the present time. It must be pointed out that improvement of the
DO levels in the Yamuna resulting from BOD load reductions via future management actions would improve the DO levels, thereby speeding up the nitrification
process in the receiving water, perhaps diluting the water quality benefits from lower
organic carbon loads in wastewater treatment. This highly possible outcome renders
a challenge to the modeling analysis in making water quality projections for future

conditions. As modelers would not have any prior knowledge on when and where this timing shift of nitrification could happen—a great uncertainty!

Another challenge the study is facing now is the derivation of CBOD deoxygenation rate in the Yamuna. As stated earlier, the simple technique in Fig. 2.26 would not work for the Yamuna. There have been many BOD/DO modeling studies of this river reported in the literature in the past five decades, but no workable solutions have been offered to this technical issue (Bhargava 1983, 1986, 2006; Paliwal et al. 2007; Sharma and Singh 2009; Mandal et al. 2010; Tyagi et al. 2014; Sharma et al. 2017). Bhargava (1983) gave a wide range from 1.5 to 5.5 day^{-1}, considered much too high for the Yamuna under the present conditions. More discussions on this CBOD deoxygenation rate can be found in Chap. 3 when the modeling analysis for the Yamuna is presented.

The DO concentration profile in Fig. 2.36 shows that the Yamuna River goes anaerobic for about 15 km downstream from Wazirabad. Gundelach and Castillo (1976) pointed out that when the DO levels reach zero in the water column for a stretch of a river, the only source of oxygen to break down carbon is from the atmosphere, thereby slowing down the decay of BOD and resulting in a lower oxygen consumption rate. Not until the water column oxygen levels recover in the downstream direction does the normal deoxygenation rate resume. The key of the analysis is to determine the deoxygenation rate in the stretch of the zero oxygen levels in the water column. Thomann and Mueller (1987) presented a quantitative relationship as the basis for this mechanism: $k_d L = k_a D$, where k_d is BOD deoxygenation rate (day^{-1}), L is BOD in mg/L, k_a is reaeration rate (day^{-1}), and D is oxygen deficit in mg/L. The deoxygenation rate will be reduced in this zero DO reach and will increase following the DO recovery. Many water quality models fail to recognize this adjustment, leading to incorrect BOD and DO concentrations in the water column. For a completion presentation of this technical issue, consult Thomann and Meuller (1987) and Chapra (1997). It is important that the water quality-modeling framework used for anaerobic rivers have this adjustment built in.

A 2020 article summarized the water quality of the Yamuna River. CPCB stated, "Based on the data, it can be interpreted that the water quality of Yamuna in Delhi stretch continues in poor state with respect to BOD during 2015–2020 (The Times of India, December 13, 2020)." High fecal levels contribute to the deteriorated water quality, indicating high fecal contamination. The CPCB report continues, "Downstream Palla, the quality degenerates. Depletion of DO level in Yamuna till Okhla was due to the demand for DO from organic matter of discharge of untreated sewage or other effluents into water." Interestingly, the sudden drop in BOD concentrations from 57 mg/L in March of this year to 5.6 mg/L in April at Nizamuddin are attributed to the Covid-19 lockdown, which led to shutdown of industrial activities.

2.11.1 Data to Support Modeling Analysis

Data needed for water quality modeling covers a wide range of information depending on the study objective, management needs, etc. For example, in a modeling study of the Washington Shipping Channel and Tidal Basin (see Sect. 6.4), the data gathering effort included the following:

1. Map showing the watershed and receiving water of the study area
2. Hydrographic/bathymetry data of tidal basin and shipping channel
3. Meteorological data
4. Freshwater inflow to the tidal basin
5. Operation data of tidal basin control gate(s)
6. Stage records of the tidal basin
7. Velocity measurements in the tidal basin and shipping channel
8. Water quality sampling station map
9. Sediment sampling station map
10. Water quality data in the tidal basin and shipping channel

    a. Conventional pollutants: pH, conductivity, alkalinity, dissolved oxygen, temperature, dissolved organic carbon, and salinity
    b. Nutrients (phosphorus and nitrogen compounds)
    c. Fecal coliform data
    d. Metals (total, dissolved, and sorbed)
    e. PAHs
    f. PCBs (total and congeners, dissolved oxygen, and sorbed)

11. Watershed data (any results from watershed modeling or available data)
12. Previous water quality studies of the tidal basin and shipping channel
13. Point source discharges, if any, into the tidal basin and shipping channel
14. Time varying water surface elevation and water quality data at the upper boundary.

Throughout this book, the importance of data to water quality modeling is emphasized. There are several categories of water quality data depending on the origination. While SOTRET is the largest database compiling water quality of natural waters in U.S., it usually lacks the spatial and temporal details for specific water quality modeling studies. A second kind of database is developed from field monitoring programs to detect water quality improvement following management actions. A third kind is also a special sampling effort to gather field data to investigate future water quality management measures. The last two types are well designed to support specific water quality modeling efforts.

The Chesapeake Bay database (http://data.chesapeakebay.net/WaterQuality) is a unique and complete database compiling water quality data and information for the Chesapeake Bay, maintained by the EPA Chesapeake Bay Program. Figure 2.39 shows the Chesapeake Bay with its over 100 surface water quality monitoring stations

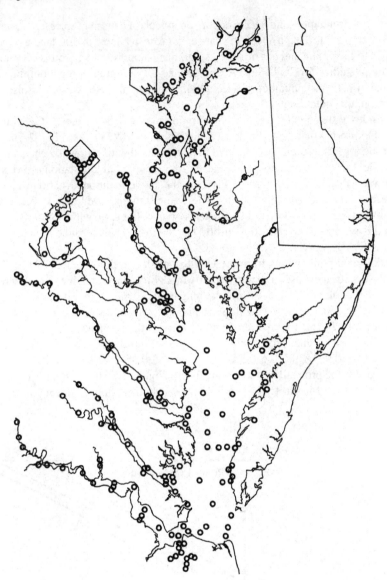

Fig. 2.39 Chesapeake Bay Tidal water monitoring stations

in the tidal water of the Bay and its major tributaries: the Patuxent, Potomac, Rappahannock, York, and James Rivers on the west bank and the Chester, Choptank,
Nanticoke, Wicomico, and Pocomoke Rivers on the eastern shore of the Bay. These
stations are visited at least once a month every year. In addition, vertical samples
are collected if warranted. Each data entry has at least 33 water quality parameters
ranging from physical/chemical (conductivity, salinity, temperature, turbidity, TSS,

VSS, light attenuation, and Secchi depth), chemical (pH, nutrient spices), and biological (phytoplankton) data. Over 30 years of data are available in this database, which is open to the public and has provided significant support to numerous water quality modeling studies for the U.S. EPA Chesapeake Bay Program throughout the years. Chapter 4 presents a number of water quality studies of the Bay estuaries using the data from this database.

Besides water quality data, another important database is the USGS flow and stage monitoring data (e.g. the flows at gaging station 02292900 in Fig. 2.7). These permanent gaging stations are located in rivers with the most downstream sampling at the fall line of tidal waters and estuaries. Flows at the fall line stations serve as the freshwater flow rates to estuaries. Flows at the above-mentioned station 02292900 represent the freshwater input to the CRE. Some stations have limited water quality data from time to time. Local state agencies often collaborate with USGS to proceed with this massive data collection effort. The database is open to the public for free information download.

Additional publications from the USGS include the development of low flows, which are extremely useful to water quality modeling analyses. For example, a report by Austin et al. (2011) documents the magnitude, frequency, and duration of low flows in Virginia streams. Low flow frequency analysis was used to determine the annual 1-, 4-, 7-, and 30-day average minimum low flow climate year time series and annual non-exceedance probabilities for long-term, continuous-record stream gaging stations. The 0.9, 0.8, 0.7, 0.6, 0.5, 0.4, 0.3, 0.2, 0.1, 0.05, 0.02, 0.01, and 0.005 annual non-exceedance probabilities were calculated. Note that a climate year is defined as April 1 through March 31. Figure 2.40 shows the low flow frequency curves for

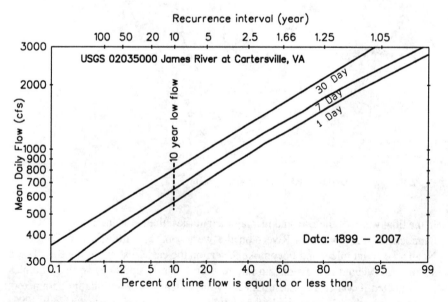

Fig. 2.40 Low flow frequency curves of James River at Cartersville, VA

1-day, 7-day, and 30-day low flows at USGS gaging station 02035000, which were derived from the analysis results by Austin et al. (2011) using the flow data from 108 climate years. The 7-day 10-year low flow at this station is 665 cfs. Other scientific and technical reports in electronic form are also available from USGS, a valuable asset providing support for water quality modeling.

References

Austin S et al (2011) Low-flow characteristics of Virginia streams. In: USGS scientific investigations report 2011–5143 prepared in cooperation with the Virginia department of environmental Quality, Richmond, VA

Bhargava DS (1983) Most rapid BOD assimilation in Ganga and Yamuna Rivers. J Environ Eng 109(1):174–188

Bhargava DS (1986) DO sag model for extremely fast river purification. J Environ Eng 112(3):572–585

Bhargava DS (2006) Revival of Mathura's ailing Yamuna River. Environmentalist 26:111–122

Blank K et al (2017) The ecological state of Lake Peipsi (Estonia/Russia): improvement, stabilization or deterioration. Proc Est Acad Sci 66(1):18–28

Brown LC, Barnwell TO (1985) Computer program documentation for the enhanced stream water quality model QUAL2E. U.S. EPA

Buhvestova et al (2011) Nitrogen and phosphorus in Estonian rivers discharging into Lake Peipsi: estimation of loads and seasonal and spatial distribution of concentrations. Estonian J Ecology 60(1):18–38. https://doi.org/10.3176/eco.2011.1.03

Buskey EJ et al (2017) Minimum flow criteria for the Caloosahatchee River Estuary. Final Peer Review Report submitted to South Florida Water Management District, 15p

Canale RP et al (2003) Case study: reduction of total phosphorus loads to big Platte Lake, MI through point source control and watershed management

Chapra SC (1997) Surface water-quality modeling. McGraw Hill, NY

Cvitanič I et al (2002) Water quality modeling of the Vrhovo Reservoir with CE-QUAL-W2. In: NATO/CCMS project report modeling nutrient loads and response in river and estuary systems

Dillon PJ, Rigler FH (1974) The phosphorus-chlorophyll relationship in lakes. Limnol Oceanogr 19(5):767–773

Di Toro DM (1978) Optics of turbid estuarine waters: approximations and applications. WR 12:1059–1068

Dyer KR (1997) Estuaries a physical introduction. Wiley, England

Edwards RE et al (2000) Final review report, Caloosahatchee minimum flow peer review panel. Report submitted to the South Florida Water Management District, West Palm Beach, FL

Gundelach JM, Castillo JE (1976) Natural stream purification under anaerobic conditions. J Water Pollut Control Fed 48(7):1753–1758

Homes RE (1970) The Secchi disk in turbid coastal waters. Limnol Oceanogr 15:668–694

Jassby A, Powell T (1975) Vertical patterns of eddy diffusion during stratification in Castle Lake, California. Limnol Oceanogr 20:530–543

Kangur K et al (2003) Phytoplankton response to changed nutrient level in Lake Peipsi (Estonia) in 1992–2001 Hydrobiologia 506–509:265–272

Kangur K, Mols T (2008) Changes in spatial distribution of phosphorus and nitrogen in the large north-temperate lowland Lake Peipsi (Estonia/Russia). Hydrobiologia 599:31–39

Lung WS (2001) Water quality modeling for wasteload allocations and TMDLs. John Wiley & Sons, NY

Lung WS, O'Connor DJ (1984) Two-dimensional mass transport in estuaries. J Hyd Eng 110(10):1340–1357

Mandal P et al (2010) Seasonal and spatial variation of Yamuna River water quality in Delhi, India. Environ Monit Assess 170:661–670

Nice AJ (2006) Developing a fate and transport model for arsenic in estuaries. Ph.D. dissertation, University of Virginia

Noges T et al (2004) The impact of changes in nutrient loading on cyanobacterial dominance in Lake Peipsi (Estonia/Russia). Arch Hydrobiol 160(2):261–279

O'Connor DJ, Lung WS (1981) Suspended solids analysis of estuarine systems. J Environ Eng 107(1):101–120

Paliwal R et al (2007) Water quality modelling of the river Yamuna (India) using QUAL2E-UNVCAS. J Environ Manage 83:131–144

Prichard DW (1969) Dispersion and flushing of pollutants in estuaries. J Hyd Eng 95(1):115–124

Reidel et al (2000) Temporal and spatial patterns of trace elements in the Patuxent River: a whole watershed approach. Estuaries 23:521–535

Sanders JG, Riedel GF (1993) Trace elements transformation during the development of an estuarine algal bloom. Estuaries 16(3A)

Sharma D, Singh RK (2009) DO-BOD modeling of River Yamuna for national capital territory, India using STREAM II, a 2D water quality model. Environ Monit Assess 159:231–240

Sharma D et al (2017) Water quality modeling for urban reach of Yamuna River, India (1999–2009), using QUAL2Kw. Appl Water Sci 7:1535–1559

Thomann RV (1977) Comparison of lake phytoplankton models with loading plots. Limnol Oceanog 22(2):370–373

Thomann RV, Mueller JA (1987) Principles of water quality modeling and control. Harper & Row, NY

Thurston et al (1979) Aqueous ammonia equilibrium-tabulation of percent un-ionized ammonia. EPA-600/3-79-091, environmental research laboratory, Duluth, MN

Tyagi B et al (2014) Analysis of DO sag for multiple point sources. Math Theory Modeling 4(3):1–7

Vollenweider RA (1968) Scientific fundamentals of the eutrophication of lakes and flowing waters, with particular references to nitrogen and phosphorus as factors in eutrophication. In: Organization for economic co-operation and development, directorate for scientific affairs, Paris, France, 159p

Wright RM et al (1998) Blackstone River initiative: water quality analysis of the Blackstone River under wet and dry weather conditions. Department of Civil and Environmental Engineering, University of Rhode Island, Kingston, RI

Chapter 3
River BOD/DO Modeling at New Normal

When considering BOD/DO modeling, we must ask the following: Are river BOD/DO problems solved? More specifically, is the point source-related BOD/DO problem eliminated? The answer is "Yes", at least for most of the streams and rivers in the U.S. and many developed countries. Vice President Gore's remarks in October 1997 on celebrating the 25th anniversary of the Clean Water Act (CWA) signaled that we had successfully mitigated this problem in the U.S. The cleanup of the upper Mississippi River in the Twin City area of St. Paul/Minneapolis is one of the best examples demonstrating the success of point source BOD control (see the data analysis in Chap. 2 and the modeling analysis later in this chapter). With the centennial of the 1925 BOD/DO modeling of the Ohio River using the Streeter-Phelps equation approaching,.the BOD/DO issue continues to evolve and presents new modeling challenges.

The technology-based control strategy to upgrade treatment plants and/or to build new treatment plants in the U.S. in the 1970s is the first step to address the point source related BOD/DO problem. When the technology-based control is not sufficient to eradicate the low DO problem, then modeling is needed to develop water quality-based effluent limits (WQBELs) for inputs along the river. Water quality standards are then translated into WQBELs for effluent permits (called National Pollutant Discharge Elimination System [NPDES] permits in the U.S.) for the dischargers. Water quality modeling is the recognized approach of translating water quality standards into effluent limitations.

Compared with the BOD/DO modeling in the U.S. during the 1970s (NACASI 1982), modelers now face new issues in low DO streams and rivers. Single point source discharges have been replaced by multiple BOD sources with varying degrees of oxygen-consuming potential along the river, rendering the traditional DO sag curve for a single discharge invalid. Single DO sags no longer exist. Varying strengths of BOD inputs have created technical issues in BOD-DO modeling for the twenty-first century, requiring a new approach and additional data support for the modeling analysis.

© The Author(s), under exclusive license to Springer Nature Switzerland AG 2022
W.-S. Lung, *Water Quality Modeling That Works*,
https://doi.org/10.1007/978-3-030-90483-8_3

3.1 BOD in Wastewater

BOD is not a substance which physically exists. Instead, it is a surrogate measuring the amount of oxygen (in mg/L) consumed by bacteria to break down organic carbon (and ammonia) in the water. Historically, BOD is reported as a 5-day value (BOD_5) and originated from the water pollution control study of the Thames River, United Kingdom in early years of the twentieth century. Wastewater (called sewage in those days) from the city of London took about 5 days to reach the North Sea, thereby out of concern at that time. The Royal Commission on Sewage Disposal recommended and adopted the BOD_5 test in 1908 with the understanding that organic carbon is completely broken down by the end of the 5 days of the test. In the U.S., regulatory agencies included BOD_5 as one of the key water quality constituents in NPDES permits in 1970s. Although it is convenient to use BOD_5 in discharge permits, modeling BOD_5 in receiving waters for permitting is becoming more challenging as discussed in the following paragraphs.

First, let's review the two underlying assumptions in the original BOD_5 concept:

1. The organic carbon is completely oxidized by the end of 5 days from the beginning of the test, i.e. $BOD_5 = BOD_u$ (ultimate BOD).
2. Nitrification does not occur during the BOD_5 test.

These assumptions were more or less valid for raw sewage or marginally treated wastewater prior to 1970s. In significantly polluted waters with very low dissolved oxygen levels such as raw sewage, the amount of dissolved oxygen consumed in 5 days is primarily from the deoxygenation of carbonaceous BOD (CBOD), as nitrification does not take place within the first 5 days. Therefore, there was little difference between BOD_5 and $CBOD_5$ in untreated wastewaters and highly polluted waters. However, this assumption has been violated many times during the past three decades. Water pollution control to improve the DO levels in the receiving water has consequently accelerated the nitrification start time, i.e. taking place prior to 5 days, resulting in $CBOD_5$ less than BOD_5. During the implementation of the CWA in the U.S., the measured effluent BOD_5 following treatment upgrade was found to be greater than the pre-upgrade BOD_5 levels in certain cases. Hall and Foxen (1983) looked into this problem and pointed out that favorable conditions occur in the effluent and provide nitrification an early start, i.e., prior to 5 days. The BOD_5 values of the well treated wastewater actually consisted of the amount of oxygen consumed by nitrifying bacteria in addition to that used to oxidize carbon. It was therefore necessary to split $CBOD_5$ from nitrification. The U.S. EPA (1984) was quick to act by modifying the secondary treatment effluent limits to either 30 mg/L BOD_5 or 25 mg/L $CBOD_5$. However, this stopgap measure was not perfect until both organic carbon and ammonia were separately accounted for and specified in the discharge permits.

Effluent permits for almost all domestic WWTPs in the U.S. are now written in terms of $CBOD_5$. For example, the key numeric limits of BOD for the Proctors Creek domestic WWTP, an up-to-date treatment facility in the suburb of Richmond, VA are written into their discharge permit. Their 2020 draft permit calls for seasonal $CBOD_5$ limits of 7 mg/L and 11 mg/L (monthly average) in summer and winter, respectively. Seasonal ammonia limits are also included in the permit as a result of nitrification at the WWTP. Note that these $CBOD_5$ limits are much lower than 25 mg/L of the 1980's, thanks to the CWA's anti-backsliding policy. The shift from BOD_5 to $CBOD_5$ also revolutionized the BOD tests (5-day or long-term) in Standard Methods.

The approach to receiving water BOD/DO modeling have also changed in response to the above actions:

1. CBOD and ammonia must be accounted for separately in the modeling analysis.
2. $CBOD_u$ (the ultimate amount of oxygen to stabilize the organic carbon) instead of $CBOD_5$ should be quantified in the model.

It may take 15 days to completely break down the organic carbon in a wastewater sample and 40 days in another, depending on the strength of the samples. Thus, the $CBOD_5$ of these two different effluents cannot be compared equally. Quantifying $CBOD_u$ instead would therefore put different effluents on a leveled playing field, a proper way to compare the ultimate strength of all the wastewaters.

The next step is converting the $CBOD_5$ loading rate to $CBOD_u$ loading rate of wastewater for the receiving water modeling analysis. The $CBOD_u$ to $CBOD_5$ ratio can be obtained from the following relationship given the k_1 (lab CBOD deoxygenation rate) value:

$$\frac{CBOD_u}{CBOD_5} = \frac{1}{1 - e^{-5k_1}} \tag{3.1}$$

Note that k_1 in Eq. 3.1 is referred to as base e coefficient, whereas older textbooks use base 10. [Multiplying the k_1 value of base 10 by 2.30 yields the k_1 of base e]. Although independent derivation of k_1 and the $CBOD_u/CBOD_5$ ratio should come from a long-term effluent BOD test, such tests require lengthy time and significant effort. Leo et al. (1984) compiled the data of 144 domestic WWTPs with treatment types ranging from primary, trickling filters, secondary, secondary with phosphorus removal, secondary with nitrification, secondary with phosphorus removal and nitrification, and secondary with nitrification and filters in the U.S. to calculate their $CBOD_u$ to $CBOD_5$ ratios. A summary of their data is presented in Fig. 3.1 starting a minimum $CBOD_u$ to $CBOD_5$ ratio of 1.0. The mean $CBOD_u$ to $CBOD_5$ ratio is 2.84 with a standard deviation of 1.17 out of the 144 treatment plant effluents, respectively, due to the variability of effluent characteristics associated with treatment levels (Leo et al. 1984). Using the $CBOD_u$ to $CBOD_5$ ratio instead of the $CBOD_u$ to BOD_5 ratio removes the effect of nitrification, thereby differentiating the attenuation of carbon and nitrogen in the modeling analysis.

Fig. 3.1 $CBOD_u$ to $CBOD_5$ ratio of domestic WWTPs

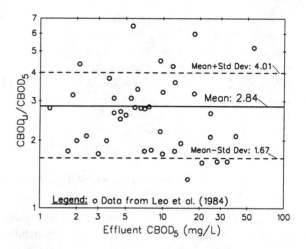

The value of the standard deviation in Fig. 3.1 suggests that the $CBOD_u$ to BOD_5 ratio is highly variable. The variability comes from not only between different treatment levels but also between different sites with the same treatment levels. For example, the $CBOD_u$ to $CBOD_5$ ratios for the Metro Plant WWPT in the twin cities of Minneapolis/St. Paul, Minnesota are 1.00, 2.50, and 3.60 for primary, secondary, and secondary with nitrification treatment, respectively (Lung 2001). One factor contributing to the standard deviation is that the BOD test itself was an inaccurate test prior to 1990. Lab procedures for long-term BOD test have been improved in the past two decades as described in a case study in the next section of the Danshui River, Taiwan. Another use of the $CBOD_u$ to $CBOD_5$ ratio is to convert the model calculated $CBOD_u$ loads back to $CBOD_5$ loads to set the $CBOD_5$ effluent limits for the NPDES permits.

Current understanding of BOD in wastewater and receiving waters can be summarized as follows:

(a) Marginally treated wastewater is mostly characterized by a $CBOD_u$ to $CBOD_5$ ratio close to 1.0 and high k_1 values above 0.5 day^{-1}.

(b) Primarily treatment (not designed to remove dissolved CBOD) would only slightly increase the $CBOD_u$ to $CBOD_5$ ratio, yet with a wide range of k_1 values.

(c) Secondary treatment would significantly increase the $CBOD_u$ to $CBOD_5$ ratio and lower the k_1 rate (to below 0.1 day^{-1}) as dissolved CBOD is highly stabilized via the treatment process.

(d) Ambient waters are usually represented by very low k_1 values (Lung 2001) but with a range of much higher $CBOD_u$ to $CBOD_5$ ratios.

Note that k_1 is referred as BOD bottle rate (Thomann and Mueller 1987) derived from BOD tests, which is different from the CBOD deoxygenation rate, k_d in the receiving water. Discussions associated with Fig. 3.1 in this section are therefore for wastewaters only and not applicable to receiving waters.

Fig. 3.2 Long-Term BOD data of the Guilford Mills effluent

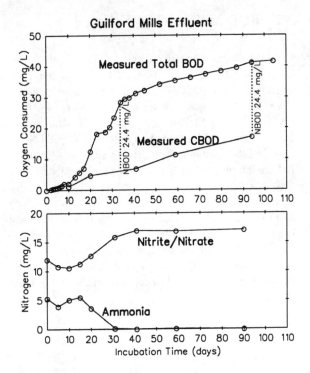

The ratios presented in Fig. 3.1 are reasonable to use when effluent data are not available. Industrial wastewaters, however, are much different from domestic wastewaters and site-specific data are crucial to developing the $CBOD_u$ to $CBOD_5$ ratios. Figure 3.2 shows long-term BOD analyses of samples from the Guilford Mills (a textile manufacturer) effluent in Kenansville, NC. First, the cumulative oxygen consumption reaches 42 mg/L at the end of the 104-day incubation period (see the top panel of Fig. 3.2). Also shown is the independently measured CBOD, reaching an ultimate level of 17.5 mg/L. Concurrent measurements of ammonia and nitrite/nitrate are presented in the bottom panel of Fig. 3.2, illustrating that nitrification started around day 8 and was completed by day 32 when ammonia is depleted. The total amount of ammonia lost is 5.33 mg/L, which is then multiplied by a factor of 4.57 (based on stoichiometry for nitrification) to yield 24.4 mg/L as nitrogenous biochemical oxygen demand (NBOD) shown in Fig. 3.2. The breakdown of dissolved carbon continued in the remaining period of the incubation. The long-term BOD data from the Guilford Mills effluent is somewhat atypical as ammonia was added during the manufacturing process, resulting in the breakdown of ammonia prior to the completion of carbon oxidation in the long-term BOD test. [Note that nitrification cannot be achieved without sufficient oxygen supply]. Applying Eq. 3.1 to the measured CBOD data yields a bottle rate, $k_1 = 0.015$ day^{-1}, indicating a well stabilized effluent in terms of organic carbon. In this case, CBOD plays a minor role in the effluent.

3.2 BOD in Receiving Water

Once the wastewater effluent enters the receiving water, bacteria consume oxygen to break down the organic carbon. Interactions between the BOD in a single wastewater discharge and the receiving water can best be demonstrated in the upper Mississippi River in a 35-mile stretch from Lock & Dam No. 1 to Lock & Dam No. 2 near the Twin City area of St. Paul and Minneapolis, Minnesota (see Fig. 3.3). The major BOD load is from the Metro Plant, a domestic WWTP with a capacity of 250 mgd (13.5 cms) in 1988. The long-term BOD results for the filtered effluent are presented in the bottom left plot of Fig. 3.3, showing the cumulative oxygen consumption (in squares linked by solid lines) in terms of dissolved oxygen in mg/L throughout the 70-day lab incubation period. Nitrogen inhibition chemicals were not used in the tests. Instead, ammonia and nitrite/nitrate concentrations in the samples were measured concurrently with the total oxygen consumption measurements. Subtracting the amount of oxygen used in ammonia conversion to nitrite/nitrate gives the calculated CBOD for the filtered samples (in squares connected with dashed lines). The difference between the solid and dashed lines in the bottom left plot of Fig. 3.3 is NBOD,

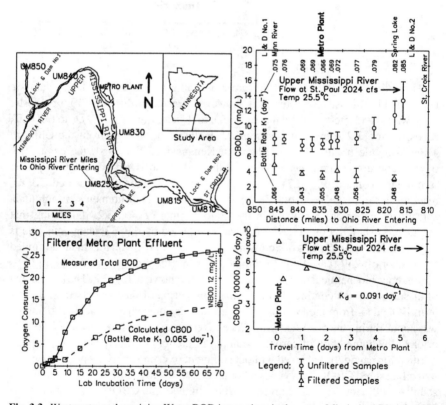

Fig. 3.3 Wastewater and receiving Water BOD interactions in the upper Mississippi River in 1988

reaching 12 mg/L by the end of the incubation period. The calculated CBOD (filtered $CBOD_u$) data points are then fitted with Eq. 3.1 to come up with a bottle rate, $k_1 = 0.065$ day^{-1} for the Metro Plant effluent (see the bottom left plot of Fig. 3.3).

Filtered and unfiltered samples were collected at eleven locations between Lock & Dam No.1 and Lock & Dam No.2 in the upper Mississippi River during a two-week period from June 17, 1988 to July 1, 1988. Each sample was run through a 70-day long-term BOD test with replicates. The long-term BOD data were fitted to Eq. 3.1 to yield the average $CBOD_u$ and range shown in the top right plot of Fig. 3.3. Measured $CBOD_u$ concentrations from the filtered river samples were shown as the mean values and ranges in triangles with bars, displaying a slight downward trend. The bottle rates, k_1 generated from fitting the filtered samples to Eq. 3.1 are displayed in numbers (from 0.055 day^{-1} to 0.048 day^{-1}), showing a slow decay of organic carbon in the water column after receiving the Metro Plant input.

Results of regressing the long-term data from unfiltered river samples with Eq. 3.1 are also shown in mean values and ranges of $CBOD_u$ in circles with bars in the top right plot of Fig. 3.3. A slight downward trend from Lock & Dam No. 1 to Metro Plant of unfiltered $CBOD_u$ concentrations is followed by a sharp rise toward Lock & Dam No. 2 due to significant algal growth in the pool behind the dam. The difference between unfiltered and filtered $CBOD_u$ represents the carbon in the algal biomass. [Additional examples of algae in BOD are presented in Figs. 3.17 and 3.18]. The algal carbon also makes a strong showing in the bottle rates,k_1 displaying a sharp rise (from 0.066 day^{-1} at immediately below the Metro Plant to 0.085 day^{-1} toward Lock & Dam No. 2). As expected, the k_1 rates for the unfiltered samples are higher than the filtered ones as the unfiltered samples contain algal biomass subject to higher degradation rates.

The filtered CBOD concentrations show a steady decrease from Lock & Dam No. 1 to Lock & Dam No. 2, a textbook spatial pattern. Decreasing k_1 rates indicate the strength of CBOD is continuing to decrease due to attenuation. The k_1 rates for filtered samples, decreasing from 0.066 day^{-1} to 0.048 day^{-1} along the river flow (from UM845 to UM818) in the downstream direction, are noticeably lower than those associated with the unfiltered samples. Again, algal biomass in the unfiltered samples contributes to this difference. Note that the bottle k_1 rates are well below 0.1 day^{-1}, indicating relatively slow degradation rates of CBOD.

Finally, the filtered $CBOD_u$ data from the river are used to obtain the CBOD deoxygenation rate, k_d in the upper Mississippi River. As suggested in Sect. 2.8, mass loading rates of $CBOD_u$ (in log scale) are plotted versus the travel time to generate a slope of 0.091 day^{-1} as the k_d rate (at water temperature of 25.5°C) in the bottom right plot of Fig. 3.3.

Having different meanings, the three data plots in Fig. 3.3 are closely related. The plot of cumulative DO consumption versus incubation time for the filtered Metro Plant effluent samples describes the long-term characteristics of the point source $CBOD_u$, revealing the strength of both carbonaceous and nitrogenous BOD. The second plot of river $CBOD_u$ vs. distance presents the spatial attenuation of unfiltered and filtered $CBOD_u$ in the receiving water. Carbon in the algal biomass is captured in the unfiltered samples, causing the increase of river $CBOD_u$ concentrations toward the pool of Lock & Dam No. 2. The third data plot of Fig. 3.3 caps the discussion by showing the independent derivation of in-stream CBOD deoxygenation rate, k_d using field data.

It is important to differentiate the receiving water bottle rate, k_1 and the deoxygenation rate, k_d for filtered $CBOD_u$. These two rates are different because the oxidation of CBOD in a natural water body includes phenomena that are not part of the BOD bottle rate (Thomann and Mueller 1987). Although not shown in Fig. 3.3, each of the bottle rates from the upper Mississippi River has an oxygen consumption versus incubation time plot, very similar to the one for the Metro Plant effluent. However, the incubation time (in days) in the long-term BOD plot has no relationship with the time of travel (also in days) in the CBOD loading plot. Having different physical meanings, these two characteristic times are not mathematically related (Thomann and Mueller 1987). The bottle rates, k_1 cannot be used for the in-stream deoxygenation rate, k_d.

The data presented in Fig. 3.3 are extracted from a comprehensive water quality survey of the upper Mississippi River during June 17 to July 1, 1988 (Metropolitan Waste Control Commission 1989) as part of the effort to evaluate the success of upgrading the Metro Plant from secondary treatment to secondary with nitrification. Over 100 technical personnel were involved in the planning, execution, and reporting for the two-week monitoring program. By its own account, the 1988 survey was claimed as a great success, providing a huge volume of valuable data to significantly improve our understanding of BOD and DO interactions in receiving waters. We are grateful for the contribution of such an immense effort to this topic.

The above analysis demonstrates that the $CBOD_u$ deoxygenation rate, k_d is strongly affected by the characteristics of BOD loading rates and therefore could change over time. The Metro Plant has been upgraded from primary treatment in 1960's to secondary treatment in 1970s, and then to secondary with nitrification in the 1980's. Figure 3.4 shows the spatial profiles of BOD, ammonia, nitrite/nitrate, and DO concentrations in a 30-mile stretch of the upper Mississippi River from upstream (at UM840) to downstream (at UM810) in 1964, 1976, and 1988, respectively. The elimination of low DO in the river is the major water quality improvement. The data show that in 1964 when the Metro Plant had only primary treatment, the low DO in the river stretched to a distance of 20 miles downstream from the Metro Plant. No CBOD, ammonia, and nitrite/nitrate were measured at that time. Thanks to the CWA, significant CBOD reductions took place in the early 1970s to raise the DO levels in the river. By 1976, the Metro Plant was upgraded to secondary treatment but the minimum DO was below 5 mg/L, still violating the water quality standard (see the second column of plots in Fig. 3.4). Nitrification at the Metro Plant was initiated in 1985 (see Fig. 2.11 for the changes in ammonia and nitrite/nitrate concentrations in the Metro Plant effluent). By 1988, the minimum DO in the upper Mississippi River below the Metro Plant was above 6 mg/L (see the plots the right-hand column of Fig. 3.4). The ammonia concentration profile in 1976 was switched to the nitrite/nitrate profile in 1988, a clear indication that the nitrification process at the Metro Plant was working. A slight, additional reduction of CBOD load from the Metro Plant was a secondary benefit of nitrification. The change of the in-stream

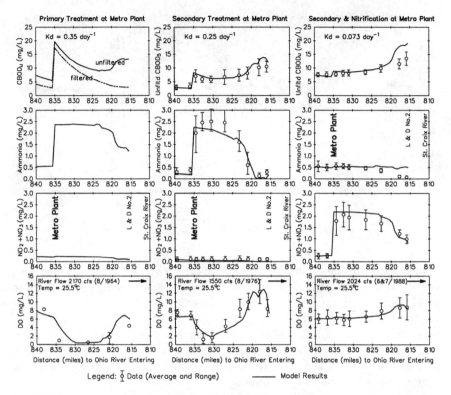

Fig. 3.4 Progressive change of in-stream BOD Kinetics following the metro plant upgrade

BOD kinetics responding to the treatment upgrade is clear: at 20°C, the CBOD deoxygenation coefficient, k_d in the upper Mississippi River is reduced from 0.35 day^{-1} (primary treatment at the Metro Plant) in 1964 to 0.25 day^{-1} (secondary treatment) in 1976, and finally to 0.073 day^{-1} (secondary with nitrification) in 1988 (Fig. 3.4). Independent derivation of k_d using the 1988 filtered $CBOD_u$ loads (in Fig. 3.3) further substantiates this rate as 0.073 day^{-1} at 20°C (equivalent to 0.093 day^{-1} at 25.5°C).

Long-term BOD tests provide useful insights into understanding the BOD/DO kinetics in receiving water systems. Figure 3.5 presents the 1996 data from the 70-day BOD tests of filtered samples at Lock & Dam No. 1 of the upper Mississippi River and its two tributaries: the Minnesota River and St. Croix River. [Refer to Fig. 3.3 for their locations]. The 70-day incubation time is necessary to achieve complete breakdown of organic carbon in the water samples. Total BOD, CBOD, and nitrite/nitrate are shown in Fig. 3.5 along with the Metro Plant effluent for a comparison. Data at Lock & Dam No. 1 represent nonpoint BOD loads from the watershed up to this point. BOD levels at the mouth of the Minnesota River are slightly higher due to the nonpoint input from the Minnesota River Watershed. The St. Croix River has very insignificant BOD levels. The BOD levels in the Metro

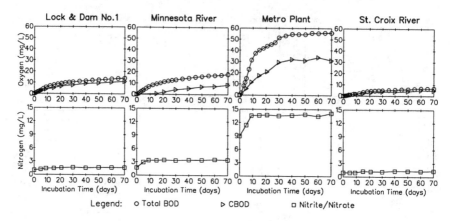

Fig. 3.5 Long-Term BOD test results of the upper Mississippi River in 1996

Plant effluent are significantly higher than its receiving water and the other two tributaries. The increase of nitrite/nitrate concentrations from 9 mg/L to 14 mg/L in the Metro Plant effluent indicate nitrification at work. The Metro Plant effluent yields a $TBOD_u$ close to 56 mg/L and $CBOD_u$ slightly above 30 mg/L. The difference of about 26 mg/L (56 mg/L less 30 mg/L) is the amount of oxygen needed to convert ammonia to nitrate at the plant. Nitrification amounts to less than 10 mg/L of oxygen consumption in the Minnesota River and is insignificant in the ambient water at Lock & Dam No. 1 and St. Croix River.

Figure 3.6 shows total BOD ($TBOD_u$) and $CBOD_u$ concentrations, and the bottle rate, k_1 of long-term BOD tests of samples taken from the upper Mississippi River (UM880 to UM800) in June 2010, presenting a historical perspective. The $TBOD_u$ and $CBOD_u$ concentrations in the river are below 10 mg/L, about the same levels observed in 1988 (see Fig. 3.3), indicating good water quality in the upper Mississippi River. Also shown are the Metro Plant effluent $TBOD_u$ and $CBOD_u$ concentrations. The Metro Plant effluent's impact on the river is minimum. Note that the bottle rate k_1 of filtered $CBOD_u$ is around 0.01 day^{-1} along the river in 2010, much lower than the 1988 values (all above 0.043 day^{-1} in Fig. 3.3). The in-stream deoxygenation rate, k_d should not be much higher than 0.01 day^{-1}, demonstrating excellent water quality with minimum oxygen consumption. Based on these data, long-term BOD tests of the river samples were discontinued in 2016 as the organic carbon is completely converted from labile to refractory in the upper Mississippi River. The Metro Plant has no impact on the river DO.

3.3   Multiple BOD Discharges

The BOD/DO modeling of the upper Mississippi River with a single discharger is considered a textbook case. Present day BOD/DO problems, in contrast, have

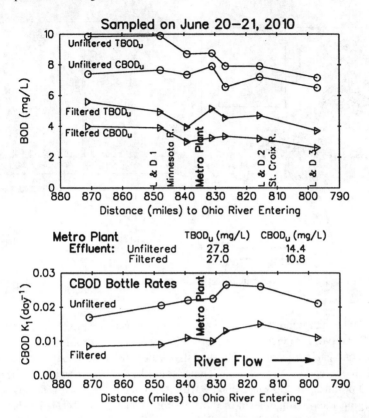

Fig. 3.6 BOD concentrations and CBOD Bottle rates in upper Mississippi River (2010)

multiple discharges of varying BOD strengths. The Danshui River in Taiwan (Fig. 3.7) is a good example. The river system originates in the mountains with an elevation of 3529 m above sea level and a watershed area of 2726 km². As the third largest river in Taiwan, the Danshui River is formed by three major tributaries: the Dahan Creek in the south, the Xindian Creek in the middle, and the Keelung River in the north. The river flows in a south-to-north/northwest direction before entering the Taiwan Strait. Rapid urban development and industrialization in the watershed in the past three decades have generated significant pollutant loads into the river system.

Figure 3.7 also presents the CBOD and ammonia loads into the Danshui River. Loads from the Dahan Creek, Xindian Creek, and Keelung River dominate the input to the Danshui River. In addition, there are point source loads directly discharged into the main stem. The top two BOD loads are from the Keelung River and Xindian River, which receive WWTP effluents receiving primary treatment. Since no long-term BOD tests were performed for these loads, the $CBOD_u$ loads from these two tributaries are calculated by multiplying the $CBOD_5$ loads by a $CBOD_u$ to $CBOD_5$ ratio of 1.75, an approximation reflecting the degree of stabilization of BOD loads

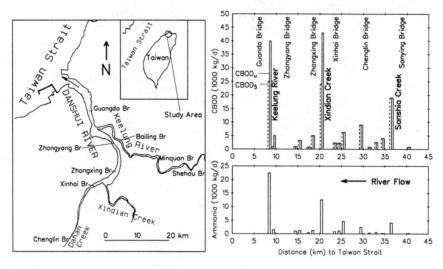

Fig. 3.7 The Danshui River (in Taiwan) with point Source CBOD and Ammonia Loads

(equivalent to marginally treated secondary effluents based on Fig. 3.1). A $CBOD_u$ to $CBOD_5$ ratio of 1.75 yields a bottle rate of 0.169 day^{-1} (according to Eq. 3.1). [Note that this rate is significantly higher than the bottle rate of 0.065 day^{-1} of the Metro Plant effluent reported in Sect. 3.2. Comparisons like these offer an idea of the degree of BOD pollution in Taiwan around 2010]. All domestic WWTP effluents directly discharging into the main stem of the Danshui River have $CBOD_5$ levels ranging from 40 mg/L to 80 mg/L. Other smaller point source loads of poorly treated effluents have $CBOD_u$ loads equal to $CBOD_5$ loads (a ratio of 1). Having various degrees of BOD potency dictates the use of $CBOD_u$ as the common currency in quantifying the DO budget along the river. Ammonia concentrations in the wastewater effluents are typically around 16 mg/L to 25 mg/L. Nitrification is inhibited due to low DO in the wastewater. Therefore, $CBOD_5$ may be construed as BOD_5 to derive the CBOD loads from these effluents.

A 1-D tidally averaged BOD/DO model based on the U.S. EPA's WASP/EUTRO framework was developed for the river system by Chen et al. (2012) for water quality management. Modeling results along with the field data are presented in multiple plots of Fig. 3.8. The top left plot shows the longitudinal profile of river flows, increasing in the downstream direction. The bottom left plot presents the model-calculated $CBOD_u$ concentration profiles in the Danshui River associated with two in-stream deoxygenation rates,k_d of 0.1 day^{-1} and 0.5 day^{-1}, respectively. As expected, the lower k_d rate (0.1 day^{-1}) in the Danshui River yields higher $CBOD_u$ levels in the water column than those with a higher k_d rate (0.5 day^{-1}). In addition, the difference in the $CBOD_u$ results between the two k_d rates is insignificant in the upstream portion of the Danshui River due to strong mass transport (i.e. relatively short residence time) in this portion of the river. This gap in $CBOD_u$ concentrations widens significantly in the downstream portion of the river where tidal action tends to

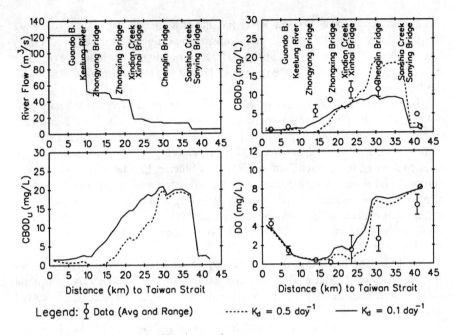

Fig. 3.8 Danshui River model calibration results

increase the residence time, thereby resulting in much lower $CBOD_u$ concentrations under the higher k_d rate (0.5 day^{-1}).

Since no $CBOD_u$ measurements are available, the calculated $CBOD_u$ must be converted to $CBOD_5$ for comparison with the $CBOD_5$ data via a $CBOD_u$ to $CBOD_5$ ratio, an unknown without the long-term BOD test. Assuming the ratio is close 1.0 for k_d of 0.50 day^{-1} (based on Eq. 3.1), the model calculated $CBOD_5$ is almost the same as the $CBOD_u$, particularly in the upstream portion of the Danshui River (see the top right plot of Fig. 3.8). The lower k_d rate of 0.1 day^{-1} (at a $CBOD_u$ to $CBOD_5$ ratio of 2.54) yields $CBOD_5$ concentrations much lower than the $CBOD_u$ concentrations, resulting in a flip between the $CBOD_5$ and $CBOD_u$ curves in the estuarine portion of the Danshui River starting at Xinhai Bridge (see the top right plot of Fig. 3.8).

The bottom right plot of Fig. 3.8 shows the model calculated DO concentrations with two different deoxygenation rates in the Danshui River. The model results associated with these different rates behave as expected as the higher deoxygenation rate yields a more depressed DO profile, particularly in the estuarine portion of the Danshui River where mass transport is significantly retarded. Results of the above modeling analysis suggest that the DO budget in the Danshui River reflect a delicate balance of mass transport and kinetics in the water column. Mass transport in the upstream portion of the Danshui River dominates the mass balance of CBOD to the extent that the effect of in-stream deoxygenation of CBOD is less significant. On the other hand, the CBOD deoxygenation process becomes a much more important mechanism in the lower Danshui River (i.e. the estuarine portion). Thus, following the dashed line of the DO profiles in the upstream portion of the river (at a high k_d rate) to about 27 km (to Taiwan Strait) and then switching to the solid line (at a low k_d rate) to reach the mouth of the river would result in a much improved match with the DO data.

The above analysis suggests that k_d rates in the Danshui River are spatially variable. Model calibration can be improved by using higher k_d rates in the upstream portion and lower k_d rates in the lower portion of the Danshui River. Such an adjustment is justified as the point source BOD loads entering the upstream portion of the river come from less treated domestic WWTP effluents, having more labile organic carbon. Point source BOD loads entering the downstream (estuarine) portion of the river come from treated effluents, thereby reflecting lower k_d rates. The analysis to derive the in-stream BOD deoxygenation rate based on a single wastewater discharge for the upper Mississippi River in Fig. 3.3 is therefore not applicable to the Danshui River.

The BOD/DO modeling of the Delaware River case offers a perspective for spatially variable k_d rates. The Delaware River (Fig. 3.9) received over 30 point sources from domestic WWTPs below Trenton in the 1970s, dropping the dissolved oxygen level to below 2 mg/L immediately downstream of the Philadelphia area. Averages and ranges of the in-stream $CBOD_{30}$ to $CBOD_5$ ratios (shown in the top panel of Fig. 3.9) were measured from 30-day BOD tests conducted in 1975–1976

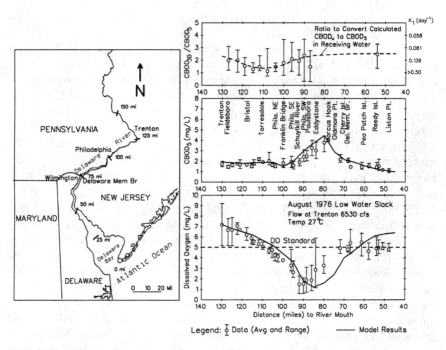

Fig. 3.9 The Delaware estuary BOD and DO model calibration results

for multiple locations along the river. They were used to convert the model calculated $CBOD_u$ to $CBOD_5$ for comparison with the measured $CBOD_5$ concentrations (in the second panel of Fig. 3.9), resulting in good matches between the model results and data for $CBOD_5$ and DO (Fig. 3.9). Note that the $CBOD_u$ values were generated from the model using a spatially constant k_d instead of spatially variable k_d reflecting site-specific BOD/DO kinetics in the river. An improved approach using spatially variable k_d rates for the Danshui River BOD/DO model is therefore recommended.

3.4 Spatially Variable Deoxygenation Rates

The approach calls for sampling at multiple stations in the main stem and its tributaries of the Danshui River for site-specific long-term BOD (filtered) tests. Results from the tests were used to derive spatially variable k_d rates for the model. While a detailed presentation of the field and lab procedures can be found in Chen et al. (2013), a succinct description of the lab procedure can be found in a later section of this chapter. A total of 4 comprehensive surveys were conducted in 2009–2010. Figure 3.10 shows the long-term BOD results from samples collected at the Xinhai Bridge station (with the highest BOD) in November 2009. The combined oxygen consumption by CBOD deoxygenation and nitrification is shown in circles connected with a solid line in the left panel of Fig. 3.10.

The increase of nitrite/nitrate concentrations with time (Fig. 3.10) suggests nitrification. Multiplying the difference in nitrite/nitrate concentrations at day 1 and day 20 by 4.57 (based on stoichiometry for nitrification) yields the amount of oxygen as NBOD (45.74 mg/L) as the NBOD of the Guilford Mills effluent in Fig. 4.2. Subtracting the amount from the total BOD yields the calculated CBOD, shown in a dashed line in Fig. 3.10. All data are reported in mg/L of DO. Also, the independently measured amount of ammonia of 8.85 mg/L is close to the amount of nitrite/nitrate produced (9.83 mg/L), maintaining the nitrogen mass balance in the test. This long-term BOD test procedure tracks concurrent ammonia consumption and

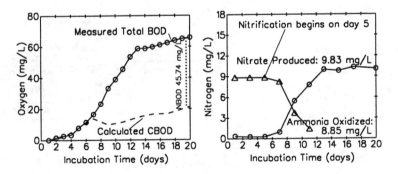

Fig. 3.10 Long-Term BOD results of Xinhai Bridge, Danshui River

nitrate production without nitrification suppressors. Figure 3.10 presents a complete picture of the long-term test to generate the calculated CBOD.

Results of the long-term BOD tests from eight sites in one of the four surveys are summarized in Fig. 3.11. Each site is shown with total BOD, calculated CBOD, and nitrate concentrations for a period of 20 days when all the organic carbon is exhausted and conversion of ammonia to nitrate is complete. Note that total BOD and CBOD data are plotted on the same scale for all stations. The same scale is also used for the nitrate data at all stations. These plots give a good perspective in the

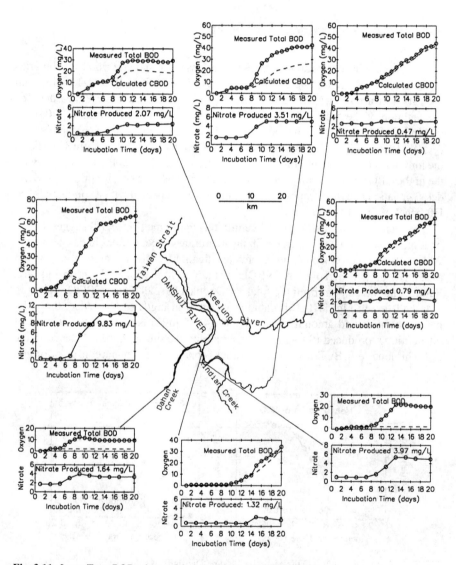

Fig. 3.11 Long-Term BODs data at 8 locations in the Danshui River system

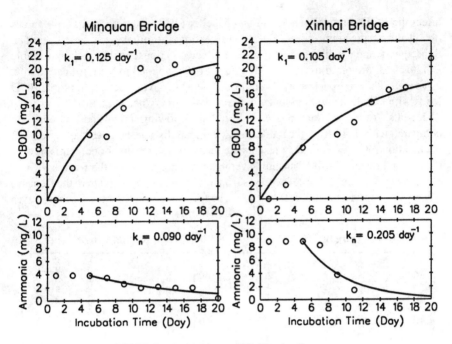

Fig. 3.12 Derivation of CBOD Deoxygenation and Nitrification Rates

spatial pattern of BOD strength in the Danshui River system. Note that nitrification starts prior to 5 days at some stations during the incubation period.

The data points of the calculated CBOD from two stations, Minquan Bridge and Xinhai Bridge, are re-plotted in the top portion of Fig. 3.12 to fit Eq. 3.1, yielding the CBOD bottle rates of $k_1 = 0.125$ day^{-1} and 0.105 day^{-1} at Minquan Bridge and Xinhai Bridge, respectively. The ammonia concentrations at the Minquan and Xinhai bridge stations are presented in the bottom two panels of Fig. 3.12, showing a decline in concentrations due to nitrification starting approximately on day 5. A simple regression analysis is performed to fit the ammonia data points to the model: $C(t) = C_0 e^{-k_n t}$, where $C(t)$ is the ammonia concentration at any given time t during the long-term BOD test, C_0 is the initial ammonia concentration at the onset of the nitrification process (8.85 mg/L in the water sample at the Xinhai Bridge), and k_n is the nitrification rate. Figure 3.12 shows fitted solid lines with nitrification rates, k_n of 0.09 day^{-1} and 0.205 day^{-1} at the Minquan and Xinhai Bridge stations, respectively.

Thomann and Mueller (1987) pointed out that the above k_1 rate derived from the lab tests of the field sample is not the in-stream deoxygenation rate, k_d used in the BOD/DO model. One big factor for this difference lies in the organic carbon sorbed on particles which settle in the water column. All water samples from the Danshui River were filtered for the long-term tests, thereby removing this factor. Still, the long-term tests do not include bio-sorption by biological slimes on river bottoms. Stream turbulence and roughness, and the density attached organisms could also affect the BOD removal (Thomann and Mueller 1987). Therefore, these k_1 values obtained from the long-term BOD tests were assigned as initial k_d rates for the model. Subsequent model calibration runs were conducted to fine tune the spatially variable k_d rates with minor adjustments. The same strategy applied to the nitrification rate, k_n as well. This approach works well for the Danshui River receiving multiple BOD loads with a wide range of potency in the river system (Chen et al. 2013).

Results from the final model calibration following minor adjustments are compared with those from the model using the spatially constant deoxygenation and nitrification rates. Figure 3.13 presents the recalibration results for comparison with the original results for the Danshui River and its major tributary, the Keelung River. First, the freshwater flow rate increases from upstream to downstream in the Danshui River (in the top left panel of Fig. 3.13). The tributary inflows of the Keelung River

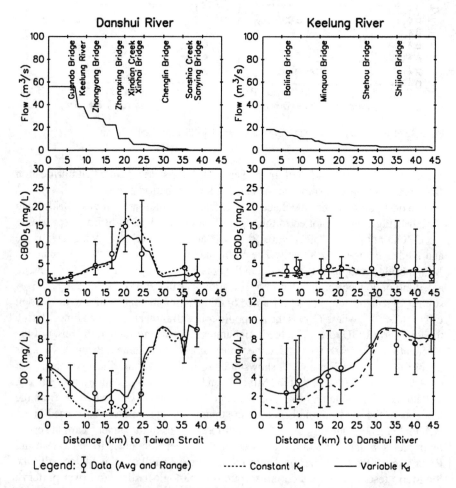

Fig. 3.13 Recalibration with spatially variable Deoxygenation rates in the Danshui River

and the Xindian Creek cause the sharp increases in the freshwater flow rates. Model results demonstrate considerable improvement of model calibration for $CBOD_5$ and DO, but only small improvement of ammonia (not shown) with the spatially variable kinetic coefficients. The improved $CBOD_5$ calibration partially benefits from the more precise quantification of the CBOD loads for model input. Also shown in Fig. 3.13 are the results for the Keelung River, a tributary to the Danshui River. Appreciable differences in results between original and improved models are demonstrated upon a comparison with the field data. This further substantiates the need of spatially variable deoxygenation and nitrification rates for the Danshui River and its tributaries.

Figure 3.14 shows the calibrated $CBOD_u$ deoxygenation and nitrification rates, k_d and k_n, respectively, in the Danshui River and its tributaries. The CBOD deoxygenation rate, k_d, shows a gradual decreasing trend in the downstream direction, while the nitrification rate, k_n, increases from upstream to downstream within the Keelung River. A combination of spatial variations from the CBOD and NBOD kinetic rates reflect the characteristics of the river: the stabilization of CBOD, accompanied by the initiation of nitrification within the downstream direction. A similar trend of higher k_d in the upstream portion of the Danshui River reflects the greater potency of point source CBOD loads. This warrants a higher deoxygenation rate in the upstream area and a lower deoxygenation rate in the downstream, estuarine portion of the river

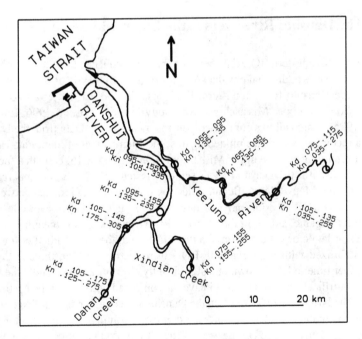

Fig. 3.14 In-stream CBOD Deoxygenation (k_d) and Nitrification (k_n) rates in the Danshui River system

system. Previous model calibration results are based on constant k_d and k_n values of 0.1 day^{-1}.

The Danshui River study demonstrates another challenge in BOD/DO modeling: quantifying the in-stream CBOD deoxygenation rate, k_d, and nitrification rate, k_n. While multiple CBOD components with varying deoxygenation rates can be configured in current water quality modeling frameworks, the key issue (i.e., quantifying the individual k_d rates) still exists. Assigning three CBOD components instead of one for the modeling analysis without data support and justification would be quite counterproductive. Collecting water samples to analyze the BOD kinetics and to generate k_1, k_d, and k_n rates is a simpler and recommended approach, particularly for rivers receiving multiple BOD inputs with different strengths.

The calculated DO level in the Danshui River is very close zero in the area approximately 20 km from the mouth (see the bottom left panel of Fig. 3.13), requiring special attention. As pointed out by Thomann and Mueller (1987) and Chapra (1997) for anaerobic rivers with DO = 0, decomposition of CBOD is slowed. That is, the k_d rate is lowered to respond to the lack of oxygen supply in the water column while the air is the only source via reaeration. Since the nitrifying bacteria are aerobic organisms, the nitrification rate, k_n is set to zero until the DO level recovers to above 1 mg/L. These safety thresholds are included in the Danshui River model.

3.5   The Danshui River Versus the Upper Mississippi River

Comparing the long-tern BOD data between the Danshui River and the upper Mississippi River reveals additional physical insights into the BOD kinetics. The lab procedure for the Danshui River long-term BOD tests in 2009 (Chen et al. 2013) was adopted from the upper Mississippi River study in 1990's (Lung 1996; Lung and Larson 1995), which allows comparison on an equal footing. Data from the Danshui River (a receiving water) are plotted against the Metro Plant effluent data (a point source discharging into the upper Mississippi River) in Fig. 3.15. Note that the long-term BOD tests for the Danshui River samples lasted only 20 days compared with 70 days for the Metro Plant effluent, reflecting different BOD kinetics between the two sets of samples. It took only 20 days for the microbial action to completely stabilize the organic carbon in the Danshui River water samples, meaning substantial levels of labile organic carbon. On the other hand, the Metro Plant effluent was mostly stabilized with the remaining BOD in refractory carbon, thereby taking a much longer time to break down at a significantly slower rate. Also, note the strong in-stream nitrification in the Danshui River as compared with that in the Metro Plant effluent. The NBOD of 45.74 mg/L in the Danshui River demands significant oxygen consumption in the river. On the other hand, nitrification in the Metro Plant wastewater is about 11 mg/L. The comparison of the nitrate production between these two cases further indicates that nitrification at the Metro Plant is successful, virtually eliminating oxygen consumption by ammonia in the receiving water. In the Danshui

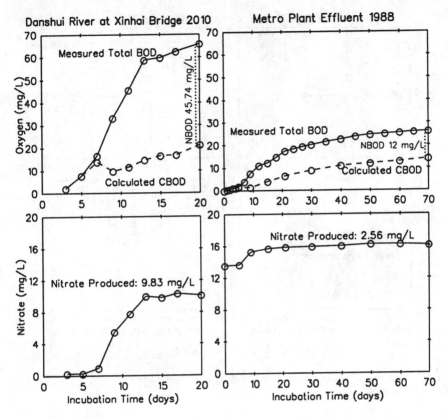

Fig. 3.15 Comparing BOD in Danshui River and metro plant effluent

River, in-stream nitrification takes a big chunk of DO due to the significant ammonia loads from the WWTPs.

Another striking difference between these two long-term BOD tests is the total BOD level (65 mg/L) in the Danshui River (a receiving water) is significantly higher than that in the Metro Plant effluent (26 mg/L), a wastewater. This difference is even more dramatic, considering that the Metro Plant effluent samples were collected two decades earlier than the Danshui River in-stream samples. The benefit of the CWA in the U.S. for water pollution control speaks for itself. By the time the water quality improves following significant wastewater treatment for the Danshui River, the long-term BOD tests of the water samples would expect to take just as much time (i.e. 70 days) to complete as the upper Mississippi River. The regulatory agencies and/or water quality managers should be made aware of the lengthened test period time prior to this transition. Lowering the CBOD loads to raise DO would move up the start time of in-stream nitrification (as in the Danshui River), thereby mandating concurrent CBOD and NBOD load reductions for water quality management.

The above discussion leads to a comparison of ammonia levels between the upper Mississippi River and the Danshui River and provides additional physical insights.

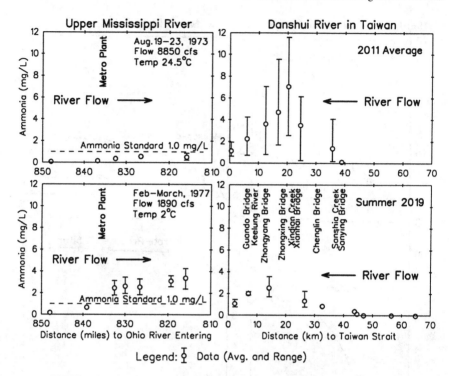

Fig. 3.16 Ammonia in the upper Mississippi River and Danshui River

The two left plots in Fig. 3.16 show ammonia concentrations in the upper Mississippi River from 1973 to 1977, prior to the implementation of nitrification at the Metro Plant. Also shown is the ammonia standard for the river, violated during the low flow period. By 1988 when the nitrification process was put online at the Metro Plant, ammonia toxicity violation was eliminated with concentrations below 0.5 mg/L meeting the standard (see Fig. 3.4). A genuine effort to reduce the ammonia loads to the Danshui River started in 2005 by adopting the Gravel Contact Oxidation Process. This process applies a fixed biofilm with the gravel contact oxidation purification tank divided into an aerated zone and a non-aerated zone. The wastewater is oxidized in the aerated zone and becomes a habitable environment for various creatures purifying water by predation and adsorption of water pollution. Sufficient oxygen is constantly supplied in the aerated zone, so that oxidation of organic substances and nitrogen can be achieved by the microorganisms. More than a dozen facilities of this type have been constructed along the river from 2005 to 2010 and they have achieved measurable success in reducing the ammonia loads to the Danshui River reflecting in significant drops of ammonia concentrations in the Danshui River between 2011 and 2019 (see plots of the right column in Fig. 3.16). By late 2019, ammonia concentrations in the Danshui River have been significantly reduced to levels not having the un-ionized ammonia toxicity problem—the onsite gravel contact oxidation process has paid off.

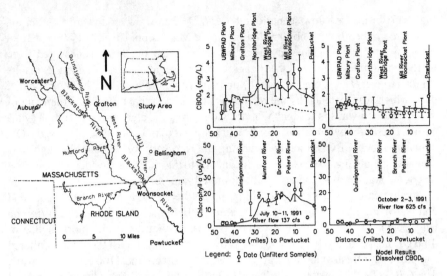

Fig. 3.17 Model results versus data of the Blackstone River

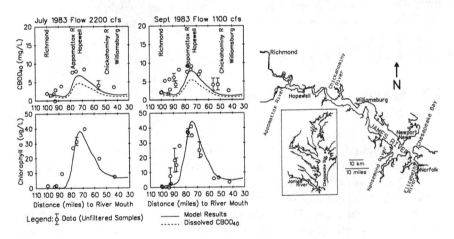

Fig. 3.18 Model results versus data of the James River Estuary

3.6 Algal Biomass in BOD

The unfiltered CBOD concentration data in the upper Mississippi River in 1976 and 1988 show an upward trend following the Metro Plant BOD load approaching the pool behind Lock and Dam No. 2 (see map in Fig. 3.3) due to the algal carbon in the samples. The Blackstone River in Massachusetts (Fig. 3.17) is another example showing the same effect (Lung 2001). The model results vs. data for unfiltered $CBOD_5$ and chlorophyll a in the Blackstone River during two flow conditions (low flow of 137 cfs and high flow of 625 cfs) at Woonsocket, MA are presented in

Fig. 3.17. In the low flow scenario, the $CBOD_5$ data show an upward trend in the downstream direction as observed in the upper Mississippi River in 1976 and 1988 (as seen in Fig. 3.4). This upward trend reflects that the BOD_5 measured from unfiltered samples contained the organic carbon in algal biomass. The model results (in filtered $CBOD_5$) do not match the unfiltered data in Fig. 3.17. The chlorophyll a data, which was matched by the model results, suggest that a significant amount of organic carbon was in the algal biomass. Converting the algal biomass carbon (via a carbon to chlorophyll a ratio) to BOD and adding it to the calculated dissolved BOD equals the total CBOD, thereby matching the data very well. Under the high flow scenario, the algal growth is not significant due to the shortened residence time. As a result, the algal carbon content in the unfiltered samples are minimal and their share of organic carbon plays a small role in the unfiltered BOD results (the right hand column of Fig. 3.18). Another case is seen in the James River Estuary in Virginia (Fig. 3.18). The effect of algal carbon is clear in that the peak chlorophyll a of 40 μg/L near Hopewell contributes approximately 2 mg/L CBOD to the unfiltered samples under freshwater flows at Richmond of 2200 cfs and 1100 cfs, respectively. The $CBOD_{40}$ data is considered as $CBOD_{40}$ in this case. The carbon in the algal biomass shows up in the unfiltered water samples, demonstrating a link between BOD and algal biomass. In rivers with significant algae growth, it is important to have BOD measurements from both filtered and unfiltered samples to complete the picture.

3.7   Amount of DO Consumption

Quantifying the amount of oxygen consumption from different pollutant sources provides useful information for water quality management. To do this calculation, the model must have been well calibrated and verified with field data. In this analysis, the BOD load of a particular source of a DO consumer is set to zero (i.e. removing the component) in the model. Next the resulting DO curve along the river is compared with the model calibration DO curve. The difference between these two DO curves is the oxygen consumption by this BOD load or DO consumer. This procedure can be repeated for any other oxygen consumption component in the model. These oxygen consumption components add up to the total amount of the consumption - superposition as a result of the linear kinetics in the BOD/DO model.

Results of the component analysis of the Delaware River (see Fig. 3.9 for the study area map) BOD-DO model are summarized in Fig. 3.19, showing the DO deficits for a number of BOD sources: (1) the upstream river flow above Trenton, NJ; (2) the point source BOD loads from wastewater treatment plants in the Philadelphia area; (3) the DO deficit from the in-stream nitrification process; (4) the sediment oxygen demand (SOD); and (5) net algal photosynthesis. The same concentration scale is used for these components for an easy magnitude comparison. The last plot of Fig. 3.19 presents the model calculated DO profile vs. data, certifying that the model has been calibrated for the August 1975 high slack water condition in the Delaware

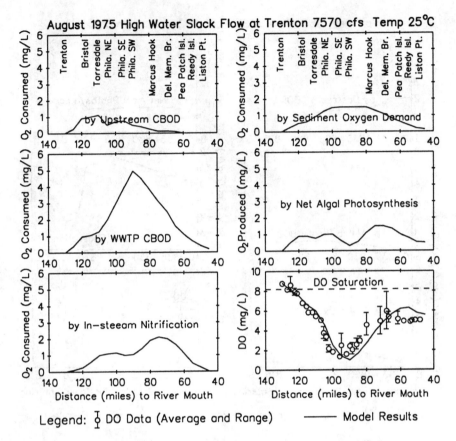

Fig. 3.19 BOD-DO component analysis of the Delaware River Estuary in 1975

River Estuary. The primary oxygen consumer is the wastewater discharges from the domestic WWTPs near Philadelphia, delivering significant CBOD and ammonia loads. SOD also consumes a significant amount of oxygen along the river while the algal photosynthesis process produces a net gain of oxygen (negative DO deficit). The low DO stretch of the Delaware River Estuary was primarily produced by the point source CBOD and NBOD in 1975 prior to achieving the full benefit of the CWA. By the 1980's, the Delaware River Estuary was cleaned up with total elimination of the WWTP DO deficits (both CBOD and in-stream nitrification).

Another DO deficit component analysis is the upper Mississippi River (see map in Fig. 3.3). Figure 3.20 shows the results for summer 1988 (a decade after the above Delaware River case). By 1988 the major BOD discharger, the Metro Plant, had a high removal rate of CBOD and nitrification in full operation. As a result, the DO consumption from the Metro Plant CBOD and in-stream nitrification was insignificant. The SOD taking oxygen from the water column reaches a high level of 3 mg/L oxygen deficit behind Lock & Dam No. 2. The net DO production of

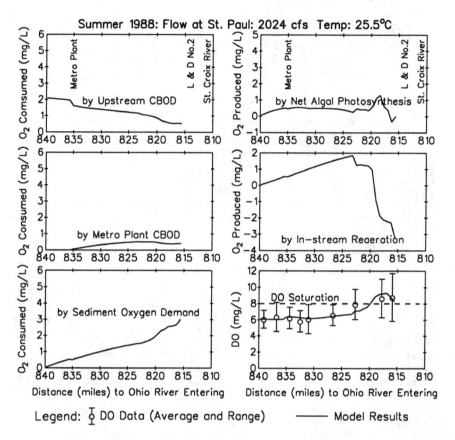

Fig. 3.20 BOD-DO component analysis of the upper Mississippi River in 1988

algal photosynthesis reaches a peak at UM818 and then decreases due to settling in the water column behind the dam. An interesting physical insight is that reaeration, providing oxygen from the atmosphere to the river, reverses it direction at UM820 because the excessive algal growth in the Lock & Dam No. 2 pool is over-saturating the DO in the water column (see the DO panel in Fig. 3.20), thereby releasing oxygen to the air. The reaeration curve in Fig. 3.20 shifts to negative at UM820. The top DO consumers in the upper Mississippi River are the SOD and upstream input, implying that additional DO level improvement in the upper Mississippi River would come from SOD mitigation and nonpoint controls for future regulatory water quality management of the upper Mississippi River.

Another demonstration of the CWA success and confirmation of the Metro Plant treatment upgrade is reviewing the historical DO deficits between the top three oxygen consumption sources: the Metro Plant, upstream nonpoint source, and sediment oxygen demand. Figure 3.21 shows the Metro Plant's diminishing role as an oxygen consumer via results of the 1964, 1976, and 1988 modeling analysis. The sharp decline of DO deficit due to the Metro Plant CBOD loads is clear from 1964 to

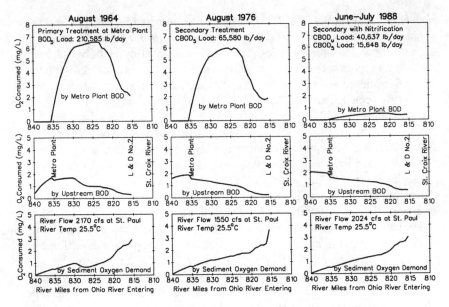

Fig. 3.21 DO deficits in the upper Mississippi River responding to treatment upgrade

1988 as the Metro Plant treatment level progressed during that 25-year period. The BOD load reduction during this time span is clear: from 210,585 lb/day of BOD_5 (note: nitrification was not present, resulting in $BOD_5 = CBOD_5$), to 65,580 lb/day of $CBOD_5$ in 1976, and to 15,648 lb/day of $CBOD_5$ in 1988. Even the $CBOD_u$ load in 1988 is less than the $CBOD_5$ load of 1976! As a result, the DO consumption by the Metro Plant effluent in 1988 is sharply reduced to well below 1 mg/L in the river. As pointed out in the previous discussion, SOD and upstream nonpoint sources are now the main oxygen consumers. It should be noted that these results are from model runs under summer low flow conditions, thereby further augmenting the significance of the DO deficits in the upper Mississippi River.

3.8 Long-Term BOD Test: Sample Collection and Lab Procedures

As the wastewater treatment level increases, differentiation of carbonaceous BOD (CBOD) from nitrogenous BOD (due to nitrification) is required (Lung 2001). The work by Hall and Foxen (1983) explains the necessity of measuring CBOD, as well as ammonia and nitrate concentrations, in the wastewater. Under previous practices after CBOD measurements were obtained, numerous problems were reported with regard to the accuracy of determining its concentration during laboratory analyses. One particular difficulty, nitrification suppression, has since been eliminated. The other anomalous results of $CBOD$ analysis often reported included circumstances where $CBOD_u$ exceeded BOD_u values and where $CBOD$ versus incubation time did not follow first-order kinetics. The study conducted by Haffely and Johnson (1994) suggests that the source of such errors was the nitrification inhibitor, 2-chloro-6-(trichloromethyl) pyridine (TCMP), used in the laboratory analysis. In their studies, they concluded that TCMP is biodegradable and can contribute significant oxygen demand in CBOD tests. Furthermore, TCMP degraders (bacteria) can be transferred between BOD bottles during routine $CBOD_u$ analyses. In many cases TCMP degrades were present in the wastewater and water samples, and the high TCMP dose and long incubation period enhanced the acclimation of the microbial population to TCMP. This potential for biodegradation casts doubt on the integrity of results obtained when TCMP, or perhaps any chemical inhibitor, is used in a long-term BOD test (Haffely and Johnson 1994).

The laboratory protocol used to quantify the $CBOD_u$ of wastewaters and ambient waters has changed significantly since the nitrification suppressor days. The current practice of determining $CBOD_u$ does not call for the use of nitrification suppressors (NCASI 1982). Instead, the total amount of oxygen consumption is recorded along with concurrent measurements of ammonium, nitrite, and nitrate concentrations, which ensures an accurate mass balance of the nitrogen components. The CBOD is then derived by subtracting the amount of oxygen used in the nitrification process from the measured total oxygen consumption. Using this protocol, Haffely (1997) has obtained excellent long-term BOD test results for the Metro Plant final effluent developed in St. Paul, Minnesota and the ambient water samples collected from the Upper Mississippi River (Lung 2001). The procedure of field sampling, water sample preparation, and lab analysis developed by the Metropolitan Council (Haffely 2009) in St. Paul, Minnesota was closely followed in the Danshui River study (see Sect. 3.3) with minor modifications to suit the local conditions. A succinct presentation of the lab procedure can be found in Chen et al. (2013).

In the Danshui River study, water samples were collected in 2-L containers within 30 min before or after the low slack of the tidal cycle at each of the sampling stations. Samples were kept at 4°C prior to being shipped to the lab. At the lab, samples were filtered using a glass fiber filter of 0.45 μm to fill the 2-L test container for the experiment. Initial concentrations of dissolved oxygen, ammonia, nitrite/nitrate nitrogen, and total Kjeldahl nitrogen (TKN) were measured as the initial conditions of the sample. The pH levels of the ambient water samples were constantly monitored within the range of 6–8. Usually, no special measures were needed to maintain the pH levels within the range.

Dissolved oxygen levels in the 2-L bottle were measured using a DO probe on a daily basis. When the DO level dropped below 4 mg/L, aeration was used to increase the DO level, but efforts were made to ensure that the concentration did not exceed 7 mg/L in order to avoid oversaturation. Operating within such a narrow range of DO—between 4 mg/L and 7 mg/L—was a precautionary measure designed to minimize unnecessary gain or loss of oxygen during the procedure. In addition, a 50-mL sample was taken from the 2-L bottle for the nitrite/nitrate analysis. Glass beads were immediately dropped into the 2-L bottle to account for the loss of water and to fill the volume up to the top of the bottle, which eliminated the possibility of air pockets near the top of the bottle. These precautionary measures were necessary to prevent any oxygen exchange across the air–water interface.

The current practice of long-term BOD tests, including not adding nitrification suppressors and taking cumulative DO and concurrent ammonia/nitrite/nitrate measurements on both unfiltered and filtered samples, has essentially rendered the textbook plots of BOD curves with nitrification initiation on day 8 obsolete. Again, the difference between the filtered and unfiltered samples would indicate whether algal biomass carbon is presents or not. The sample collection and lab test procedures developed from these recent studies should be followed for future modeling analysis.

3.9 Dissolved Oxygen Modeling of the Yamuna River

Following the data analysis of BOD and COD in the Yamuna River presented in Sect. 2.11, a BOD/DO model was calibrated using the May 2018 data from CPCB. The month of May is usually the warm and dry month before the monsoon season begins in Delhi, India. Model results vs. BOD_5 and DO data from May 2018 for the 25-km stretch of the Yamuna River near Delhi is presented in Fig. 3.22, showing the strong impact of the Najargarh Drain on the DO in this section of the Yamuna. Model results to raise the DO levels based on BOD load reductions show that the minimum DO of 4 mg/L in this stretch of the river can be achieved with a 90% reduction of the BOD loads. The control strategy points to curbing the BOD load of the Najafgarh drain, the most significant source of waste loads. Since no data were available to derive the in-stream deoxygenation rate, k_d for the Yamuna River, model projection analysis must include this as one of the variables to address the future conditions of the river. Note that the mechanisms for anaerobic conditions as presented in Sect. 3.4 are turned on in this modeling analysis. The nitrification rate, k_n is set to zero while the CBOD deoxygenation rate, k_d is reduced under diminishing oxygen supply in the river as discussed in Sect. 3.4.

Results in Fig. 3.22 can only be viewed as a first-cut analysis to project DO levels in the Yamuna as there are many uncertainties associated with this result. In addition to the BOD loads to the Yamuna and the background DO levels in the river, the BOD deoxygenation coefficient, k_d, which in turn is related to the in-stream conditions, is the key to the model calculations. Since the Najafgarh Drain provides the single largest BOD loads to the Yamuna near Delhi, its $CBOD_5$ and DO levels are also included in the model projections. Low DO levels in the Najafgarh Drain have strong impact on the water quality of the Yamuna. DO levels in the drain ranging from 1 mg/L to 4 mg/L are also included as another key variable in the projection analysis. Past modeling studies (Bhargava 1983; Paliwal et al. 2007; Sharma and Singh 2009; Mandal et al. 2010; Parmar and Bhardwaj 2015; Sharma et al. 2017; and Tyagi 2014) did not include the Najargarh Drain. Figure 3.23 summarizes the calculated DO at Nizamuddin in a 3-D plot. The range of k_d values in Fig. 3.23 is from 0.1 day^{-1} to 0.4 day^{-1} and is low compared with the current assimilative speed

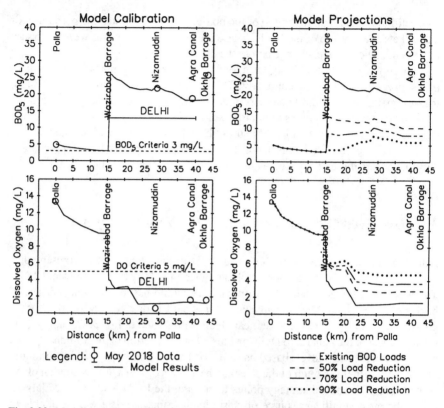

Fig. 3.22 Modeling analysis of the Yamuna River near Delhi, India

of BOD. This range of low k_d rates is designed for the Yamuna River under future conditions following the cleanup of the Najafgarh Drain.

Modeling results show that the Najargarh Drain (after receiving loads from the Supplementary Drain) provides the single largest COD and BOD load (over 60%) to the Yamuna. A water-quality monitoring program was launched in April 2017 to collect water quality data from the Drain. This is the first time the Najargarh Drain has been sampled. Over 30 sampling locations along the 50-km Najafgarh Drain (Fig. 3.24) were identified to collect water quality samples under dry weather conditions. Water quality parameters at locations immediately upstream and downstream of key tributary drains were also sampled. In addition, water flows and hydraulic geometry (velocity, depth, width, etc.) were measured along the Drain to establish the hydrodynamic conditions for the modeling analysis. The physical parameters include the following: water temperature, pH, conductivity, turbidity, total suspended solids, and alkalinity. The chemical parameters to be analyzed from the water samples include 5-day BOD (filtered and unfiltered), DO, TKN, ammonia, nitrate, total phosphorus, particulate phosphorus, ortho phosphorus, fecal coliforms, and total coliforms. The sampling program was carried out from April–June, 2017.

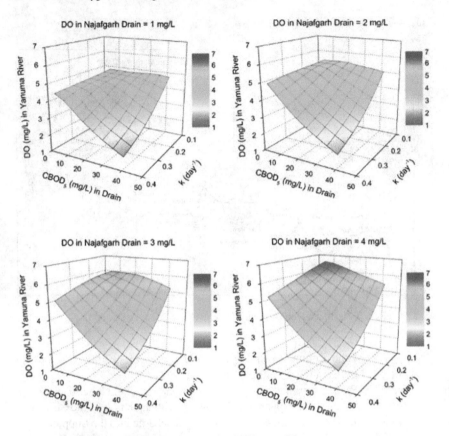

Fig. 3.23 Factors affecting model projections for the Yamuna River

Figure 3.24 shows the measured flows along the Najafgarh Drain, increasing from zero at the beginning of the drain to about 19 cms just upstream of the confluence with the Supplementary Drain, adding another 15 cms. By the time the Najafgarh Drain enters the Yamuna, the total flow is 34 cms.

This flow carries a significant amount of pollutants to the Yamuna. Figure 3.25 shows the longitudinal concentration profiles of TSS, BOD_5, TKN, ammonia, nitrate, Ortho-P, total P, and DO along this 57-km drain. These high concentrations provide significant TSS, BOD_5, and nutrient loads into the Yamuna. The spatial pattern of BOD_5 matches TSS quite well, suggesting that the bulk of BOD_5 samples (not filtered) is attached to suspended particles in the water column. The nitrogen components are plotted in the same concentration scale with ammonia matching TKN closely, indicating that the majority of TKN is ammonia. The combination of BOD_5 and TKN rises sharply in the final 20 km of the Najafgarh Drain, reducing DO to anaerobic conditions. Although the nitrate levels are much lower than TKN, they are all above 10 mg/L, another indication of a highly polluted river. The spatial patterns of high Ortho-P and total P concentrations are similar with total P reaching

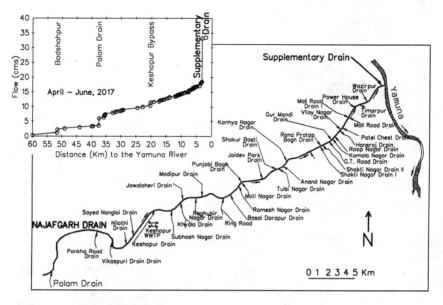

Fig. 3.24 Flows in Najafgarh Drain with supplementary and Tributary Drains

60 mg/L entering the Yamuna River. In general, DO levels in the drain are quite low, mostly below 2 mg/L in the water column during the survey.

Major WWTPs discharging directly to the Najafgarh Drain include Keshopur (72 mgd), Nilothi (60 mgd), and Pappankalan (40 mgd). Rithala (80 mgd) and Coronation Pillar (40 mgd) are the two main WWTPs directly discharge into the Supplementary Drain, which enters the Najafgarh Drain before reaching the Yamuna River. It is evident that these capacities are far below the needs to serve the Delhi area. The most logical approach to improving the water quality of the Yamuna River (i.e. raising the DO levels) is technology-based control as a first step via adding new and upgrading existing WWTPs. These WWTPs would significantly reduce the BOD and ammonia loads to the Najafgarh Drain via setting pollutant concentration limits in the WWTP effluents. Then BOD/DO modeling would come in to set WQBELs to complete the clean-up process.

3.10 In-Stream CBOD Deoxygenation Rate Summary

A number of BOD/DO modeling studies have been presented in this chapter, ranging from river stretches receiving a single point source discharge to multiple point source discharges along the river. One of the key model coefficients in model configuration is the receiving water deoxygenation rate, k_d. Determining this rate is a critical task requiring full knowledge of the characteristics of the wastewater as well as the receiving water. Independent derivation of the k_d rate from field samples of CBOD

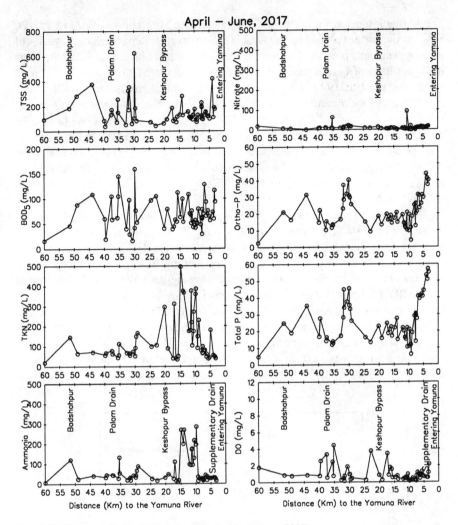

Fig. 3.25 Water quality in the Najafgarh Drain (April-June, 2017)

along the river is the most efficient approach. Field work needs to be expanded for multiple sources of variable strength to develop spatially variable k_d rates along the river. This chapter also demonstrated field and lab studies to perform long-term BOD tests, in which the resulting data were used to determine the site-specific, spatially variable k_d rates for the modeling analysis. In lieu of field data, the k_d rates can be obtained via the model calibration analysis. Given the importance of this kinetics coefficient, however, there is still not a physically based methodology to quantify this rate a priori.

A summary of k_d rates by U.S. EPA (1995) offered a plot of k_d versus river depth. An update of that plot can be found in Lung (2001) with the addition of the Shirtee

Creek, Blackstone River, Roanoke River, and upper Mississippi River. Figure 3.26 presents the current plot of k_d values at 20°C versus river depth including studies of rivers and streams presented in this Chapter. The dashed line in Fig. 3.26 represents the lower limit of k_d as a function of water depth, derived from Lung (2001) based on studies prior to 1990. The k_d rates from individual studies are added in this plot to expand the summary. First, the k_d rates in the Hudson River and Delaware River are well above 0.1 day^{-1} in the 1970s as secondary treatment had just begun at that time. The ranges shown indicate spatially variable deoxygenation rates in different stretches of the rivers derived from model calibration runs. Their k_d rates fall above the pre-1990 dashed line. Note that the 1975–1976 rates from the Hudson and Delaware are presented to give a historical perspective. Their rates are below 0.1 day^{-1} today.

The k_d rate in Shirtee Creek, Alabama in the 1991 study is around 0.3 day^{-1}, typical for very small streams during that time. The two studies of the Blackstone River, MA and Roanoke River, VA show the k_d rates around 0.1 day^{-1} by late 1990's, representing the prevailing conditions of rivers receiving wastewaters from secondary treatment in the latter part of that decade. Following the CWA of 1970s, the point source BOD related DO problems have subsided in the U.S. as demonstrated in the upper Mississippi River in the Minneapolis and St. Paul area of Minnesota. The progressive treatment plant upgrades from primary treatment, to secondary treatment, and secondary with nitrification offered significant water quality benefits in the upper Mississippi River, resulting in a downward trend for the k_d rate, dropping from 0.35 day^{-1} in 1964 to 0.073 day^{-1} in 1988. The more recent data reveal that it drops

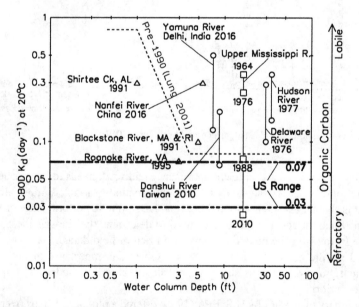

Fig. 3.26 Summary of CBOD Deoxygenation rates

to below 0.03 day^{-1} by 2010. There is not much labile organic carbon left in the CBOD of the river samples. The measurable CBOD is all in the refractory form. The rates from the upper Mississippi River, albeit from a single point source, were derived from the receiving water CBOD loading plot substantiated by independent long-term BOD tests of the receiving water samples.

The DO levels in the 20-km stretch of the Yamuna River near Delhi, India have been extremely low during the past two decades, a result of unabated pollution from domestic BOD discharges around the capital city. The calibrated range of k_d values from 0.15 day^{-1} to 0.5 day^{-1} is high. The modeling study of the Nanfei River, an urban river in China by Huang et al. (2017), yields a k_d rate of 0.3 day^{-1}. The extensive modeling study of the Danshui River in Taiwan, consisting of field determination of spatially variable k_d rates reported in this chapter, shows a range of rates below those of the Yamuna and Nanfei Rivers. Note that the rates reported in the Yamuna River and Nanfei River were developed from model calibration while the range of spatially variable rates for the Danshui River was obtained directly from very comprehensive field sampling and lab measurements.

In conclusion, 0.1 day^{-1} is now considered the upper bound of CBOD deoxygenation rates in systems receiving well treated (secondary and beyond) wastewaters as the CBOD gradually shifts from labile organic carbon to refractory organic carbon. For the majority of the receiving waters in the U.S., k_d would fall in the low range of 0.03 day^{-1} to 0.07 day^{-1} (see the band in Fig. 3.26) regardless of water depth and highlights an achievement of the CWA. These low rates further demonstrate the strong dependence of k_d on the characteristics of wastewater BOD loads. For rivers receiving marginally treated wastewaters, 0.5 day^{-1} may be used as the upper bound for the in-stream k_d rates. For a perspective, the deoxygenation rates of stormwater runoff BOD reported by McCabe et al. (2021) fall in a range between 0.03 day^{-1} to 0.07 day^{-1}, slightly below 0.1 day^{-1} of the present day rate for receiving waters. It should be pointed out that this plot not be construed as a replacement for deriving the rates using field data, as in the Danshui River (Sect. 3.4).

3.11 BOD/DO Modeling Recap

BOD modeling has come a long way for water quality management since the Streeter-Phelps equation in 1925 albeit still playing a surrogate role in the analysis. It is indispensable in quantifying a DO budget for the receiving water system as DO is the water quality endpoint for water quality management. BOD rarely is picked as the water quality endpoint with few exceptions. In water quality management of some small, fast moving streams in Taiwan, healthy DO levels sustained by strong stream reaeration from high velocities easily disqualify DO as the endpoint. Instead, a BOD standard is assigned by the regulatory agencies for these streams.

Even as a surrogate, BOD has been broken into multiple components in BOD/DO modeling by mandate: unfiltered and filtered BOD, CBOD and NBOD, ultimate CBOD and 5-day CBOD. For example, carbon in algal biomass would show up in unfiltered water samples (see Figs. 3.17 and 3.18 in Sect. 3.6). Obviously, differentiating these components require data, resulting in substantial expansion of the data gathering effort. Even today, not every permit is required to filter their samples for their 5-day BOD tests, a situation creating problems for the modelers who need to differentiate between the dissolved and particulate BOD for model configuration. In addition, SOD plays an important role in the DO budgets, as seen in the Delaware River and the upper Mississippi River, respectively (see Figs. 3.19 and 3.20 in Sect. 3.7). The DO problems, connected intimately with primary productivity of algae and sediment effects, in spite of the long history, tend to be considerably more complex than generally believed (Thomann 1987). Although not serving as an endpoint, the role of BOD in DO modeling cannot be underestimated.

It is clear that the dilemma of assigning the in-stream deoxygenation coefficient, k_d will remain. Putting effort and resources into field data collection to generate this coefficient instead of manipulating the modeling framework is the preferred path since k_d is a highly parameterized coefficient. Some modeling frameworks offering multiple BOD slots with different k_d rates (slow and fast) is counter-productive as it is difficult enough to derive a single, accurate k_d rate to begin with. Trying to assign two k_d rates in the same river reach without data support is not justified. In addition, differentiating concentrations of slow and fast BOD in the water samples (for model calibration) is an impossible task, let alone dealing with multiple discharges of variable strength. After all, since BOD is only a surrogate and not a water quality endpoint, it is better to keep the modeling exercise as simple as possible and to focus the effort on data collection. The approach adopted for the Danshui River in Taiwan (Sect. 3.4) is therefore highly recommended. Finally, a model post-audit to confirm or update the k_d rates (see Fig. 3.4) would complete the process.

References

Bhargava DS (1983) Most rapid BOD assimilation in Ganga and Yamuna Rivers. J Environ Engrg 109(1):174–188

Chapra SC (1997) Surface water-quality modeling. McGraw-Hill, NY

Chen CH et al (2012) Technical challenges with BOD/DO modeling of rivers in Taiwan. J Hydro-Environ Research 6:3–8

Chen CH et al (2013) Spatially variable deoxygenation in the Danshui River: improvement in model calibration. Water Environ Res 85(12):2243–2253. https://doi.org/10.2175/106143013X13596524516301

Epa US (1984) Handbook of advanced treatment review issues. Office of Water Program Operation, Washington DC

Haffely G (1997) Long-term BOD Data on the Upper Mississippi River, Minnesota River, St. Croix, and the Metro Plant. Personal communication, Metropolitan Council, Minneapolis, MN

Haffely G (2009) Personal communications, Metropolitan Council, Minneapolis, MN

Haffely G, Johnson L (1994) Biodegradation of TCMP (N-Serve) nitrification inhibitor in the ultimate BOD test. In: Proc WEFTEC 1994, 64th Ann Conf., Chicago, IL.

Hall JC, Foxen RJ (1983) Nitrification in BOD test increases POTW noncompliance. J Water Pollut Control Fed 55(12):1461–1469

Huang J et al (2017) (2017) Modelling dissolved oxygen depression in an urban river in China. Water 9:520. https://doi.org/10.3390/w9070520

Leo MW et al (1984) Before and after case studies: comparison of water quality following municipal treatment plant improvements. Report submitted by HydroQual, Inc. to U.S. EPA Office of Water Program Operations, Washington DC

Lung WS (1996) Post audit of the Upper Mississippi River BOD/DO model. J Environ Eng 122(5):350–358

Lung WS (2001) Water quality modeling for wasteload allocations and TMDLs. Wiley, New York, NY

Lung WS, Larson CE (1995) Water quality modeling of Upper Mississippi River and Lake Pepin. J Environ Eng 121(10):691–699

Mandal P et al (2010) Seasonal and spatial variation of Yamuna River water quality in Delhi, India. Environ Monit Assess 170:661–670

McCabe KM et al (2021) Particulate and Dissolved Organic Matter in Stormwater Runoff Influences Oxygen Demand in Urbanized Headwater Catchments. Environ. Sci. Technol. https://dx.doi.org/ https://doi.org/10.1021/acs.est.0c04502

Metropolitan Waste Control Commission (1989) 1988 Mississippi River low flow survey report. Saint Paul, MN

NCASI (1982) A review of ultimate BOD estimation and its kinetic formulation for pulp and paper mill effluents. National Council of the Paper Industry for Air and Stream Improvement, Inc. Technical Bulletin 382, Medford, MA

Paliwal R et al (2007) Water quality modelling of the river Yamuna (India) using QUAL2E-UNVCAS. J Environ Manage 83:131–144

Parmar KS, Bhardwaj R (2015) Statistical, tie series, and fractal analysis of full stretch of river Yamuna (India) for water quality management. Environ Sci Pollut Res 22:397–414

Sharma D, Singh RK (2009) DO-BOD modeling of River Yamuna for national capital territory, India using STREAM II, a 2D water quality model. Environ Monit Assess 159:231–240

Sharma D et al (2017) Water quality modeling for urban reach of Yamuna River, India (1999–2009), using QUAL2Kw. Appl Water Sci 7:1535–1559

Thomann RV (1987) System analysis in water quality management – a 25 year retrospect. In: Beck MB (ed) System analysis in water quality management. Pergamon Press, Tarrytown, NY, pp 1–14

Thomann RV, Mueller JA (1987) Principles of surface water quality modeling and control. Harper & Row, New York

Tyagi B et al (2014) Analysis of DO sag for Multiple Point Sources. Math Theory and Modeling 4(3):1–7

US EPA (1995) Technical guidance manual for developing total maximum daily loads, Book 2: streams and rivers, part 1: biochemical oxygen demand/dissolved oxygen and nutrients/eutrophication. EPA 823-B-95–007, Washington DC

Chapter 4
Nutrients, Algae, and DO

Water quality management to restore eutrophication is an ongoing effort in the twenty-first century. While point source related BOD-DO problems are eliminated in developed countries, the nutrient-related DO problem has surfaced, making the assessment of eutrophication much more complex and difficult. Mitigating eutrophication points to nutrient control. Nutrient control has been adopted to curb eutrophication in lakes, reservoirs, and estuaries for over 70 years. Phosphorus reductions from point sources were implemented to manage the blue-green algal blooms in the Potomac Estuary in 1970 and the river has not seen any significant algal outbreaks since the summer of 1983 algal bloom (see Chap. 1)—a success story.

Nutrient control still generates heated debates as to which nutrient(s) should be selected for control and perhaps more importantly, how to implement the control with respect to spatial and temporal patterns of nutrient reductions. For example, Conley et al. (2009) suggested that for controlling eutrophication, nitrogen and phosphorus should be selected. They demonstrated that improvements in the water quality of many freshwater and most coastal marine ecosystems requires reductions in both nitrogen and phosphorus inputs. On the other hand, Schindler et al. (2008) strongly stated that lake eutrophication could not be controlled by reducing nitrogen input based on the results of a 37-year study of Lake 227, a small lake in the Precambrian Shield, Canada. Bryhn and Håkanson (2009) suggested using models to guide the analysis.

More recently, Schindler et al. (2016) reported that reducing inputs of a single nutrient, phosphorus, led to the success of curbing eutrophication in a number of lakes in Europe and North America. They credited Vollenweider's work in 1960s as the foundation for adopting phosphorus control for the Laurentian Great Lakes, leading to the subsequent success of mitigating eutrophication of these lakes in the following two decades. However, Vollenweider's loading plot was highly empirical. Thomann (1977) reviewed Vollenweider's loading plot (see Sect. 2.8) and provided a solid scientific base for the analysis, laying the groundwork for eutrophication modeling of lakes and estuaries for water quality management.

© The Author(s), under exclusive license to Springer Nature Switzerland AG 2022   107
W.-S. Lung, *Water Quality Modeling That Works*,
https://doi.org/10.1007/978-3-030-90483-8_4

This chapter focuses on interplay of the modeling results and field data to extract physical insights into nutrient limitations for controlling eutrophication. Results of such analyses are used to address environmental mangers' questions on nutrient control. Instead of discussing eutrophication modeling frameworks and their configuration, we employ a massive amount of data to support the modeling analysis. Field data prove critical to eutrophication analysis.

4.1   Modeling for Nutrient Control

Thomann (1977) made the following assessment following an evaluation of the simple nutrient plot model by Vollenweider (1968) and the complex (mechanistic) models:

1   They are not mutually exclusive
2   They must be derivable from each other
3   More complex models are needed to obtain a better estimate of parameters which also exert an influence on loading tolerance, and
4   Simple models are of value in predicting the effects of reduction of nutrient load on algal biomass.

Further, Thomann pointed out that using complex modeling would be a reasonable option in developing management strategies for eutrophication control. His modeling work on Lake Ontario (Thomann et al. 1976) demonstrated this thinking. Bryhn and Hákanson (2009) also suggested model before acting. Their only concern was the validity of the model results. Fortunately, many modeling frameworks (the so-called complex models by Thomann) incorporating both nitrogen and phosphorus are readily available for the analysis. The remaining roadblocks would be to collect the observation data, which are needed to develop the site-specific model for the system, to calibrate the model (Bryhn and Hakanson's main concern), and to identify data gaps for refining the model. The focus of this chapter is to demonstrate the interplay of modeling results and field data to develop sound management strategy for eutrophication control. This is accomplished by using a number of case studies of lakes and estuaries with sufficient data for model calibration and verification.

Environmental scientists favor the use of the Redfield ratio in assessing nutrient control. While this ratio is useful to estimate relative amounts of nutrients in the water column, it has many drawbacks:

1.   The Redfield ratio is not a universal law, rather it is an empirical relationship (established about 80 years ago) which is subject to local deviations and alternations under changing environmental conditions. Therefore, there is no surprise in that it does not hold universally.
2.   The Redfield ratio more or less works well on average and when neither chemical component is a limiting factor in the local ecosystem. Site-specific data may come from a system that is limited by either P or N content.

It should be pointed out that nitrogen and phosphorus are rarely considered as water quality endpoints in eutrophication control. Instead, dissolved oxygen (DO) and chlorophyll *a* are the common endpoints in eutrophication management. Nutrient criteria in terms of nitrogen and phosphorus levels are developed to meet the chlorophyll *a* and/or DO endpoints for the receiving water quality management. Chlorophyll *a* and/or DO are first evaluated for water systems to see if the water quality is impaired or compromised. If so, reduction of nutrient inputs (nitrogen and/or phosphorus) can be quantified via eutrophication modeling, supported by site-specific data, to meet the water quality standards for specific endpoints of chlorophyll *a* or DO, i.e. modeling before acting (Bryhn and Håkanson 2009). Adopting this strategy would avoid arbitrarily reducing nutrient levels in the receiving water simply because of high nutrient levels without evaluating whether the water system is impaired or not. Such actions have resulted in much wasteful effort in eutrophication control using DO and chlorophyll *a* as water quality endpoints.

Another key factor in eutrophication control and management is sediment nutrient release, which could easily cancel out the effort in reducing external nutrient inputs. Further, reducing external nitrogen inputs may not be effective in eutrophication control for some water systems, which are known to have nitrogen fixing blue-green algae (cyanobacteria) such as *Microcystis aeruginosa*. In addition to the biological aspects of eutrophication, physical characteristics (e.g. hydrodynamics) associated with the water system may play a key role in eutrophication analysis. All these factors make arbitrary reductions of nutrient inputs to lower the nutrient concentrations in the water system extremely risky and potentially wasteful.

Logical steps to take in a eutrophication modeling analysis would be:

a.   Water quality impairment assessment
b.   Establishing water quality endpoints (chlorophyll *a* and/or DO)
c.   Configuring a eutrophication model with substantial field data support
d.   Identifying limiting nutrient(s) with respect to temporal and spatial distribution
e.   Calibrating and verifying the model with quantified goodness of fit
f.   Developing dynamic nutrient budgets with the model results
g.   Back-calculating the allowable external nutrient inputs to restore eutrophication
h.   Assessing the response time to nutrient reductions.

4.2   Nonlinear Nutrient Kinetics

In many eutrophication models, nutrient limitation on algal growth is formulated with the Michaelis-Menton kinetics relationship (Thomann and Mueller 1987; Chapra 1997):

$$V(S) = \frac{S}{K_m + S} \tag{4.1}$$

Fig. 4.1 Michaelis-Menton
kinetics for nutrient
limitation

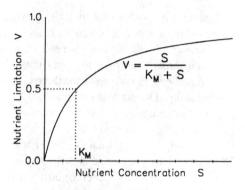

where S is the concentration of the nutrient (nitrogen or phosphorus) needed for
phytoplankton growth and K_m is the concentration of the nutrient at which the nutrient
effect $V(S) = \frac{1}{2}$. K_m is called the Michaelis constant on the uptake kinetics of organisms on substrates. Figure 4.1 shows the plot of $V(S)$ versus S. The curve of $V(S)$
versus S consists of two portions. The first portion with the nutrient concentration
less than K_m is close to a linear relationship, where the nutrient limiting effect is
clear and obvious. In the second portion, when a present nutrient input lands on the
nonlinear part of the curve with S much greater than K_m, the effect of nutrient limiting
the phytoplankton growth is minimal. This later, nonlinear relationship explains the
many unsuccessful nutrient reduction programs for eutrophication control.

A good example to demonstrate this nonlinearity effect on eutrophication control
is the investigation of phosphorus load reductions to Lake Pepin in Minnesota.
Figure 4.2 shows the study area from the upstream end at Lock and Dam No. 1
to the downstream end at the outlet of Lake Pepin for a distance of 100 miles.
The Metro Plant, a domestic WWTP serving the twin city area of St. Paul and
Minneapolis, has long been blamed as the cause of eutrophication in Lake Pepin
which exhibits elevated chlorophyll a levels. Phosphorus removal at the Metro Plant
was suggested. However, Lung and Larson (1995) and Lung (2001) showed that
phosphorus removal at the Metro Plant alone on the upper Mississippi River has a
limited effect on reducing the chlorophyll a levels in Lake Pepin. That conclusion
has raised a number of questions from the regulatory agency regarding the fate of
phosphorus loads from the Metro Plant in Lake Pepin. To what extent is phosphorus
from the Metro Plant transported to Lake Pepin under both existing and potentially
reduced phosphorus loading conditions? Perhaps a more specific question is how
much phosphorus in the Lake Pepin algal biomass is originated from the Metro Plant
effluent?

Recalling the component analysis performed with the BOD-DO model of the
upper Mississippi River (see Fig. 3.20), the sum of the DO deficits due to individual
oxygen consumption components is equal to the total DO deficit calculated from a
model run with all individual components being turned off. The analysis procedure
is straightforward: a certain BOD load is removed from the model, the model is
rerun, and the resulting DO concentrations are compared to the original results. The

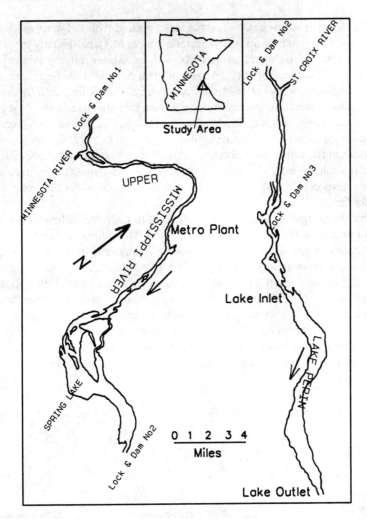

Fig. 4.2 Upper Mississippi River and Lake Pepin

difference in DO concentrations between these two model results represents the portion of the overall DO consumption in the receiving water attributable to this specific component of the BOD loadings.

However, this procedure would fail on eutrophication modeling. Removing individual phosphorus sources would result in proportional reductions in river phosphorus concentrations but unproportioned reductions in chlorophyll *a* concentrations. In other words, if the sources of phosphorus were removed one at a time and the resulting chlorophyll *a* concentrations were added up, the total chlorophyll *a* concentrations would be much higher than the chlorophyll *a* concentrations resulting from the model run in which all phosphorus sources are included. This superposition principle works for the BOD/DO model component analysis because the BOD and

DO deficit kinetics is linear. On the other hand, algal growth and nutrient limitation are in a nonlinear relationship as demonstrated in Eq. (4.1), particularly in the range of high phosphorus concentrations (see Fig. 4.1). In addition, phytoplankton growth is characterized with nonlinear light and nutrient dynamics in the model. Under existing conditions in the upper Mississippi River, nutrients are generally in excess and do not exert much control over phytoplankton growth. Instead, the algal growth is light limited (see the light extinction coefficients in Figs. 2.30 and 2.31). Removing one phosphorus source may not be significant enough to affect the phytoplankton growth rate and thereby would have little impact on the chlorophyll a concentrations. A more rigorous approach is needed to address the above question by tracking the fate and transport of the Metro Plant phosphorus along the upper Mississippi River and Lake Pepin.

Instead of using $^{32}PO_4$ (a radioactive tracer) to track the different phosphorus inputs, a numerical tracer is used for Lake Pepin. The calibrated model was modified to include a numerical tracer representing phosphorus from a given origin to perform the numerical tagging analysis. For a comprehensive presentation of this tagging technique, consult Lung and Testerman (1989) and Lung (1996). Results of this numerical tagging analysis for the upper Mississippi River and Lake Pepin are summarized in Fig. 4.3 showing orthophosphorus and chlorophyll a concentration profiles in the study area from Lock and Dam No. 1 to the outlet of Lake Pepin under the 1988 summer low flow conditions. The left column of Fig. 4.3 presents the model calibration results matching the measured orthophosphorus and chlorophyll a in the water column.

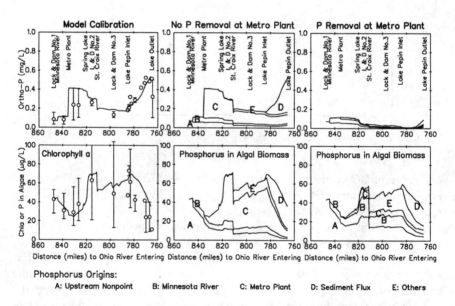

Fig. 4.3 Tagging phosphorus in upper Mississippi River and Lake Pepin (Summer of 1988)

In the middle column of Fig. 4.3, the numerical tagging results associated with the Metro Plant effluent without phosphorus removal are presented. First, results are plotted from the model run in which phosphorus from the upstream nonpoint source (i.e., the Mississippi River at Lock and Dam No. 1) is tagged. Next, results from the model runs with the Minnesota River loads tagged are added. Subsequently, results from the model runs with phosphorus loads from the Metro Plant, other sources, and sediment flux are superimposed. The key aspect is that the *total* concentration profiles match the original model calibration results (in the left column of Fig. 4.3) for orthophosphorus and chlorophyll *a*, confirming the successful operation of the tagging model (Lung 1996).

Result in the middle column of Fig. 4.3 shows the dominating effect of the Metro Plant (without phosphorus removal) discharge on the orthophosphorus concentrations in the water column during the summer months of 1988. Under extremely low flow conditions in 1988, the relatively stable flow from the Metro Plant delivers significant phosphorus loads to the upper Mississippi River compared with nonpoint origins from the upstream and tributaries. As a consequence, phosphorus loads from the Metro Plant have a greater influence on ambient river concentrations in low flow years than in higher flow years, as will be demonstrated in further discussion later. Although the Metro Plant effluent contains little or no phytoplankton biomass, phosphorus from the effluent is gradually taken up by the phytoplankton in the river, leading to the wide band of viable chlorophyll *a* attributed to the Metro Plant phosphorus origin in the middle bottom plot of Fig. 4.3. It should be pointed out that the numerical tagging model is designed to proportion algal uptake of phosphorus according to the proportion of various sources of orthophosphate in the water column. Because the Metro Plant contributes the largest portion of orthophosphorus in summer of 1988, it is not surprising to see a significant portion of the algal biomass below the plant incorporating this tagged nutrient source. The model uses a one-to-one ratio of phosphorus to chlorophyll *a* in the algal biomass. Figure 4.3 shows that roughly 50 μg/L of net algal phosphorus at the end of pool behind Lock and Dam No. 2 and 40 μg/L of net algal phosphorus at the inlet of Lake Pepin came from Metro Plant phosphorus in summer of 1988. On the other hand, the role of phosphorus from the headwaters (i.e., at Lock and Dam No. 1) is greatly diminished, reducing from 40 μg/L of algal phosphorus at the upstream end to only about 4 μg/L at the outlet of Lake Pepin (see phosphorus origin A in middle column of Fig. 4.3). This is in response to its decreasing portion of orthophosphate in the water column. However, it should not be interpreted from Fig. 4.3 that the Metro Plant phosphorus is the *cause* of the large increase in chlorophyll *a* behind Lock and Dam No. 2. In fact, flow and light are the primary controlling factors of algal growth in the upper Mississippi River. High growth in the pool behind Lock and Dam No. 2 are attributable to reduced flows and increased light in the vicinity of Spring Lake. Also, given the nonlinear nature of algal dynamics, it cannot be assumed that reductions in Metro Plant phosphorus will result in similar reductions in algal biomass, as will be seen in the tagging results under the reduced Metro Plant load.

The right column of Fig. 4.3 presents the model results under a much reduced Metro Plant load with phosphorus removal in operation, thereby minimizing algal uptake of phosphorus. In fact, the Metro Plant (phosphorus origin C) would be responsible for only 10 µg/L of chlorophyll *a* behind Lock and Dam No. 2 and even less (about 5 µg/L) in Lake Pepin (see the right hand column of Fig. 4.3). Least but not last, a noticeable amount of algal phosphorus under this scenario is from the sediment origin (i.e. phosphate flux in Lake Pepin). Phosphorus origins D and E together make a formidable contribution to the phosphorus in Lake Pepin, more than compensating for the reduced contribution from the Metro Plant phosphorus. Despite the significant reduction of phosphorus loads from the Metro Plant, the overall peak phytoplankton biomass reduction in Lake Pepin amounts to only 10 µg/L (reduced from 70 to 60 µg/L) of chlorophyll *a*, in agreement with the finding by Lung and Larson (1995).

While the concentrations of various phosphorus components in Fig. 4.3 show the relative strengths of these sources, the effects of phytoplankton nutrient dynamics, settling to the sediment, and dilution from other flows cannot be differentiated. Perhaps additional physical insights into the fate of individual phosphorus origins can best be illustrated by mass rates along the river. That is, for a particular phosphorus origin, the mass rates of a phosphorus component along the main channel is calculated by multiplying the phosphorus concentration by the advective outflow. Lateral flows between the main channel and the shallow area in Spring Lake are not included in this calculation.

The left hand plots of Fig. 4.4 show the mass rates in Mg/day (1 Mg = 1000 kg) for total P, ortho-P, total organic P, and P in algal biomass under the summer of 1988 condition, associated with two sources: Metro Plant and upstream river flow, respectively. The total organic P curve includes the P in algal biomass. The sum of the total organic P and ortho-P makes up the total P. The amount of total P released by

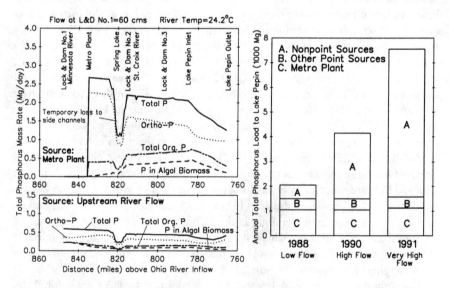

Fig. 4.4 Phosphorus components in upper Mississippi River and phosphorus loads to Lake Pepin

the Metro Plant is about 2.7 Mg/day, with the majority (about 84%) in the ortho-P form. The sharp decreases of ortho-P and total P mass rates in Spring Lake are due to lateral flows and dispersion transporting mass out of the main channel. The main channel later regains this mass from the lateral area near Lock and Dam No. 2.

The top left plot of Fig. 4.4 follows the fate of phosphorus output from the Metro Plant alone. P in the algal biomass mass rate (the dashed line) increases following the Metro Plan input to about 0.45 Mg/day by Lock and Dam No. 2, with the sharpest rise in the Spring Lake area. This mass rate continues to grow, reaching a peak of 0.5 Mg/day at the entrance of Lake Pepin, and then reduces throughout the lake to a level of 0.09 Mg/day at the outlet of Lake Pepin. The shallow depths in the Spring Lake area offer a favorable condition for algal growth, while the significant depths in Lake Pepin, particularly in the lower portion, drastically reduce the available light for algal growth as the algae are mixed to greater depths. Such a difference in the P in algal biomass mass rates between Spring Lake and Lake Pepin is particularly pronounced during a low flow condition with extremely shallow waters in the Spring Lake area.

There is a net loss of about 0.68 Mg/day (= 2.25 − 1.57) of Metro Plant ortho-P between river miles UM825 and UM815 (see the dotted line), of which 0.27 Mg/day is converted to P in algal biomass. The balance of 0.41 Mg/day represents a net loss of phosphorus via the settling of algae and nonliving detritus. While there is tremendous algal growth in Spring Lake, more algal biomass settles into the sediment behind Lock and Dam No. 2 than makes it over the dam. As a result, the pooled area behind Lock and Dam No. 2 served as a phosphorus sink in the summer of 1988. Similarly there is a net loss of approximately 0.41 Mg/day (= 1.37 − 0.96) Metro Plant ortho-P between the inlet and outlet of Lake Pepin (the dotted line). Like Spring Lake, there is also a net loss in P in algal biomass (about 0.36 Mg/day) due to settling in Lake Pepin. Thus, a total loss of 0.77 Mg/day phosphorus is seen for Lake Pepin. In summer, the two major losses of phosphorus from the Metro Plant are in pooled areas behind Lock and Dam No. 2 (about 0.41 Mg/day) and in Lake Pepin (about 0.77 Mg/day).

Results from a similar analysis for phosphorus origin of the upstream river flow are shown in the bottom left plot of Fig. 4.4. The total P mass rate from Lock and Dam No. 1 is much lower than that of the Metro Plant in summer of 1988. A similar, sharp decline and regain of ortho-P and total P between river miles UM825 and UM815 also occur for this phosphorus source—an artifact of the lateral mass exchanges. P in algal biomass from this source continues to decrease in the downstream direction due to settling and recycling to total organic P and ortho-P. In fact, recycling minimizes the loss of phosphorus throughout the study area for this phosphorus source. The total reduction of the upstream phosphorus mass rate over the study reach is roughly 0.11 (= 0.60 − 0.49) Mg/day in summer of 1988, representing about 20% of the total P mass rate at Lock and Dam No. 1. At the outlet of Lake Pepin, the majority of phosphorus is ortho-P, as a result of this recycling. Interactions between these phosphorus components in the upper Mississippi River and Lake Pepin quantified by this mass rate tracking analysis reveal important insights into the fate and transport of individual phosphorus sources in the system, thereby offering valuable information for decision makers in water quality management.

The above discussions can be put into a perspective by examining the annual phosphorus loads of different origins to Lake Pepin. The right-hand plot in Fig. 4.4 shows annual total phosphorus loads from the upper Mississippi River Watershed to Lake Pepin in 1988, 1990, and 1991. The phosphorus loads are grouped into three origins: the Metro Plant, other point sources, and nonpoint sources. While the Metro Plant and other point source loads remain relatively constant over these 3 years, nonpoint loads increase sharply from 1988 (a low flow year) to 1991 (an extremely high flow year). Benefits of phosphorus removal at the Metro Plant may be realized in a low flow year, but will disappear in a high flow year.

This example demonstrates the nonlinearity of phosphorus and algal interactions in Lake Pepin and its effect on nutrient management. Controlling the Metro Plant phosphorus input alone would not mitigate the eutrophication problem in Lake Pepin. For years, the residents around Lake Pepin have faulted the Metro Plant for the water quality problem in lake. The above analysis suggests that reducing the phosphorus from the Metro Plant alone would not solve the problem. As demonstrated in Fig. 4.4, reductions of other phosphorus origins such as the nonpoint phosphorus loads from upstream of the upper Mississippi River and the Minnesota River must be addressed concurrently to achieve chlorophyll *a* reductions in Lake Pepin. This misconception and public concern are described in "Treatment Plant May not be the Source of Pepin Problem"—St. Paul Pioneer Press, August 3, 1992.

4.3   Chorophyll *a* Endpoint

The James River is one of the most iconic rivers of the United States. Affectionately known as "American's Founding River", the James brought early English settlers to Jamestown, the first permanent English settlement in the New World. The James River is not only Virginia's largest river, but is also the largest tributary to Virginia's portion of the Chesapeake Bay (the Bay) and the third largest (behind the Susquehanna and Potomac) of all Bay tributaries (Fig. 4.5). The James River has been having eutrophication problems with consistent, significant algal growth in the Hopewell area. Phosphorus input from major point sources along the river, particularly in the upper estuary from Richmond to Hopewell, provided the needed nutrient for significant algal growth in the 1980s. Figure 4.6 shows the orthophosphorus and chlorophyll *a* concentrations from the fall line (where tides stop) in Richmond to the mouth of the river during the summers of 1976 and 1983. The July 1976 data shows low chlorophyll *a* levels along the estuary. By September 1983 the orthophosphorus levels exceeded 0.4 mg/L immediately after receiving phosphorus input from the Richmond WWTP. This and other phosphorus input from sources in the Hopewell area supported the algal growth to reach a peak chlorophyll *a* level exceeding 40 μg/L at the location about 75 miles from the mouth. In the meantime, however, BOD and

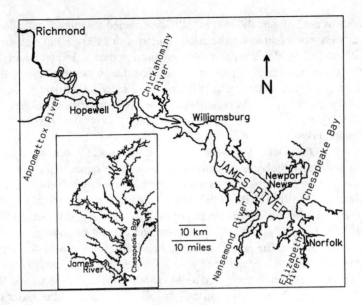

Fig. 4.5 The James River Estuary from Richmond to Chesapeake Bay

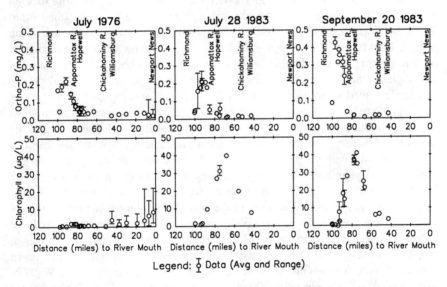

Fig. 4.6 Rising peak chlorophyll *a* levels in the James River Estuary from 1976 to 1983

nitrogen levels in the river were not high enough to cause dissolved oxygen depression (Lung 1986).

While the Bay and other major tidal tributary waters (below the fall line) may have algal-related impairments due to eutrophication, U.S. EPA and the seven sub watersheds of the Bay jurisdictions initially determined that numerical chlorophyll *a* criteria were not required for the tidal tributaries in 1990s. Following an extensive evaluation of the Bay's water quality monitoring data and water quality model simulations, the Bay Program partners concluded that implementation of the new dissolved oxygen and water clarity criteria would mandate nutrient reductions to these tributaries, thereby addressing the algal-related impairments in these waters (U.S. EPA 2003). In other words, DO and turbidity were identified as the water quality endpoints.

However, DO was later determined not to be the correct water quality endpoint for the James Estuary because the nutrient loads from the James River Watershed do not significantly impact dissolved oxygen concentrations or water clarity conditions in the water column (U.S. EPA 2003). Unlike the other major tributary systems of the Bay, the tidal James River is relatively shallow and very well mixed (Kuo and Neilson 1987; Hagy et al. 2000), offering enhanced input of atmospheric oxygen into the water column. The proximity of the James River to the Atlantic Ocean and its input of relatively well oxygenated waters also keeps the DO levels in the James Estuary at acceptable levels compared to the other Bay tributaries (Kuo and Neilson 1987). As a result, chlorophyll *a* instead of DO was picked as the water quality endpoint for eutrophication control for the James Estuary.

Control measures to curb the chlorophyll *a* levels in the James therefore focused on reducing the nutrient inputs, starting with a phosphate-detergent ban (the ban) in the Commonwealth of Virginia on January 1, 1988. [Other jurisdictions such as the State of Maryland and Washington, District of Columbia had their bans earlier for other major tributaries to the Bay.] Subsequently the total phosphorus concentration of 2 mg/L in the effluents of domestic WWTPs with flows of 1 mgd and greater was implemented in Virginia by March 1, 1991. Figure 4.7 shows the impact of these two control measures on the influent and effluent phosphorus concentrations at one of the largest domestic WWTPs (Chesapeake-Elizabeth) discharging to the James River Estuary. Phosphorus concentrations are obtained from the discharger's NPDES permit monthly reports submitted to VDEQ. The influent total phosphorus concentrations were lowered slightly following the ban as shown in the top plot of Fig. 4.7. The effluent concentration was also lowered after the ban but with additional reduction due to phosphorus removal at the plant. [Note that the ban has no impact on the effluent concentrations once phosphorus removal is online. It only reduces the plant's operating cost to remove phosphorus.] Biological phosphorus removal at WWTPs started in late 1990 and finally brought the effluent phosphorus level down to below 2 mg/L. By 1994 the effluent total phosphorus concentrations at all WWTPs along the James River reached the 2 mg/L target. The question: Did this phosphorus reduction help to lower the chlorophyll *a* levels? The answer: Not yet.

The VDEQ established numerical chlorophyll *a* criteria for the tidal James River in 2005. The criteria are set for different segments of the river under spring and summer conditions. The scientific basis for the 2005 criteria was questioned once it became apparent that the nutrient load limits set by the Bay Total Maximum Daily Loads (TMDLs) for chlorophyll *a* criteria attainment in the James River basin (U.S. EPA 2010) was much lower than an earlier estimate (U.S. EPA 2003). In 2011, VDEQ

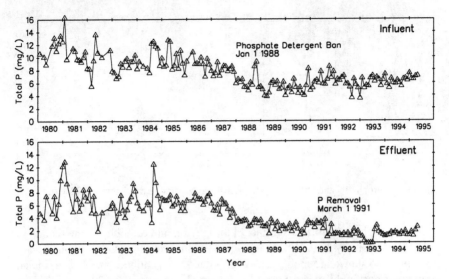

Fig. 4.7 Monthly average influent and effluent phosphorus concentrations at Chesapeake-Elizabeth WWTP from 1980 to 1995

launched a study to evaluate the 2005 criteria. Following a 5-year evaluation of the chlorophyll *a* criteria for the James River, a new set of criteria was recommended in 2019. Figure 4.8 shows the 2005 and 2019 criteria in four sections of the James River Estuary: upper tidal fresh, lower tidal fresh, oligohaline, and mesohaline/polyhaline for the summer months. The chlorophyll *a* peaks of July 2010 occur immediately below the entrance of the Appomattox River, easily exceeding the 2005 criteria of 23 μg/L and the 2019 criteria of 24 μg/L, respectively (see the right panel of Fig. 4.8).

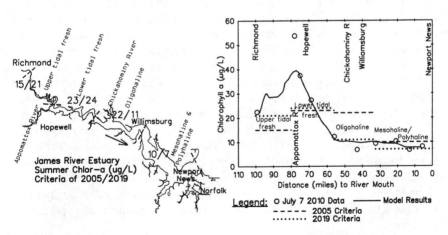

Fig. 4.8 Chlorophyll *a* Criteria for the James River Estuary set in 2005 and 2019

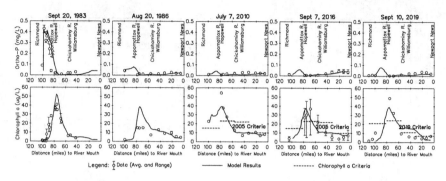

Fig. 4.9 Chlorophyll *a* and Orthophosphate Levels in the James River Estuary (1983–2019)

A review of the historical summer chlorophyll *a* and orthophosphate data in the James River Estuary from 1983 to 2019 (continuing from Fig. 4.6) is summarized in Fig. 4.9. Also shown are model results generated from calibration, verification, and continuing post audit of the James Estuary model (Lung 1986, 2001, 2017; Lung and Testerman 1989). These select data sets are from the Bay database for the James Estuary under summer low flow conditions. Although the peak orthophosphate level has been substantially reduced following nutrient control measures such as the ban and phosphorus removal (effluent TP not exceeding 2 mg/L) at WWTPs since 1983, the drop in chlorophyll *a* peaks near Hopewell is not significant, still exceeding the 2005 and 2019 criteria.

The next step of analysis is to determine the effluent total phosphorus limits at WWTPs to meet the chlorophyll *a* criteria under a very stringent condition—the summer 7Q10 low flow. The September 2019 point source loads were used for the model calculation. Model results are presented in Fig. 4.10. The total phosphorus concentrations from the point sources in 2019 are shown in the left plot of Fig. 4.10 and they are all below 2 mg/L. The model calculated total phosphorus, orthophosphate, and chlorophyll *a* concentrations in the James River Estuary are shown in the next two plots in Fig. 4.10. The TP and ortho-P concentrations in the upper tidal fresh section (between Richmond and the Appomattox River) are still dominated by the WWTP and industrial phosphorus loads, leading to a significant chlorophyll *a* peak of 55 μg/L in the Hopewell area. Chlorophyll *a* levels in the water column between river miles 85 and 55 (i.e. the entire lower tidal fresh stretch of the river) exceed the 2019 summer criteria of 24 μg/L. Note that the chlorophyll *a* peak of 55 μg/L is higher than that on September 10, 2019 in Fig. 4.9 because the summer low flow is lower than the September 2019 flow. As reported earlier, the calculated chlorophyll *a* peak is controlled by the depleting phosphorus supply (with ortho-P levels close to the Michaelis constant of 0.001 mg/L) in the Hopewell area and relatively strong light extinction in the water column of the oligohaline and mesohaline portions of the river.

To meet the 2019 chlorophyll *a* criteria for the lower tidal fresh portion of the James River Estuary under the 7Q10 low flow, the model was run under three loading

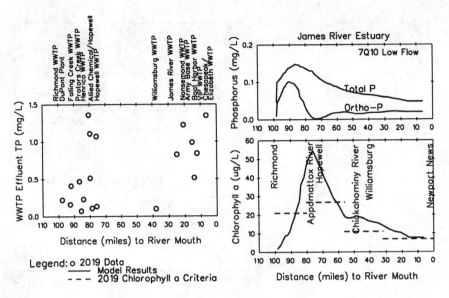

Fig. 4.10 Model prediction for 7Q10 low flow conditions

scenarios for effluent total phosphorus concentrations at 0.5, 0.4, and 0.2 mg/L, respectively. Figure 4.11 presents the model projection results. The ortho-P concentrations would reach all-time low levels in the upper tidal fresh portion, suppressing the chlorophyll *a* peak in the lower tidal fresh portion substantially. The 24 μg/L criteria of 2019 would be met with effluent total phosphorus concentrations below 0.5 mg/L. In a separate analysis based on the Bay wasteload allocations study, VDEQ developed the effluent total phosphorus concentration of 0.1 mg/L for all dischargers except the Richmond WWTP (0.4 mg/L) and Hopewell WWTP (0.3 mg/L) to meet the 2005 criteria at 23 μg/L (Brockenbrough 2016).

Chlorophyll *a* is picked as the water quality endpoint to address eutrophication control of the James Estuary by the regulatory agencies. A simple model with strong data support has been able to mimic the observed phosphorus and chlorophyll *a* levels in the water column from 1980s to the present. After the phosphate-detergent ban and subsequent phosphorus removal at WWTPs, the chlorophyll *a* levels in the James are still high under dry weather, low flow conditions, forcing VDEQ to set chlorophyll *a* criteria in 2005, amended in 2019. In the meantime, nutrient loads to the tidal portion of the James River have been fully allocated for the point source dischargers along the river. New dischargers would only be allowed via nutrient trading to acquire the loads from the unused allocations of existing dischargers. In 2016, a modeling study was launched to evaluate a potential industrial manufacturing facility with sizable nutrient loads near the Proctors Creek WWTP in the Richmond area for a NPDES permit. Nutrient trading involving shifting loads between upstream and downstream discharges resulted in no significant impact on the chlorophyll *a* and DO levels while still meeting the criteria under dry weather low flow conditions (Lung 2017).

Fig. 4.11 Point source effluent phosphorus concentrations to meet the 2019 Chlorophyll *a* Criteria

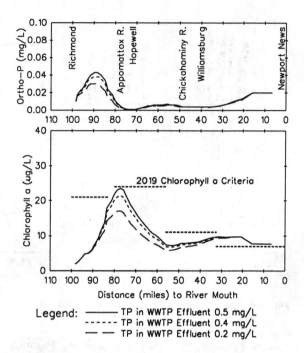

4.4   Dissolved Oxygen Endpoint

4.4.1   *Eutrophication in the Patuxent Estuary*

One of the purposes of modeling is to use model results for water quality management and control. While fully calibrated and verified models are generally used to perform model predictions associated with carefully developed simulation scenarios to provide answers for decision makers, at times lingering questions remain prior to firming up a sound strategy for decision-making. Examples include the Potomac Estuary algal bloom mystery and the numerical tagging of phosphorus for the upper Mississippi River presented in earlier sections. A similar but more intriguing case in the Patuxent River Estuary reveals interesting physical insights into the nutrient dynamics as well as interactions between mass transport and phytoplankton in the water column. Although DO is not picked as the water quality endpoint for eutrophication control in the James River Estuary, it is identified as the endpoint for the Patuxent Estuary.

The Patuxent River Estuary in Maryland (Fig. 4.12), located between Washington, DC and Baltimore, has been a focus of water quality studies for the past four decades. Persistent algal growth in the upper estuary near Nottingham in the spring have been observed (HydroQual 1981; Lung 1992; Breitburg et al. 2003). Subsequent anoxic conditions that developed around Broomes Island in the lower estuary over the summer months has been the primary water quality concern. Historical legal

Fig. 4.12 The Patuxent
River Watershed and Estuary

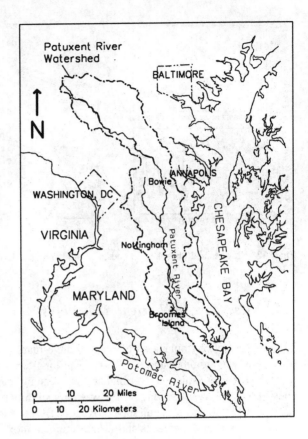

actions brought by local citizens in the lower watershed against the upper watershed
have been reported, continuing the call for control measures to curb nutrient inputs
from point sources above the fall line at Bowie. Note that all but one of the domestic
WWTPs are located above the fall line.

Subsequent point source nutrient controls in the Patuxent River Watershed have
been in effect for over three decades, implemented in different phases. The first
phase of nutrient control called for a phosphate-detergent ban, which began in 1986,
immediately resulting in a sharp reduction (over 50%) of phosphorus loads to the
Patuxent Estuary (Fig. 4.13). By 1991, nitrogen removal was installed at the major
wastewater treatment plants in the Patuxent watershed, resulting in the significant
decline of total nitrogen loads. Testa et al. (2008) displayed a similar temporal pattern
of nutrient loads to the Patuxent Estuary from 1985 to 2003.

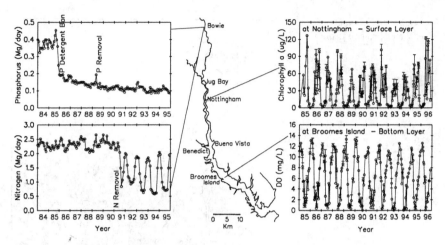

Fig. 4.13 Nitrogen and Phosphorus Loads, Chlorophyll *a*, and Dissolved Oxygen in the Patuxent Estuary

4.4.2 Response to Nutrient Reductions

The response of phytoplankton biomass in the estuary to point source nutrient load reductions was minimal and very slow. The long-term trend of the chlorophyll *a* level in the upper estuary at Nottingham, where significant spring algal blooms are consistently observed (see Fig. 4.13), was only slightly reduced from 1986 to 1996 (Lung 2001). However, the dissolved oxygen (DO) levels in the deep waters of the lower estuary near Broomes Island, where summer anoxic conditions are measured, did not show any improvement over the same period (Fig. 4.13). The point source nutrient reduction program has resulted in little improvement of the DO levels. These observations are consistent with the model prediction results by Lung (2001) and Lung and Bai (2003), based on a 38-segment model in a tidally averaged two-layer mass transport scheme. The Lung and Bai model utilized the CE-QUAL-W2 (W2) framework and simulated state variables (i.e. water quality constituents): ammonia, nitrite/nitrate, orthophosphate, chlorophyll *a*, and dissolved oxygen.

While the hydrodynamic simulations from the W2 model are generally robust and efficient in generating real time circulation patterns in the estuary, water column kinetics utilized by the model for the Patuxent Estuary had some deficiencies at that time. First, organic phosphorus and organic nitrogen are not considered as state variables; hence, the model was not capable of tracking the complete cycle of nutrients in the water column. Secondly, sediment–water interactions are not simulated by the model. Without a direct linkage of the water column and sediment, the model's capability to predict impacts of nutrient reductions was severely limited, especially for long-term model simulation runs. While the Lung and Bai model (2003) was successfully designed to provide a better understanding of eutrophication in the water column of the Patuxent, the model could not be utilized to predict the expected reduction in sediment nutrient fluxes and sediment oxygen demand (SOD) following nutrient reductions to the estuary.

A new modeling effort was launched in 2004 to explore nonpoint nutrient controls in the Patuxent River Watershed. The purpose of this work was to enhance the Patuxent Estuary model developed by Lung and Bai (2003) to address the deficiencies regarding nutrient cycling and sediment–water interactions. First, the spatial resolution was significantly increased to 163 segments in the longitudinal direction from the fall line to the mouth. Each segment has multiple vertical layers to take advantage of the multiple data sampling points, resulting in a total of 1993 segments. Next, water column kinetics were enhanced to a level matching that in the EPA's WASP/EUTRO model (Ambrose et al. 1993), adding several state variables, and providing a more complete picture of nutrient cycling in the water column. WASP/EUTRO kinetics routines were ideal for this effort because the modeled state variables closely matched the extent of available field data for this study. The water column kinetics and sediment–water interactions are shown in Fig. 4.14 and described in the following paragraphs.

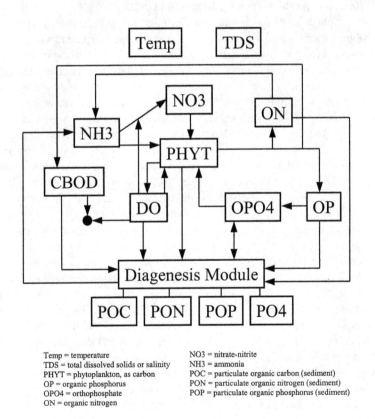

Temp = temperature
TDS = total dissolved solids or salinity
PHYT = phytoplankton, as carbon
OP = organic phosphorus
OPO4 = orthophosphate
ON = organic nitrogen

NO3 = nitrate-nitrite
NH3 = ammonia
POC = particulate organic carbon (sediment)
PON = particulate organic nitrogen (sediment)
POP = particulate organic phosphorus (sediment)

Fig. 4.14 The Patuxent Estuary model kinetics (Nice 2006)

A sediment diagenesis model based on work by Di Toro (2001) and Di Toro et al. (1990) was added to formulate a direct linkage between the water column and sediments (Lung and Nice 2007). The module performs calculations to simulate deposition of organic matter to the sediment (including phytoplankton, organic carbon, organic phosphorus, and organic nitrogen) and subsequent SOD, ammonia flux, and phosphate flux. Interaction between the sediment diagenesis module and the rest of the modeling system can be found in Nice (2006). Sediment diagenesis refers to the mineralization or conversion of particulate organic carbon (POC), particulate organic nitrogen (PON), and particulate organic phosphorus (POP) to methane, ammonia, and phosphate, respectively. Mineralization of the particulate organic matter was achieved in the model using mass balance equations and reaction rates for three different reactivity classes, as specified by Di Toro (2001). SOD and ammonia fluxes are calculated based on an empirical formulation for methane and ammonia oxidation (Di Toro et al. 1990) and are a direct link to the diagenetic fluxes generated by the mineralization of POC and PON. Similar applications of this model formulation for SOD and ammonia flux have been applied to the Anacostia River (Lung 2001) and for the Pocomoke River (Hunt 2005).

Sediment phosphorus flux is quantified using a two-layer (aerobic and anaerobic) mass balance formulation, which simulates partitioning, particle mixing, and diffusion between layers and the overlying water column. The phosphorus flux model differs from the SOD and ammonia flux model in that it tracks concentrations and mass of phosphate in the aerobic and anaerobic layers, and calculation of flux is based on these concentrations rather than the diagenesis flux generated by the mineralization of POP. The flux model is patterned after a sediment flux model developed by Hunt (2005) for the Pocomoke River, which is based on equations proposed by Di Toro (2001). Diagenetic fluxes generated by mineralization of organic phosphorus supply phosphate to the anaerobic layer. Theoretically, phosphate sorbs strongly to particles in the sediment during aerobic conditions in the overlying water column. When oxygen levels in the water column fall below a critical level, phosphate desorbs from particles and diffuses from the sediment pore water to the water column.

Because re-suspension of particulate organic matter (POM) is not simulated by the model and because sediment diagenesis and subsequent SOD and nutrient fluxes are driven by settling and deposition of POM, settling velocities for phytoplankton and POM are assigned to vary longitudinally in the model. Net settling rates vary according to the region in the estuary; lower net settling rates are assigned to the shallower, upper regions of the estuary, while higher net settling rates are used for regions in the lower estuary to reflect the net settling characteristics of deeper waters. In addition, field studies indicate that 5–10 times more POM collect in the channel and deeper portions of the Bay than in the pycnocline and euphotic zone (Kemp and Boynton 1992). To account for higher amounts of POM collecting in the deeper portions of the Patuxent Estuary, an algorithm is included in the settling routine which increases settling velocities with depth in a linear fashion. A factor is determined using a linear function based on vertical grid cell depth and total segment depth and is applied to the net settling rate for each vertical grid cell. Application of the factor results in a net settling rate for the most bottom layer, which is 10 times greater than the net settling rate for the most top layer. Such an approach is consistent with the observation that 5–10 times more POM settles in deeper portions of estuaries. While the current algorithm for settling may add to the level of parameterization, the resulting model more realistically simulates net settling (including re-suspension), thus, more realistically simulating SOD and nutrient fluxes. SOD and nutrient flux data were not available for the simulation period. However, Boynton et al. (1990) recorded sediment flux data at Buena Vista and St. Leonard's Creek in the Patuxent Estuary during 1985. The sediment flux data can be qualitatively compared to results produced by the model.

Watershed flows and nutrient loads based on data collection and watershed modeling by Weller et al. (2003) were used to calibrate the Patuxent Estuary model by Lung and Nice (2007). The first step of model calibration was confirmed with close reproduction of the salinity and temperature in both spatial and temporal distribution to confirm the mass transport (Lung and Nice 2007). In the next step, the water quality model calibration focuses on the surface layer of chlorophyll a levels, particularly near Nottingham, where persistent algal growth has been observed (see Fig. 4.13). Figure 4.15 presents longitudinal concentration profiles for chlorophyll a in the top layer (above the pynocline, where phytoplankton blooms likely occur) from a two-year period from August 1997 to July 1999 with a wet year in 1998 followed by a dry year in 1999. As shown, the model reproduces the spatial trends of phytoplankton biomass well, including peaks of growth during the late spring and summer months in the upper estuary at Nottingham. In addition, phytoplankton blooms during the winter of the wet year (e.g., February 1998) are also reproduced in the lower Estuary at Broomes Island. For reference to the figures, the monitoring

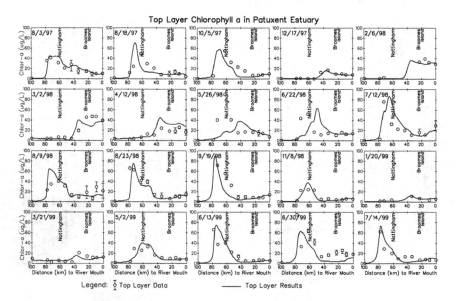

Fig. 4.15 Patuxent Estuary model calibration—top layer chlorophyll a (1997–1999 data)

station at Nottingham is located approximately 63 km from the mouth of the estuary, while the monitoring station at Broomes Island is located approximately 24 km from the mouth. Model results and data in the vertical direction are layer-averaged (above and below the pynocline) for presentation in Fig. 4.15. Model calculated chlorophyll a peaks usually occur near Nottingham, determined by the freshwater flow, matching the data closely. The very shallow water near Nottingham offers a natural, fertile ground for phytoplankton growth. The eutrophication endpoint for the Patuxent Estuary is the depressed DO levels in the deep water column of Broomes Island.

Figure 4.16 shows the model calibration results for DO of the bottom layer at Broomes Island for the same period of August 1997 to July 1999. Summer low DO concentrations are reproduced by the model. Figure 4.17 presents 40 vertical profiles for a comprehensive look of the Broomes Island DO. As shown in the plots, DO stratification in the water column during the late spring and summer months is reproduced by the model, further substantiating the validity of the model. Another phenomenon, evident in Figs. 4.15, 4.16, and 4.17, is that high phytoplankton levels and low DO levels were observed during spring and summer of 1998, a wet year. This trend of low DO levels following higher river flows (and nutrient loading) is consistent with relationships established by Hagy et al. (2004) regarding nutrient loading and river flow for the Bay. The Patuxent Estuary model has a sediment flux model to account for diagenesis of organic material in the form of carbon fluxes, nitrogen fluxes, total SOD, total benthic gas flux, and phosphate flux. The sophistication of the sediment system interacting with the overlying water deserves more in-depth discussion and is presented in Sect. 4.10.

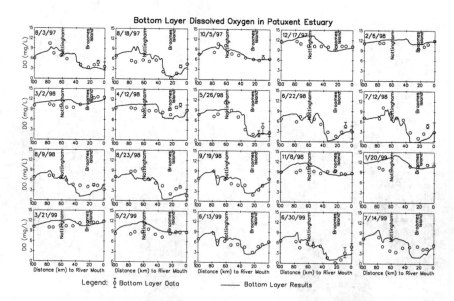

Fig. 4.16 Patuxent Estuary model calibration—bottom layer DO (1997–1999 data)

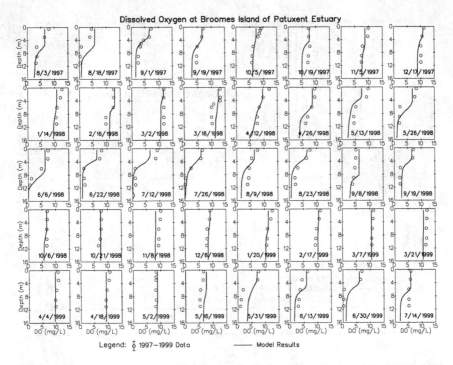

Fig. 4.17 Patuxent Estuary model calibration—vertical DO profiles at Broomes Island (1997–1999 data)

4.4.3 Watershed Nutrient Control

The calibrated model was used to assess the impact of watershed nonpoint nutrient controls. The top graph in Fig. 4.18 displays a hydrograph of daily discharge recorded at the USGS gauging station 01,594,440 at the head of the Patuxent Estuary at Bowie, Maryland from August 1997 to August 1999. The hydrograph represents flow entering the upstream segment of the model and is the largest single freshwater inflow. Flows in 1998 and 1999 show the typical temporal pattern of high flows in the spring months followed by low flows in the summer. In addition, 1998 was a very wet year with significant freshwater flows into the estuary during the spring season. On the other hand, the 1999 flows were much lower in the spring, characterizing a relatively dry year. The difference between the spring flows in 1998 and 1999 has significant implications on the anoxic conditions at Broomes Island and can reveal interesting physical insights into the estuarine system aided by the results from the calibrated model.

Anoxic water volume is a useful indicator of the eutrophication status with respect to the DO endpoint. The middle panel of Fig. 4.18 shows the model calculated percent of the total volume of the water column having a DO level below 2 mg/L in a temporal pattern from August 1997 to July 1999 under three watershed nutrient

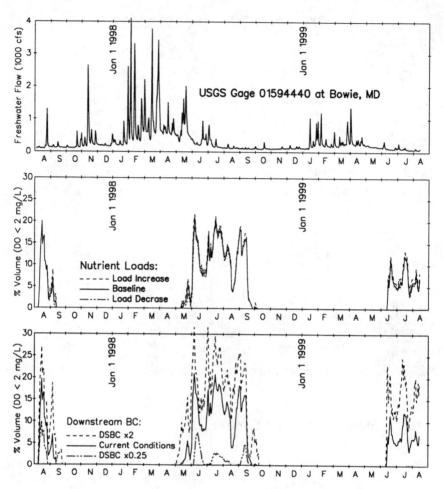

Fig. 4.18 Effects of watershed nutrient loads and open boundary conditions on DO in the Patuxent Estuary

load conditions: current, 50% load increase, and 50% load decrease, respectively. The results show that while watershed nutrient load reductions would curtail algal biomass in the water column, nutrient reductions would still have a minimal effect on improving the anoxic conditions in the lower estuary during the summer months. It is interesting to note that the start of the anoxic conditions has a three-month lag behind the spring high flow months in 1998 and 1999. Even more intriguing is that the volume of DO below 2 mg/L is significantly larger in 1998 than that in 1999. The estuarine hydrodynamics play a key role in this result in that the relatively high flows in spring 1998 form an effective blocking pynocline between the layers in the water column, thereby cutting off the oxygen supply from the surface layer to the deeper layers. On the other hand, the water column is better mixed in the vertical direction in the summer 1999 due to lower freshwater flows in the spring, producing smaller anoxic volumes in the system. The back-to-back wet and dry years offer interesting insights into the hydrodynamics of the estuary. Note the trend of lower dissolved oxygen levels following higher river flows (and nutrient loading), a phenomena that was also discussed by Hagy et al. (2004).

The bottom panel of Fig. 4.18 compares the water quality response to downstream, i.e. open boundary conditions, for model predictions. Theoretically, a boundary condition must be free from any influence associated with nutrient load changes to the system. For example, the Bay water quality model (Cerco 1995) has its downstream boundary conditions set at the continental shelf, free from any influence within the Bay. Following the same strategy, the downstream boundary conditions should be set at the continental shelf too, which is not practical for the Patuxent Estuary study. Therefore, a range of boundary conditions are chosen for the Patuxent Estuary: the existing conditions, doubling of the existing conditions, and a quarter of the existing conditions for the nutrient levels at the mouth of the Patuxent Estuary. The bottom panel of Fig. 4.18 shows significant differences in predicted percentage volume of DO below 2 mg/L associated with the open boundary conditions. This analysis indicates that the effect of nutrient control within the Patuxent watershed depends strongly on the open boundary conditions and has implications for water quality management. Nutrients running up and down the Bay from the neighboring estuaries (e.g. the Potomac, the York, the Rappahannock, and the James) could deliver high nutrient concentrations to the mouth of the Patuxent Estuary, negating the Patuxent River Watershed nutrient control effort. "The benefits of in-house cleaning could be totally offset by the neighbors dumping trash at your front door!"—A wasted effort. The outcome of this modeling analysis clearly suggests the need of a Chesapeake Bay nutrient TMDL study.

4.4.4 Water Quality Management Questions

The above discussion suggests that watershed nutrient load reductions alone were not sufficient to reverse eutrophication in the Patuxent Estuary. Ensuing questions therefore surfaced among the staff and researchers of the regulatory agencies. Is the Patuxent River Watershed a net source of nutrients to the Bay, or is it a net importer of nutrients from the Bay? How does the source/sink balance differ between phosphorus and nitrogen? Does the source/sink balance change seasonally or between wet/dry years? How would the source/sink balance change with different watershed and downstream boundary condition scenarios? Is primary production in the Patuxent Estuary limited by nitrogen or phosphorus? Where and when is nitrogen limitation likely? Where and when is phosphorus limitation? Do the spatial patterns of limiting nutrients change between wet and dry years? How do the spatial and seasonal patterns of nutrient availability, chlorophyll a, and primary production change between wet and dry years? What are the implications for managing and restoring both the Patuxent estuary and the Bay? The above discussions are grouped into a set of more precise questions for the model to explore:

1. Where and when is nitrogen or phosphorus limiting the algal growth in the water column of the Patuxent Estuary?
2. Do the spatial patterns of limiting nutrients in the upper and lower estuary change between wet and dry flow years?
3. To what extent (percentage) do nutrient loads from the watershed contribute to the total loads to the Patuxent Estuary?
4. Is the Bay a nutrient source for the Patuxent River?
5. How significant is the role of sediment in the overall nutrient budget?

The fully calibrated and verified model by Lung and Nice (2007) with data support was then used to reveal additional insights into the nutrient dynamics related to nutrient control for water quality management of the Patuxent Estuary.

Model calculated time-variable phosphorus and nitrogen limitation factors (between 0 and 1 based on Eq. (4.1) in the surface layer of the water column at Nottingham and Broomes Island for the two-year period from August 1997 to July 1999 are shown in Fig. 4.19. The left column plots show results for phosphorus and the right column plots are for nitrogen. At Nottingham, phosphorus levels are near saturation (i.e., algal growth not limited by phosphorus) throughout this two-year period (see the first plot in the left column). At Broomes Island, phosphorus is not a limiting factor for algal growth except during the winter months of 1999. In fact, significant phosphorus limitations (with a limitation factor of 0.5) are reached in

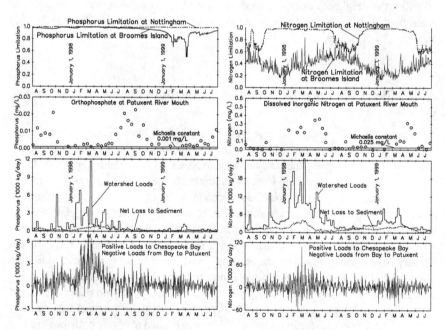

Fig. 4.19 Phosphorus and nitrogen dynamics in the Patuxent Estuary (1997–1999)

February and March 1999 at Broomes Island. This limitation prompts the following questions: Why only the winter of 1999? What triggered this phosphorus limitation?

Orthophosphate (Ortho-P) data at Station LE1.4 (just outside of the mouth of the Patuxent Estuary) are plotted in the second panel of the left column in Fig. 4.19. Orthophosphate levels at the mouth, reflecting influence from the Bay, are quite low and very close to the Michaelis constant of 0.001 mg/L, set in the model for phosphorus limitation, in both winters of 1998 and 1999. Yet, the model results show only phosphorus limitation in the winter of 1999, not the winter of 1998. These results suggest that the phosphorus limitation at Broomes Island in winter 1999 was not influenced by the seasonally low orthophosphate levels at the mouth of the Patuxent Estuary. Instead, phosphorus limitation at Broomes Island in winter 1999 was most likely affected by the nonpoint watershed loads.

Both spatial and temporal nitrogen limitation (see the first plot of the right column in Fig. 4.19) in the Patuxent Estuary is quite different from phosphorus limitation. At Nottingham, the model results show a maximum degree of nitrogen limitation during the summer months when the watershed flows are at minimum for the year. On the other hand, the Bay appears to play a major role in the seasonal pattern of nitrogen limitation at Broomes Island. A review of the DIN concentrations at Station LE1.4 (the second panel of the nitrogen column) shows a reverse seasonal pattern of the orthophosphate concentrations. Further, the nitrogen limitation at Broomes Island exhibits a temporal pattern closely following that of the DIN at Station LE1.4 with an approximate time lag of one and half months, thereby suggesting that the nitrogen limitation at Broomes Island is strongly affected by the Bay, not by the watershed loads.

The above analysis reveals complicated temporal and spatial nutrient limitations between phosphorus and nitrogen in the Patuxent Estuary, characterized by various factors at the upper and lower estuary in different times of the year. It shows that phosphorus limitation at Broomes Island in winter 1999 is controlled by the significantly low watershed phosphorus loads. On the other hand, the Bay has a strong impact on nitrogen limitation at Broomes Island. Seasonal and spatial variations between phosphorus and nitrogen limitations in the Patuxent Estuary cannot be derived from this in-depth analysis without a well calibrated eutrophication model supported by the wealth of data. This level of extracting physical insights into the model results would therefore require strong skills of the modeler supported by data. What the above analysis demonstrates is the essence of water quality modeling, which is far more valuable and interesting than simply running models (i.e. pressing the button).

It is expected that during any given year, nutrient budgets are not balanced, thereby leading to a gain or loss of nutrient mass for the estuary. Such a phenomenon simply indicates that the estuarine system is never at a steady state and that the nutrient input is not equal to the nutrient output on a yearly basis. A dynamic nutrient budget throughout an entire year or a period of years would serve as a more accurate account of all nutrient sources and sinks for the Patuxent Estuary. Nutrient budgets of total nitrogen and total phosphorus were developed for the Patuxent Estuary during a two-year period (a wet year followed by a dry year) from August 1997 to July 1999 using the field data and model results to demonstrate the interplay of model results and field data. Four groups of nutrient loads (in 10^6 g/day) were quantified for this two-year period:

1   nutrients from the watershed into the Patuxent River Estuary
2   net loss of nutrients from the water column to the sediment
3   nutrient loads transported from the Patuxent River to the Bay at the mouth of the river
4   nutrient loads received from the Bay to the Patuxent River at the mouth of the river

While the Group 1 load was input into the model for the simulation runs, the other three groups were calculated by the model. The third plot of the left column in Fig. 4.19 show Groups 1 and 2 phosphorus loads from August 1, 1997 to July 31, 1999. The total phosphorus input from the watershed varies seasonally, reaching a peak in the spring months of 1998 (the wet year). The net deposition of total phosphorus from the water column to the sediment represents a small fraction of the total watershed loads, again showing seasonal variations.

The difference between Groups 3 and 4 loads, representing the net (exchange) phosphorus loads across the open boundary between the Patuxent Estuary and the Bay, is shown in the fourth plot of the left column in Fig. 4.19. The positive loads of the exchange represent the total phosphorus loads from the Patuxent to the Bay, while the negative loads are from the Bay to the estuary. The seasonal pattern of the net exchange load closely follows that of the watershed loads, suggesting the strong influence of the watershed phosphorus loads at the mouth of the Patuxent River. However, over a 12-month period from January to December, the net phosphorus loads are positive across the open boundary, delivering net phosphorus loads from the Patuxent watershed to the Bay.

During the wet year from August 1, 1997 to July 31, 1998, the total phosphorus loads from the watershed to the Patuxent Estuary were 544 Mg (1 Mg = 1000 kg), while the estuary delivered 389 Mg of total phosphorus to the Bay, resulting in a net gain of phosphorus of 155 Mg during the wet year (Table 4.1). The magnitude of phosphorus loads during the dry year from August 1, 1998 to July 31, 1999 was much lower, showing only 85 Mg from the watershed and 36 Mg to the Bay, respectively,

Table 4.1 Nutrient budgets (Mg) for the Patuxent Estuary

	Wet year	Dry year
Total phosphorus		
Watershed input	544	85
Delivery to bay	389	36
Net gain	155	49
Total nitrogen		
Watershed input	2214	871
Delivery to bay	647	270
Net gain	1567	601

which amounts to only 48 Mg of net phosphorus gain in the estuary during that period. The net loss of total phosphorus from the water column to the sediment is insignificant compared with the other loads and therefore not included in Table 4.1. Results presented in Table 4.1 clearly show that the Patuxent Estuary is a source of phosphorus for the Bay, albeit a small amount during the dry year.

The nitrogen loads to the Patuxent Estuary during the same two-year period from August 1, 1997 to July 31, 1999 are presented in the third and fourth plots of the right column in Fig. 4.19. While the watershed nitrogen loads have the similar temporal pattern of the watershed phosphorus loads, the net nitrogen exchange loads at the open boundary of the Patuxent Estuary do not. In addition, the net nitrogen exchange loads are higher than the watershed nitrogen loads by a factor of five. Such a result suggests that the Bay, not the Patuxent River Watershed, strongly influences the nitrogen budget in the Patuxent Estuary. Nitrogen loads from the water column to the sediment account for a very small portion of the total watershed nitrogen loads. The annual nitrogen summary for the two-year period is presented in Table 4.1, showing a net delivery of total nitrogen loads to the Bay of 647 and 270 Mg for the wet year and dry year, respectively. This result of the Patuxent Estuary delivering net nutrient loads to the Bay is consistent with the finding by Boynton et al. (2008).

As expected the total nitrogen mass has a seasonal pattern closely mimicking the net exchange nitrogen loads at the open boundary (Fig. 4.19). The significantly augmented net exchange of nitrogen loads (see the 4th panel of the nitrogen column in Fig. 4.19) between the Patuxent Estuary and the Bay further substantiates this result. To check the phosphorus loading rates, the total mass of phosphorus in the water column was quantified on a daily basis throughout the two-year period. Figure 4.20 shows the total phosphorus and total nitrogen mass in metric tons, displaying seasonal fluctuations throughout this two-year period. The total mass of nitrogen and phosphorus in the Patuxent Estuary was higher in the wet year (1998) than the dry year (1999).

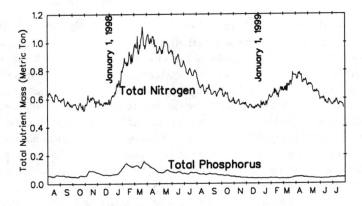

Fig. 4.20 Nutrient mass in Patuxent Estuary (Aug 1977–July 1999)

The preceding in-depth analysis highlights the interplay between model results and data and reveals insights into nutrient/eutrophication dynamics in the Patuxent Estuary. The two-year period from August 1977 to July 1999, characterized with a hydrological condition of a wet year followed by a dry, offers an excellent opportunity to explore the year-to-year variation in nutrient limitations in the Patuxent Estuary. The spatial and temporal differences of phosphorus and nitrogen limitations closely relating to hydrology and hydrodynamics (i.e. mass transport) in the Patuxent Estuary during this two-year period are demonstrated from both the model results and field data. They also display the utility of a calibrated and verified model in obtaining additional insights to address management questions. The contributing watershed phosphorus loads control the phosphorus dynamics at the mouth of the river while the Bay controls the nitrogen dynamics. Additionally, the no response to watershed nutrient load reductions in mitigating low DO levels in the water column around Broomes Island can be explained as follows: the percent volume of the water column having DO below 2 mg/L becomes more significant in a wet year than a dry year due to the unique estuarine hydrodynamics. High freshwater flows tend to form a natural barrier in the water column prohibiting atmospheric oxygen resupply from reaching the bottom waters.

Reductions of the watershed loads are mostly compensated by the nutrients from the neighboring rivers such as the Susquehanna, Potomac, Rappahannock, and the York Rivers that deliver sufficient nutrients at the mouth of the Patuxent River. Therefore, reducing nutrient loads from the Patuxent River Watershed alone would not do the job. A more comprehensive approach would be to address this issue on a Bay-wide basis, leading to the later development of the Bay TMDLs. Note that the Bay-wide approach would simply remove the open boundary issue at the mouth of the Patuxent!

4.5 Estuarine Turbidity Maximum

Turbidity maximum is closely related to salinity and phytoplankton growth in an estuary. It is a unique phenomenon in partially mixed estuaries (such as the James and Patuxent Estuaries in Sects. 4.3 and 4.4, respectively). Generally, turbidity maximum is located in the vicinity of the tail of the salinity intrusion, called the null zone, where the horizontal stress at the bottom of the water column becomes zero. River freshwater flow variations also determine the relative location of the null zone up or down the estuary. Two significant effects can be attributed to the presence of a null zone in estuaries: it is typically the area of the most rapid sediment accumulation and the highest concentrations of suspended solids occur in the vicinity. In general, the maximum concentration of suspended solids is expected to occur immediately downstream of the null zone due to prolonged advective residence time in the water column. Turbidity limits light penetration, reduces the thickness of the euphotic zone, and therefore prevents excessive primary production in the water column.

Predicting turbidity maximum has been a common practice for decades. An early work which deserves great commendation and has set the tone for turbidity maximum analyses is the sediment model proposed by Odd and Owen (1972). They formulated a two-layer model to simulate the tidal flow and transport of mud in the Thames. The model assumed uniform properties in each layer with a rectangular section. In another treatment of suspended sediment transport, Ariathurai and Krone (1976) used a two-dimensional (in a horizontal plane) depth averaged model with advection–diffusion equations of mass conservation solved by the finite element method. Since modeling estuarine circulation is an essential prerequisite in modeling the sediment transport, two-dimensional (longitudinal-vertical) models designed to calculate the tidal currents and stage in time variable fashion are usually employed. The essential point, however, is not the tidal currents and stage, but the tidally averaged values, which characterize the significant circulation features of estuarine flow: a landward flow in the lower layer and a seaward flow in the upper layer, with consistent vertical flows to maintain hydraulic continuity (Fig. 4.21). The water column is divided into two layers by the plane of no net motion, where the tidally averaged velocity is zero. The vertical profile of the longitudinal velocity in the estuary is shown with the layer-averaged seaward velocity in the top layer and the landward velocity in the bottom layer. The plane of no net motion meets the bed at the null zone. This mass transport pattern is also responsible for the occurrence of phytoplankton peaks located in the null zone, where the freshwater and saline water meet to form a net upward vertical flow, thereby countering algae settling in that region.

In addition to the velocity impacted by the typical two-dimensional estuarine circulation, suspended particles possess a vertical settling velocity. If the vertical water velocity is downward it enhances the vertical flux of solids in that direction; by contrast, if the water velocity is upward, it tends to cancel the settling velocity. Thus, for the larger and denser particles, such as sand and silts, the net vertical velocity is in the downward direction, but less than the settling velocity. For the smaller and less dense particles, such as clays and organics, the net velocity is directed upward, if the water velocity is greater than the settling velocity. This phenomenon, in conjunction with the convergence of the landward-flowing density current and the seaward-flowing surface river current at the tail of salinity intrusion, is responsible for the solids concentrations in the saline zone of the estuary, which are greater than

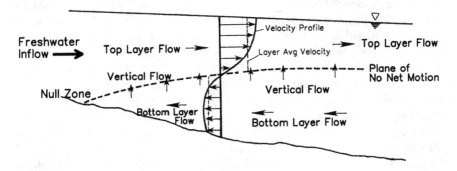

Fig. 4.21 Tidally averaged two-layer estuarine mass transport

those of the upstream freshwater inflow and downstream density current—creating a turbidity maximum.

With today's sophisticated hydrodynamic and sediment transport modeling frameworks, modeling the turbidity maximum is not a difficult task but still requires a substantial amount of effort. In addition, verifying the hydrodynamic model calculations requires strong data support, which is not always readily available. Recognizing that the key of the turbidity maximum is associated with the tidally averaged two-layer mass transport in the estuary, an alternative modeling approach is suggested, i.e., decoupling the estuarine hydrodynamics and mass transport. This approach is based on the condition that longitudinal and vertical distributions of salinity in the water column are known or may be assigned. Martin and McCutcheon (1999) called it the Lung and O'Connor method from the work by Lung and O'Connor (1984) and O'Connor and Lung (1981). A complete presentation of this analysis can be found in Lung (1993). In this analysis, salinity is used as a tracer to calibrate the mass transport model. Vertical dispersion coefficients between the top and bottom layers are assigned by an appropriate modification of the eddy viscosity by means of the Richardson Criterion, which provides a qualitative relationship for the momentum-mass transformation. To complete the loop, the derived two-layer transport pattern is used to calculate the distribution of salinity for comparison with the originally given salinity distribution on a tidally averaged basis. Repeated applications of this procedure produce a consistent set of mass transport coefficients under the specific tidal condition and freshwater flow. One key step in the analysis is to correlate vertical eddy viscosity in the water column with the longitudinal salinity gradient at the water surface. Establishment of this relationship provides an additional guide for assigning the vertical eddy viscosity in the prediction analysis. Preliminary development of estimating the vertical eddy viscosity at the null zone area is also carried out. According to the Richardson Criterion, this value should be equal to the vertical dispersion coefficient due to the neutral stability at the null zone, and therefore, becomes the lower limit on the vertical dispersion coefficient.

Based on the two-layer mass transport model for the Sacramento-San Joaquin Delta (see map in Fig. 2.25), a total suspended solids (TSS) model was developed. The Sacramento River runs from east to west, is joined by the San Joaquin River prior to reaching the San Pablo Bay (Fig. 4.22) and eventually the Golden Gate Bridge in San Francisco Bay. One of the key mechanisms responsible for the TSS distribution is the vertical settling in the water column. Equations (2.6) and (2.7) are used to derive the settling velocity of 8 ft/day (2.44 m/day) in the water column based on the particle density and sizes (see Fig. 2.33 in Chap. 2). Results of the two-layered salinity model at two freshwater flows of 4000 cfs (125 cms) and 10,000 cfs (283 cms) are presented in Fig. 4.22. As expected, salinity intrusion is further upstream under the lower freshwater flow than the higher flow. Salinity levels are also higher under the lower freshwater flow. The turbidity maximum is located slightly downstream of the null zone (at the tail of salinity intrusion) at approximately mile 40 and mile 45 for the flows of 10,000 cfs and 4000 cfs, respectively. TSS concentrations are higher in the bottom layer than those in the surface layer due to vertical settling. Further, TSS levels are higher at 10,000 cfs freshwater flow than those at 4000 cfs because

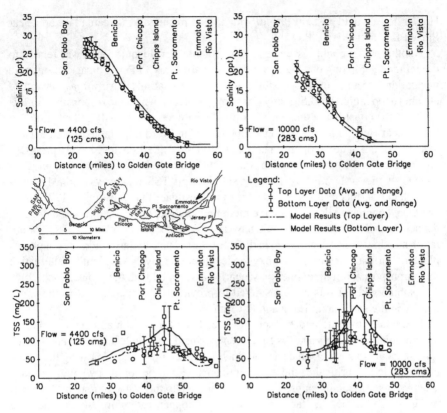

Fig. 4.22 Salinity and TSS modeling of the Sacramento-San Joaquin Delta, California

the higher flow carries greater particle loads from the upstream Sacramento River. It is noted that an earlier 1-D modeling analysis, lacking the proper mass transfer, had failed to reproduce the turbidity maximum in the estuary (O'Connor and Lung 1981).

Development of the two-layer salinity and TSS models was a response to the water quality management needs for the Sacramento-San Joaquin Delta, marked by salinity fluctuations due to the seasonal variation in freshwater flows from the Sacramento and San Joaquin Rivers. The Suisun Bay, Grizzly Bay, and Honker Bay (Fig. 4.22) are shallow waters offering favorable conditions for algal growth, which has been under control due to significant light extinction provided by high TSS levels as a result of significant solids loads from the Sacramento River. This cause-and-effect balance would be lifted when the Sacramento River flows were diverted south to provide supplementary water supply for southern California in the 1970s. The flow diversion would significantly reduce the TSS loads to these shallow bays, thereby minimizing the turbidity maximum and lightening up the water column for algal growth. Note that at a freshwater flow of 10,000 cfs, the turbidity maximum is located between Port Chicago and Chipps Island, immediately adjacent to Honker Bay (Fig. 4.22).

Higher flows would push the turbidity maximum further downstream to the area near Suisun Bay and Grizzly Bay.

Another water quality issue in the Sacramento-San Joaquin Delta was related to dredging the shipping channel from the San Pablo Bay to Port Chicago to improve navigation. Because of the possibility of increased salinity intrusion as a result of channel deepening, the U.S. Army Corps of Engineers, who were responsible for this project, was considering the construction of a submerged rock sill in the Carquinez Strait (near Benicia in Fig. 4.22) to impede the landward movement of saline water near the bottom of the shipping channel. The foremost question raised by the proposed construction is the extent to which water quality would be affected by this physical change of the system. The two-layer salinity and TSS model was linked with a eutrophication model for the study area to investigate this question in 1970s. The principal conclusion drawn from the modeling analysis is that the proposed channel deepening and sill placement would have minimal effect on the phytoplankton population and dissolved oxygen level in the study area. The simulated increases of phytoplankton biomass and dissolved oxygen levels are relatively small under alternative mass transport and light extinction scenarios associated with the project conditions. The submerged sill was not considered to provide any serious threat to the water quality in the study area (Lung and O'Connor 1984).

In tracking the fate and transport of metals and toxic contaminants, including the emerging chemicals such as EDCs (endocrine disrupting chemicals) and PPCPs (pharmaceutical and personal care products) in receiving waters, knowing the distribution of TSS is critical as these contaminants tend to be sorbed by solid particles. The two-layer modeling approach has been applied to the Patuxent Estuary (see map in Fig. 4.12) to track the fate and transport of copper, cadmium, and arsenic. The development and calibration of the Patuxent Estuary TSS model was supported by massive amounts of water quality data available from the Bay data base. Following the eutrophication modeling effort reported in Sect. 4.4, the solids model started with a time-variable, multi-layer, 1993-segment configuration for the two-year period of 1995 to 1997. Figure 4.23 presents the recalibrated salinity results and data on 20 select dates through this two-year period. For presentation purpose, the multi-layer model results and data are processed into averages of two-layers separated by the pynocline. The extent of salinity intrusion moving up and down the estuary is mandated by the freshwater flow to the estuary. Being a tributary to the Bay, salinity levels at the Patuxent Estuary mouth rarely exceed 20 ppt as the estuary is quite a distance away from the mouth of the Bay. Vertical stratification of salinity is closely related to the freshwater flow as high flows tend to intensify the vertical salinity stratification in addition to pushing the salt content further downstream. These key features are displayed by the model results and data in Fig. 4.23, confirming the validity of the mass transport model. The strong vertical stratification of salinity on January 23, 1996 demonstrates the freshwater flow effect. A freshwater flow of almost 9000 cfs (255 cms) entered the estuary on that day, exerting a strong shear effect and virtually blocking the vertical mixing in the water column, thereby creating sharp vertical salinity gradients in the water column. Based on 43 years of records, the median flow at Bowie, MD in the month of January is only 200 cfs (5.7 cms)!

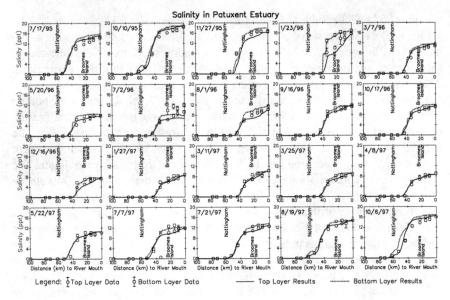

Fig. 4.23 Salinity model results versus data of the Patuxent Estuary (1995–1997)

It is not surprising that the salinity intrusion could only reach about 40 km from the mouth on that day, a much shorter distance than any other date in Fig. 4.23.

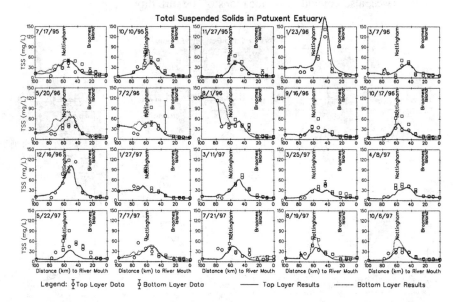

Fig. 4.24 Turbidity maximum model results versus data of the Patuxent Estuary (1995–1997)

The TSS model results and data for this same period are shown in Fig. 4.24. Again, model results match the data quite well for this two-year, time-variable model simulation. In general, the locations of the turbidity maximum match the end of salinity intrusion (compare Figs. 4.23 and 4.24) as explained in earlier discussions. The unique mass transport pattern at the null zone creates a favorable condition to keep the solids in suspension. As expected, higher solids levels are seen in the bottom layer than the surface layer. Like the salinity stratification, the vertical stratification of suspended solids is strongly influenced by the freshwater flow. In addition, the high flow on January 23, 1996 carries significant amounts of solids from the upstream watershed, thereby generating very high solids levels at the turbidity maximum. The calculated peak concentration reached 170 mg/L, higher than any peak levels during this two-year period. The effect of freshwater flow on the longitudinal movement of salinity and solids distributions in the Patuxent is also displayed in Figs. 4.23 and 4.24. Results and data of metals modeling utilizing this solids modeling framework for the Patuxent Estuary is presented in Chap. 7.

A solids modeling analysis is not complete without checking into phytoplankton as suspended solids affect available light for algal growth. In return algal biomass (organic solids) contributes to the TSS in the water column. Figure 4.25 presents the layer-averaged model results and data of chlorophyll a in the Patuxent Estuary for the period of 1995–1997. The results are consistent with those reported in Fig. 4.15 for the surface layer chlorophyll a of the Patuxent Estuary from 1997 to 1999. Like suspended solids, locations of chlorophyll a peaks, moving upstream and downstream in the estuary near Nottingham, are closely linked to the tail of salinity intrusion and in many cases, also coincide with locations of the turbidity maximum as well. As

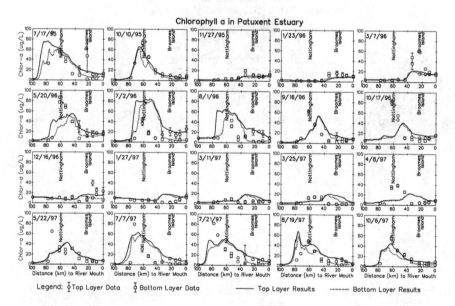

Fig. 4.25 Chlorophyll a model results versus data of the Patuxent Estuary (1995–1997)

expected, phytoplankton chlorophyll *a* levels in the surface layer (reaching 80 μg/L) are higher than those in the bottom layer.

Algal blooms and turbidity maximum are closely related in many estuaries. A turbidity maximum in the Potomac Estuary (see map in Fig. 1.1) during the 1977 algal bloom is presented in Fig. 4.26, showing the longitudinal profiles (from the Chain Bridge to the Bay) of salinity, suspended solids, and chlorophyll *a* concentrations, measured on August 23–25, 1977. The turbidity maximum is in the vicinity of between miles 40 and 50, immediately downstream of the salinity intrusion. Note that the TSS profile includes inorganic as well as organic particulate (associated with chlorophyll *a*) peaks in the vicinity of mile 30, matching the chlorophyll *a* peak of the algal bloom in Fig. 1.1. The estimated inorganic solids (in triangles with dashed lines) are obtained from chlorophyll *a*-solids relationships. The turbidity maximum of the inorganic solids is shifted downstream from the tail of the salinity intrusion to Quantico. The significant algal bloom produces a chlorophyll *a* peak reaching 320 μg/L, which contributes 20 mg/L of solids to the total solids concentration and

Fig. 4.26 Salinity, turbidity maximum, and chlorophyll *a* in the Potomac Estuary (USGS data)

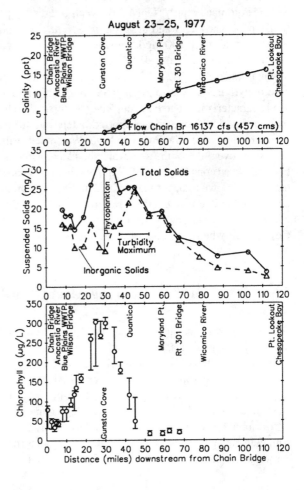

thereby accounts for a significant amount of algal self-shading in light extinction at Gunston Cove. For comparison, chlorophyll *a* peaks in the Patuxent Estuary rarely exceed 80 μg/L at Nottingham (Fig. 4.24); its contribution to TSS is substantially small when compared with the Potomac Estuary algal blooms in 1977 (Fig. 4.26) and 1983 (Fig. 1.1).

Another conversion factor affecting the chlorophyll *a* and TSS relationship is the carbon-to-chlorophyll *a* ratio. Jakobsen and Markager (2016) compiled data from 7578 coastal seawater samples collected in Danish waters from 1990 to 2014 to provide an in-depth analysis of this factor. They listed a number of factors affecting this ratio. Nitrogen limitation on algal growth tends to increase the C:Chl *a* value. The ratio decreased from 130 to 10 when temperature increased from 0 to 30°C under nutrient-replete and light-saturated conditions (Geider 1987; Thompson et al. 1992). Clearly, the interrelationships between salinity, suspended solids, and algal growth in estuaries are intriguing with respect to eutrophication and have profound implications on water quality management.

Finally, comparing the TSS compositions between the Patuxent Estuary (Fig. 4.24) and the Potomac Estuary (Fig. 4.26) reveals interesting physical insights into the relationship between TSS and algal biomass. The TSS concentrations in the Patuxent Estuary could reach 150 mg/L, of which the algal biomass accounts for 5 to 6 mg/L of TSS. In contrast, the peak TSS concentration in the 1977 Potomac Estuary algal bloom is under 35 mg/L, yet 20 mg/L of that is from the algal bloom of over 320 μg/L chlorophyll *a*. This variability in TSS composition comes from the different watershed characteristics. The Patuxent River Watershed at the fall line at Bowie, MD (see map in Fig. 4.12) amounts to approximately one third of the total watershed area while in comparison the Potomac River Watershed at the fall line (at Chain Bridge) represents over 90% of the total watershed area. Although phytoplankton growth has been observed consistently in the Patuxent Estuary near Nottingham, no algal blooms have been known to occur in the Patuxent Estuary. On the other hand, repeated algal blooms in the Potomac Estuary prior to the 1990s produced significant chlorophyll *a* concentrations, thereby contributing substantial levels of organic solids at Gunston Cove to overwhelm the TSS loads from the upstream watershed.

4.6   Mass Transport Limiting Algal Growth

With so much attention paid to nutrient limitation on algal growth in eutrophication modeling analyses, one should not overlook the effect of other factors such as physical characteristics of the receiving water on eutrophication potential. Shortly following the oil spill from the tanker EXXON Valdez in Prince William Sound (Fig. 4.27), Alaska in spring 1989, a clean-up program calling for bioremediation was launched. Fertilizer (chemicals with high contents of nutrients: phosphorus and nitrogen but low on carbon) was applied to embayments in Prince William Sound to stimulate the growth of indigenous bacteria that in turn would break down the hydrocarbon in the oil for carbon to sustain their growth. Prior to the chemical application,

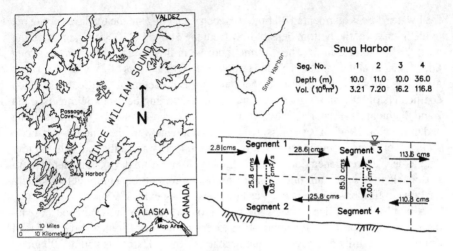

Fig. 4.27 Prince William Sound and two-layer mass transport in Snug Harbor

one of the concerns raised was whether the added nutrients would cause excessive algal growth during the growing season in the embayments. The concern was raised by the Federal and state agencies on the Shoreline Committee during the planning of the bioremediation experiments and later when EXXON proposed to apply nutrients on a wide scale. Another concern was whether there would be an adverse long-term water quality impact after the fertilizer application was completed and discontinued. Snug Harbor (Fig. 4.27) is one of the embayments chosen for the eutrophication potential study. Mixing and circulation play a major role in determining the eutrophication potential of these small embayments. Flushing time and residence time of a conservative substance are good indicators of how long a contaminant would remain in the embayment water. With a small freshwater inflow of 2.8 m^3/s, it would take 8974 tidal cycles (at 12.54 h for a typical M2 tide in Prince William Sound) to flush the system by the freshwater alone. That is, a conservative substance could spend almost 4688 days in Snug Harbor prior to reaching the Sound. On the other hand, the strong tide of 4.57 m from the Prince William Sound could flush the substance out of Snug Harbor in no time, i.e. approximately 5.76 tidal cycles (in just 3 days). Neither the flushing nor residence time estimated for Snug Harbor is reasonable for this the system. Amore robust analysis to accurately quantify the mass transport in Snug Harbor is needed.

The methodology presented in Sect. 4.5 is used for this Snug Harbor analysis. Following Fig. 4.21, the water column of the harbor is divided into two layers in a 4-segment configuration (see Fig. 4.27) for the mass transport modeling of salinity. First, vertical profiles of tidally averaged longitudinal velocities are calculated based on salinity data. Next, mass transport coefficients such as longitudinal and vertical advective flows and vertical dispersion coefficients are determined. Figure 4.27 shows the mass transport coefficients for Snug Harbor. Note that the freshwater flow rate of 2.8 m^3/s is small compared with the two-layer flows. At the open boundary, the freshwater flow is augmented 40 times to reach 113.5 m^3/s in the top layer going out to the Sound; yet the bottom density flow from the Sound is 110.8 m^3/s, maintaining a net outflow of 2.8 m^3/s. The 40-time increase in flow is generated by the strong tide of 4.57 m at the mouth of Snug Harbor. The vertical flows of 25.8 and 85.0 m^3/s are needed to maintain hydraulic continuity for each segment. Vertical dispersion coefficients of 0.87 and 2.0 cm^2/s, responsible for mixing between Segments 1 and 2 and Segments 3 and 4, respectively, are part of the two-layer mass transport (Lung and O'Connor, 1984). These mass transport coefficients were then incorporated into a two-layer, four-segment salinity model. The model was calibrated with salinity data to derive a residence time of 28.35 tidal cycles (= 14.8 days), considered more reasonable to describe mixing in Snug Harbor and for subsequent model sensitivity analyses to further substantiate the calculations (Lung et al. 1993).

A number of fertilizer types and application procedures are evaluated next by the bioremediation team for application to Prince William Sound. A typical load estimate for fertilizers sprayed on the beaches of Snug Harbor is about 7.0 kg/day and 0.7 kg/day for nitrogen and phosphorus, respectively. For the modeling analysis, successive loading rates at factors of 10, 100, and 1000 are considered to cover all possible loading conditions. For a conservative analysis, no fertilizer losses are considered during their transport from the beaches to the water column, nor during any uptake by organisms in the beach or nearshore zone. Model simulation results indicate that the maximum chlorophyll a concentrations could reach a negligible level of 2.9–3.3 μg/L in Snug Harbor—practically no impact.

To determine how long it would take the embayments to recover after the nutrient treatment stops, the model is run with constant nutrient loads for about 100 days so the algal biomass can approach equilibrium with the loads prior to discontinuing the application. The model results show that it would take approximately an additional 20–60 days for the nutrient concentrations return to their pretreatment levels in Snug Harbor. Subsequent tracking of the system showed that Snug Harbor returned to its pristine conditions in less than 6 months. Although high ammonia concentrations are predicted following the nutrient treatment, Snug Harbor is not expected to exceed the 0.02 mg/L un-ionized ammonia criterion even under the maximum nutrient loading rate of 1000 times of the base rate. The eutrophication potential in the Prince William Sound embayments is controlled by their hydrodynamics instead of the nutrient loads as demonstrated in this modeling analysis (Lung et al. 1993). In fact, the Prince William Sound recovered shortly after the incident with minimum damage on the water quality, thereby confirming the model predictions.

4.7   Lake Phosphorus Modeling and Response Time

While the Vollenweider phosphorus loading plot presented in Fig. 2.15 is useful with limited data support, the model framework of phosphorus and oxygen for stratified lakes by Chapra and Canale (1991) is well suited for the next level of analysis. Their model consists of two components: a total phosphorus budget and a model of hypolimnetic oxygen deficit. Total phosphorus in the water column and sediment is tracked. Dissolved oxygen in the hypolimnion is calculated based on the areal hypolimnetic oxygen demand (AHOD), which is dependent on the total phosphorus concentration in the hypolimnion during the stratification period. Platte Lake in Michigan is selected to demonstrate this analysis. Vertical profiles of measured temperature (see Fig. 2.23) show the thermal structure of this temperate climate lake. The spring overturn mixes the water column well in April. The warming season gradually raises the surface temperature to start the summer stratification in June. By July, a well-defined thermocline is formed to separate the epilimnion and hypolimnion. The cooling season starts in late September to lower the surface temperature and to break up the stratification. The fall overturn in October brings a fully mixed lake with uniform temperature from surface to bottom. A mild reverse stratification is observed in January and February as the water density is highest at 4°C.

While the analysis by Chapra and Canale (1991) consists of a single water column compartment, the water column of Platte Lake is divided into two layers. Figure 4.28 shows the average and range of total phosphorus concentrations in the epilimnion (upper layer) and hypolimnion (lower layer) of Platte Lake from 1990 to 1997, respectively. The total phosphorus levels in the hypolimnion are slightly higher than the epilimnion. Oxygen transfer between the two layers diminishes in the summer months due to vertical stratification, making oxygen in the epilimnion not available for the hypolimnion. AHOD starts to work its way in to deplete oxygen in the hypolimnion. The oxygen data clearly show that a steep slope of oxygen decline (i.e.,

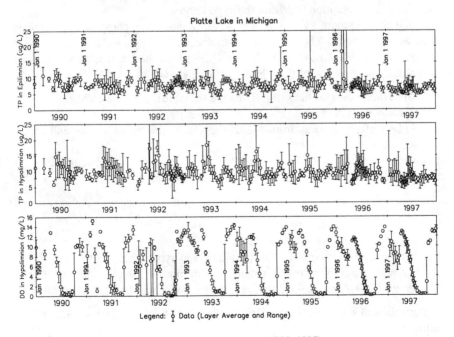

Fig. 4.28 Platte Lake phosphorus and dissolved oxygen (1990–1997)

Fig. 4.29   Areal
hypolimnetic oxygen
demand in Platte Lake

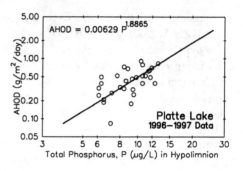

AHOD) repeated every summer in the hypolimnion (Fig. 4.28). During the summer months of 1996 and 1997, oxygen was measured on a weekly basis, more frequently than previous years, thereby providing more data to develop the relationship between AHOD and total phosphorus concentration in the hypolimnion. Figure 4.29 shows the data points from 1996 to 1997 in a LOG–LOG scale plot. The AHOD values ranging from 0.1 to 1.0 g/m^2/day are consistent with literature data (Chapra and Canale 1991). The regression line in Fig. 4.29 is defined as:

$$AHOD = 0.00629 P^{1.8865}$$

where $AHOD$ is areal hypolimnetic oxygen demand (g/m^2/day) and P is total phosphorus concentration (μg/L) in the hypolimnion.

While the sediment system is not modeled, sediment phosphorus release is included in the model. The two system variables in the Platte Lake model are total phosphorus and dissolved oxygen. The total phosphorus budget in the water column is tracked by accounting for the following flow and mass rates:

1.   Watershed flow (via the Platte River and the contributing watershed)
2.   Outflow from the lake
3.   Setting of phosphorus from the epilimnion to the hypolimnion, and to the sediment
4.   Vertical exchange of phosphorus between the epilimnion and hypolimnion
5.   Sediment release from the sediment to the hypolimnion.

The dissolved oxygen is tracked by the following processes:

1.   Inflow from the Platte River and the watershed
2.   Reaeration at the air–water interface
3.   Vertical exchange between the epilimnion and hypolimnion
4.   Oxygen consumption in the hypolimnion during the summer months, characterized by $AHOD$, linking the phosphorus model and the oxygen model.

Derivation of key model coefficients for the Platte Lake model is summarized as follows:

1 Flow data from the USGS gaging station 04,126,740 were used to quantify the flows and total phosphorus loads to the lake.

2 Settling velocity of total phosphorus in the water column was derived from the study by Walker (1998), yielding seasonally variable settling velocities ranging from 8 to 37 m/yr. Parameterizing time-variable settling velocity is justified because the composition of total phosphorus, i.e., organic and inorganic phosphorus, changes throughout the year. Organic phosphorus in algal biomass contributes to a significant portion of the total phosphorus concentration in the growing season, implying a higher settling rate.

3 The seasonally variable vertical dispersion coefficients were derived using the temperature data in Fig. 2.23 and Eq. (2.3) from Chap. 2. This vertical exchange of phosphorus and oxygen is significantly reduced during the summer stratification period.

4 A constant mass transfer rate of 5 m/day was used for oxygen exchange at the air–water interface.

Model results of total phosphorus and dissolved oxygen for 1997 are shown in Fig. 4.30 and match the data well in the epilimnion and hypolimnion, respectively. Seasonal trends of total phosphorus and dissolved oxygen levels in both layers are reproduced by the model. The phosphorus concentration in the epilimnion follows the seasonal trend of the Platte River input closely. It reaches a low level of 6 μg/L by the second half of August as the summer low flow in the Platte River becomes more pronounced. This concentration rises following the fall overturn with phosphorus supplied from the hypolimnion. Total phosphorus concentration in the hypolimnion is influenced by settling from the epilimnion and vertical dispersion (i.e. mixing) between the two layers. During the summer months when the hypolimnion becomes anaerobic, sediment release plays a major role in the hypolimnetic phosphorus budget.

Dissolved oxygen concentrations in the epilimnion during the spring months are mainly controlled by the saturated levels via reaeration at the surface. The concentrations in both layers start to drop as the warming season begins in April and the rate of decline in the hypolimnion follows AHOD closely. Although the hypolimnetic oxygen levels drop to almost zero in August, temperature stratification in the water column prevents the oxygen levels in the epilimnion from further decrease. Following the fall overturn, the water column is well mixed and the dissolved oxygen levels rise to saturation.

Figure 4.31 shows the dynamic phosphorus budget in terms of these categories: input (phosphorus loads entering the lake), output (leaving the lake), settling from the epilimnion to the hypolimnion, settling from the hypolimnion to the sediment, release from the sediment, and vertical exchange between the two layers. The loads from the hypolimnion to the sediment are an integral component of the overall budget. Phosphorus release from the sediment reaches a maximum of about 6 kg/day during the summer anoxic months at the sediment–water interface. As expected, the vertical dispersion is high during the spring and fall overturns supplying phosphorus from the hypolimnion to the epilimnion. Additional insights into Platte Lake are provided

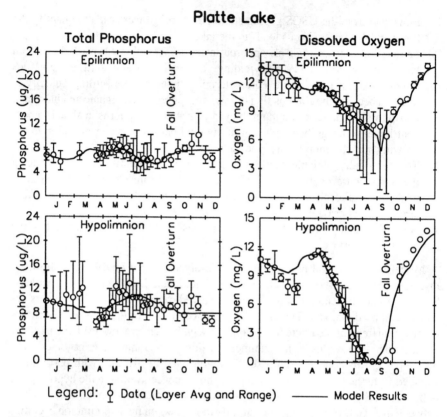

Fig. 4.30 Model results and data of total phosphorus and DO in Platte Lake

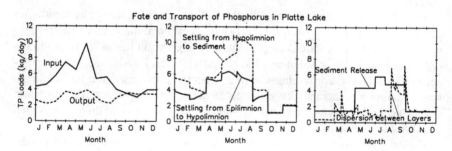

Fig. 4.31 Total phosphorus budget in Platte Lake

by these modeling results, which offer useful input for water quality management of the lake.

A frequently asked question in lake water quality management is: how long does it take to see a positive response in the lake due to nutrient reductions (e.g., lowered nutrient concentrations reaching the lake)? Obviously, the simple Vollen-weider model (in Sect. 2.6) cannot provide the answer as the analysis looks at only at steady-state conditions, i.e. when the lake phosphorus concentrations are in equilibrium with the constant, external phosphorus loads. A related critical question to the above raised in water quality management includes: how long does it take for the lake to respond to nutrient reductions? Understandably, it takes time for a lake to respond to nutrient load changes because the lake concentrations are rarely in equilibrium with the external loads. Such a condition becomes even more pronounced in large lakes with long residence time. In theory, it takes an infinite length of time to reach an equilibrium state following nutrient load changes. In other words, a sharp reduction in the external nutrient load would not necessarily produce an immediate reduction in the lake nutrient concentration for a large lake. For Lake Ontario (one of the Great Lakes between Canada and the U.S.), it would take about 20 years to reach an equilibrium between the constant, external nutrient load and the lake concentration. This length of time is about 2.5 times the detention time (8 years for Lake Ontario) to reach about 95% of a new steady state, i.e. equilibrium. Imagine how long it would take for Lake Superior (with a retention time of 180 years) to reach equilibrium. It would take centuries for Lake Superior to approach a new steady state.

Since large lakes seldom reach a true equilibrium between the lake concentration and external loads, changes in lake water quality are therefore difficult to perceive on a year-to-year basis, thereby creating a problem for the modeler: the estimation of long-term system response based upon a short observational period. The problem is somewhat similar for attempting to estimate the occurrence frequency of a one in ten-year drought flow based upon one or two years of record (Thomann et al. 1976). Long-term data are therefore needed to post audit changes in large lakes.

The theoretical time-variable responses for a lake can be computed from the following equation (Thomann and Mueller 1987):

$$C = \underbrace{\frac{W}{Q + KV}\left[1 - e^{-\left(\frac{1}{t_o}+k\right)t}\right]}_{External\ Loads} + \underbrace{C_o e^{-\left(\frac{1}{t_o}+k\right)t}}_{Initial\ Condition} \qquad (4.2)$$

where C = the whole lake nutrient concentration, W is the external nutrient loads, V is the lake volume, Q is the flow through the lake, t_o is the hydraulic detention time $(= V/Q)$, k is the overall attenuation rate of the nutrient due to settling or chemical reactions, and t is time. For example, a large lake with a hydraulic detention time of 20 years has a current total phosphorus concentration of 0.045 mg/L and is subject to an instantaneous load reduction to decrease the in-lake concentration. Assuming phosphorus is a conservative substance, Fig. 4.32 shows the dominant effect of the initial condition during the first year. At the end of the first year, the lake is responsive primarily to the specified initial conditions rather the external inputs—the lake has a memory! This rather long memory is one of the reasons that many nutrient reduction programs would take an extended time to achieve the new equilibrium. In this case, it takes about two and half times the hydraulic retention time (i.e. 50 years) to reach

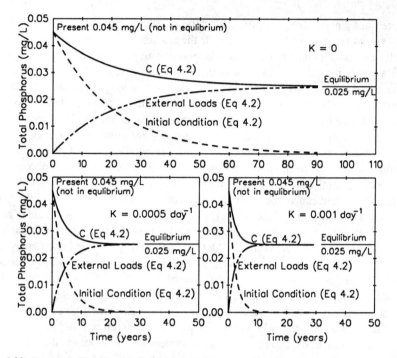

Fig. 4.32 Response time to nutrient control in Large Lakes

a concentration within 5% of the new equilibrium (see Fig. 4.32). The time to reach the new equilibrium of 0.025 mg/L would be infinity theoretically.

For non-conservative substances such as nutrients, one could make estimates for the external nutrient load, W and overall loss rate, k using the above equation. W is estimated from a variety of data sources but usually for a given year or group of years close to the sampling timeframe for the nutrients. The overall decay rate is estimated as part of the settling of the phytoplankton (as in the dynamic phytoplankton model) or is estimated from the nutrient data itself. However, the latter course of action assumes the lake to be in equilibrium with the present load, an assumption that cannot be checked until the lake has actually been observed for a period of at least one to two detention times (Thomann et al. 1976). The dilemma is made clear in Lake Ontario which would take over 16 years to reach a new equilibrium. Figure 4.32 shows that the total phosphorus concentrations at the end of one year are 0.041 mg/L and 0.038 mg/L for $k = 0.0005$ and 0.001, respectively. The difference is 0.003 mg/L (or 3 µg/L), too small to precisely determine the k value.

The above analysis points out that long-term data are crucial in eutrophication modeling of large lakes. Several decades have passed since Thomann's analysis; many lakes have gathered long-term data including sediments to support eutrophication modeling with phytoplankton and detrital settling, nutrient—algal dynamics, and nutrient cycling. Another insight from this analysis is that long response times are expected for nutrient control of large lakes. Another more dramatic element in determining the response time resides in the sediment system, which has even greater time scales than the overlying waters. We will explore this insight into sediment–water interactions later in this Chapter (see Sect. 4.9).

4.8 Reservoir Eutrophication Modeling

4.8.1 Vrhovo Reservoir in Slovenia

This section demonstrates eutrophication modeling and key physical insights of reservoirs for water quality management. As part of the NATO CCMP study, the Vrhovo Reservoir (see Fig. 2.2) was selected for a modeling study to support the water quality management. Some physical, chemical, and water quality data for this reservoir has already been presented in Chap. 2.

The W2 model was configured for the Vrhovo Reservoir with 1-m layers using temperature data in 2001 to calibrate the mass transport. Figure 4.33 shows the model calculated temperature profiles matching measured data at Radeče Bridge, Vrhovo Bridge, and Vrhovo Barrier from January 24 to August 23. Like the field data, the model results show no vertical stratification in the water column at these locations from January through May. Only slight temperature vertical stratification in the water column is seen during the summer surveys: June 28, July 12, July 26, August 9, and August 23. The most significant temperature stratification observed on August 23 is well reproduced by the model results. These results demonstrate

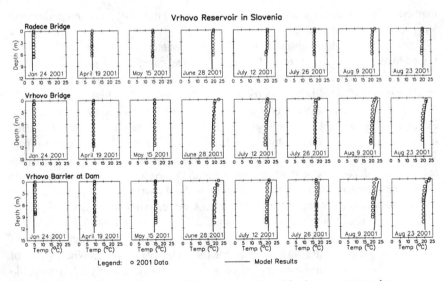

Fig. 4.33 Hydrodynamic calibration of Vrhovo reservoir model using temperature data

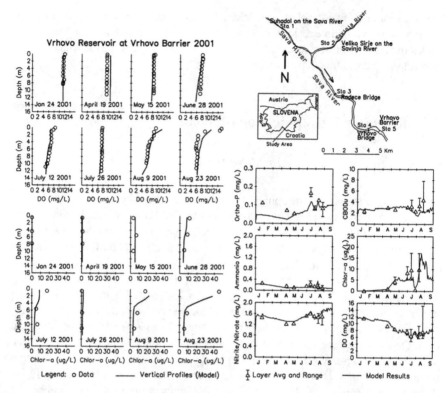

Fig. 4.34 Model results versus data for DO, nutrients, and Chlorophyll *a* at Vrhovo Barrier

that the hydrodynamic model is reproducing the 2-D distribution of temperature in Vrhovo Reservoir very well.

The next step of the analysis was to calibrate the water quality model using the calibrated hydrodynamics. Figure 4.34 shows the model calculated vertical DO profiles versus collected data at Vrhovo Barrier, the deep station near the dam. Summer DO stratification in August is also reproduced by the model. The chlorophyll *a* results also match the data, particularly the strong algal growth close to the water surface in August. Figure 4.34 also presents temporal plots of orthophosphate, ammonia, nitrite/nitrate, CBOD, chlorophyll *a*, and DO vs. data over a 9-month period from January 1, 2001 to September 30, 2001 at Vrhovo Barrier. Both the model results and data are depth-averaged values for comparison. Note that the DO levels drop during the summer months. The temporal patterns of nutrients in the water column follow the watershed nonpoint loads closely. A more complete presentation of the water quality modelling results can be found in Cvitanic et al. (2002).

The next question is how to implement dissolved oxygen as the eutrophication endpoint in water quality management for the Vrhovo Reservoir. When (time of the year) and where (in the reservoir) does the DO standard apply? A reasonable alternative is to apply the water quality standard (i.e. DO above 4 mg/L) for the

Fig. 4.35 Vrhovo Reservoir volumes of DO below 4 mg/L

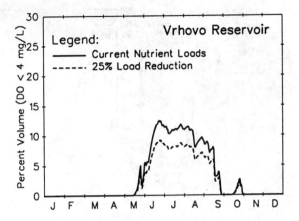

entire water column. Figure 4.35 shows the calculated reservoir volume percentage with DO below 4 mg/L in a solid line over the year. Obviously, this volume is most significant during the summer months. With 25% nutrient reduction, the volume is reduced (see the dashed line in Fig. 4.35).

4.8.2 Taiwan Reservoirs

Three of the top four largest Taiwan reservoirs in terms of volume are: Tsengwen (659×10^6 m^3), Feitsui (406×10^6 m^3), and Techi (183×10^6 m^3). They are located in southern, northern, and central Taiwan, respectively. See Fig. 4.36 for their relative locations (from north to south) in Taiwan. The Tsengwen Reservoir is located immediately below the Tropic of Cancer in the tropical climate zone while the other two reservoirs are sub-tropical. Primary water use of these three reservoirs are as follows: Feitsui for domestic water supply in Taipei City, Techi for power generation, and Tsengwen for agricultural irrigation. Taiwan reservoirs are characterized with strong fluctuations of the water surface elevations and corresponding storage volumes caused by torrential rainfall and massive runoff from typhoons (hurricanes) in the summer months. Therefore, hydrodynamics is a critical factor in modeling of these reservoirs (Kuo et al. 2006).

Figure 4.36 shows the calculated water surface elevations from the W2 model and measured surface elevations for these three reservoirs in two-year runs using the same elevation scale for ease of comparison. The Techi Reservoir is located over 1400 m above sea level, significantly higher than the other two reservoirs. The 44 m drop in the Feitsui Reservoir during the spring drought of 2000 is followed by a rapid recovery from hurricanes in the summer. Not a single hurricane made landfall during all of 2020, which created a severe drought in Taiwan by spring 2021 and forced restrictions of domestic water supply to 5 days a week. Although there were still no hurricanes hitting Taiwan by August 2021, the drought was relieved by sufficient

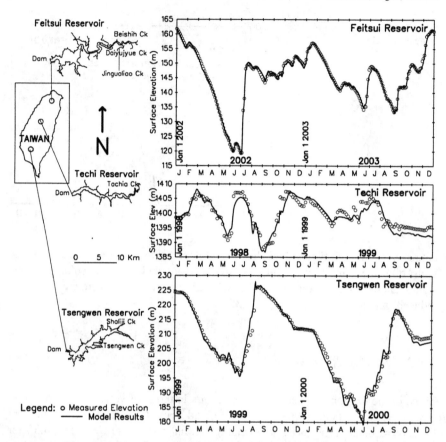

Fig. 4.36 Temporal variation of elevations of Taiwan Reservoirs: Feitsui, Techi, and Tsengwen

rainfall in late spring and early summer to replenish the reservoirs and to resume 100% domestic and industrial water supply.

Although control measures are in place, such as disallowing new development in the Feitsui Reservoir Watershed, eutrophication (with algal growth in the surface and depressed DO levels in deep water) is the top concern for these reservoirs. The first step in the modeling analysis is quantifying hydrodynamics, reflected in accurately calculating the water surface elevations of the reservoirs (Fig. 4.36), matching available data for model calibration. Reservoir temperature is strongly influenced by hydrodynamics. The left plot of Fig. 4.37 presents the model results versus temperature data in the Feitsui Reservoir from 2002 to 2003 and Tsengwen Reservoir in 1999 to 2000. Similar results for the Techi Reservoir in 1998–1999 are also shown in the right plot of Fig. 4.37. Note that the depth and temperature scales are kept the same for all three reservoirs so one can compare the vertical temperature profiles between the reservoirs closely as their climatic conditions vary. Seasonal variations of temperature as well as vertical stratification are clearly displayed in the Feitsui Reservoir,

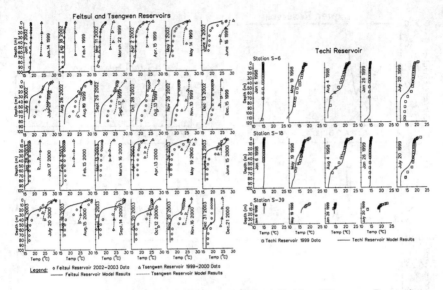

Fig. 4.37 Seasonal variation of water temperature in Feitsui, Techi, and Tsengwen Reservoirs

the most northern reservoir of the three. In addition, water temperature levels are higher in Tsengwen Reservoir than Feitsui Reservoir as Tsengwen Reservoir located to the south of the Tropic of Cancer is the most southern water of the three. The seasonal temperature variation is also less pronounced in Tsengwen Reservoir. The thermocline is vividly displayed in the Feitsui Reservoir in the summer months with a clear epilimnion in the surface. On the right, water temperatures in Techi Reservoir are shown at three reservoir stations: S-39, upstream and shallow; S-18, mid-section of the reservoir; and S-6, deep and close to the reservoir. Note that surface water temperatures are slightly lower in Techi Reservoir than Feitsui Reservoir despite the fact that the Techi is located south of the Feitsui. This temperature difference is because the Techi Reservoir is situated over 1200 m higher than the Feitsui Reservoir. Model results match the temperature data in these three reservoirs in Fig. 4.37. The reservoir hydrodynamics, which are verified with the model calculated water elevations and temperatures, are the first step of the reservoir modeling analysis.

Inflows and nutrient loads to the Tsengwen and Techi Reservoirs are plotted in Fig. 4.38. Comparing the water elevations and inflows in Fig. 4.38 reveals that temporal variation of the elevations closely follow the inflows in the reservoirs. For example, the dry spell in the spring of 1999 drove the water level down about 47 m in the Tsengwen Reservoir by the end of June. Hurricanes brought significant inflows starting in July 1999 (see Fig. 4.38) and refilled the reservoir quickly in July and August of that year. With lesser scale, a similar seasonal trend in the inflows and water elevations is also observed in the Techi Reservoir in 1998 and 1999. Nutrient loads in terms of total phosphorus, ammonia, and nitrie/nitrate concentrations also follow the hurricanes closely, especially in the Tsengwen Reservoir.

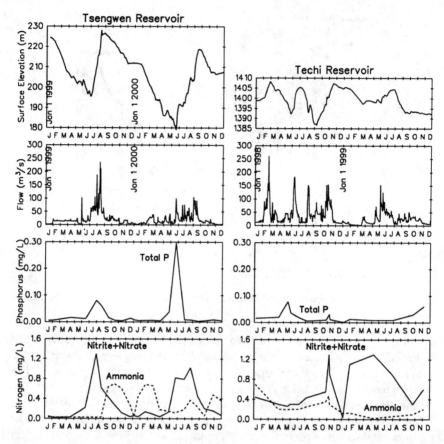

Fig. 4.38 Inflow and watershed nutrient loads to Tsengwen and Techi Reservoirs

Figure 4.39 presents the model results of DO versus data in vertical profiles for the Feitsui Reservoir (2002–2003) and Tsengwen Reservoir (1998–1999). The seasonal variations of DO in the Feitsui Reservoir are well reproduced for this two-year run (Ciou et al. 2012). DO data for the Tsengwen Reservoir are less sufficient compared with the Feitsui Reservoir. Nevertheless, the seasonal trend of DO is also reproduced by the model. As the Tsengwen is located in the tropical climate zone, the vertical stratification of DO is not as pronounced as the Feitsui Reservoir. The W2 model worked well for these reservoirs. Figure 4.40 shows the water quality results of ammonia, nitrite/nitrate, total phosphorus, chlorophyll *a*, and DO versus data at the dam site for these two reservoirs. In general, ammonia concentrations are quite low compared with nitrate concentrations. Significant nutrient loads delivered by significant inflow to the Tsengwen Reservoir during the hurricane seasons in 1998–1999 reflect on the rise of nutrient concentrations. Water quality in the Tsengwen Reservoir was closely controled by the hydrodynamics to the extent that chlorophyll

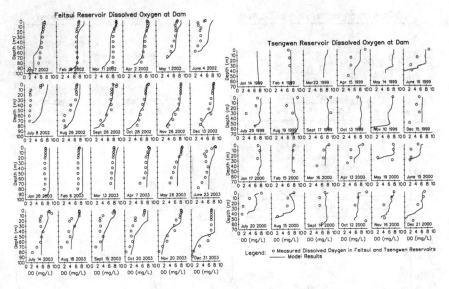

Fig. 4.39 Model calculated DO versus data in Feitsui (2002–2003) and Tsengwen (1999–2000) Reservoirs

a levels were not alarming, nor were the DO levels too low to be a concern in 1998–1999. The chlorophyll *a* concentration peak in the Feitsui Reservoir during June and July of 2002 was caused by the significant drawdown of the reservoir during that period, providing a favorable hydrodynamic condition for algal growth supported by sufficient nutrients. It is clear that the water quality of these three Taiwan reservoirs is strongly controled by hydrodynamics during the hurricane dominated seasons which can easily deliver precipitation over 1000 mm in a single day!

4.8.3 Maryland Reservoirs

Figure 4.41 shows four reservoirs in Maryland: Triadelphia (23×10^6 m^3), Rocky Gorge (21×10^6 m^3), Pretty Boy (72×10^6 m^3), and Loch Raven (87×10^6 m^3). Compared with the three Taiwan reservoirs in previous section, these reservoirs are small in volume. The Triadelphia and Rocky Gorge Reservoirs, operated by the Washington Suburban Sanitary Commission, provide drinking water to Montgomery and Prince George Counties while the Pretty Boy and Loch Raven Reservoirs are the drinking water source for the City of Baltimore and its surrounding counties in Maryland. The Triadelphia and Rocky Gorge Reservoirs are in the Patuxent River watershed while the Pretty Boy and Loch Raven Reservoirs are in the Gunpowder River watershed. Results of the hydrodynamic modeling of these four reservoirs match the water surface elevations quite well (Fig. 4.41). Note the same elevation scale is used in the plots for all four reservoirs, not only showing the altitudes of their

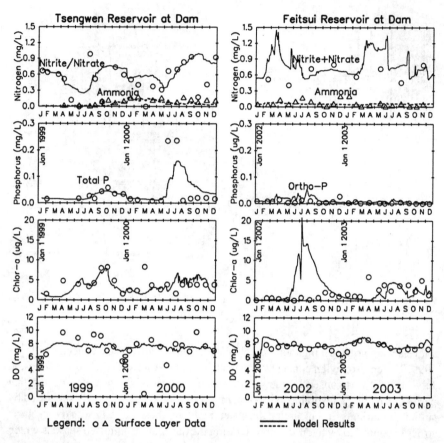

Fig. 4.40 Model calculated nutrient, chlorophyll *a*, and DO versus data Tsengwen (1999–2000) and Feitsui (2002–2003) reservoirs

respective locations, but also giving a good perspective of water surface seasonal variations. The significant drops in surface elevations in the last four months of 1997 in all four reservoirs were caused by water supply withdraws because of the summer drought. The seasonal variations of water elevations in these Maryland reservoirs are nowhere near those in the Taiwan Reservoirs, which are closely controlled by hurricanes. A unique water quality characteristic of these Maryland reservoirs is oversaturation of dissolved oxygen in the surface layer (limited to the top 3 m) during the summer months, which is the topic covered in this section.

While a common eutrophication concern with reservoirs is low DO in the deep portion of the water column, the DO levels close to the water surface reach saturation with continuing oxygen supply from the air. However, over-saturation of dissolved oxygen in the surface layer has been observed in many reservoirs. In their modeling studies, Cole and Wells (1999) reported super-saturation in a number of reservoirs: Monroe Reservoir, West Point Reservoir, Neely Henry Reservoir, and Walter F.

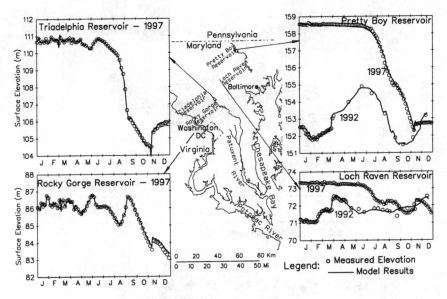

Fig. 4.41 Hydrodynamic model calibration—surface elevations in Triadelphia, Rocky Gorge, Pretty Boy, and Loch Raven reservoirs

George Reservoir. Measured DO and temperature values in the top 3-m of the water column of Triadelphia, Rocky Gorge, and Loch Raven Reservoirs are presented in Fig. 4.42. The solid line is the temperature dependent DO saturation curve for freshwater. It is clear that measured DO levels in the top 3-m of the water column are consistently above the saturation when water temperature is above 22°C (during the summer months). Such a phenomenon is even more pronounced in Loch Raven Reservoir than Triadelphia and Rocky Gorge Reservoirs.

Figure 4.43 shows the model results of dissolved oxygen versus data along with temperature dependent saturation in Triadelphia Reservoir during 1997. While the bottom DO levels in the summer months are well reproduced by the model, the calculated DO values in the top 3–4 m are consistently below, albeit coming close to saturation, the measured values from June to mid-September. In addition, the measured DO values are higher than the DO saturation levels during this summer period.

Modeling of the Loch Raven Reservoir with the W2 model was performed using the 1992 data (Lung and Zuo 2001). The vertical profiles of chlorophyll *a* and dissolved oxygen at the dam from February 3, 1992 to December 14, 1992 are presented in Fig. 4.44. The model results of chlorophyll *a* match the data quite well, particularly during the summer months. While the model reproduces the low DO levels in the bottom water of the reservoir closely, its results under-predict the DO in the over-supersaturated surface layer, primarily in the top 3 m (Fig. 4.44). The comparison shows that starting with the vertical profile of May 4, dissolved oxygen levels measured in the surface 3-m depth are greater than the saturation levels. Such

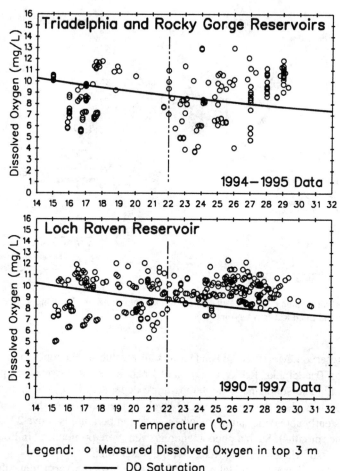

Fig. 4.42 Dissolved oxygen versus water temperature in Maryland Reservoirs

a phenomenon continues until late September/early October when the fall overturn begins in the water column. Note that the model calculated chlorophyll *a* levels match the vertical profiles of the measured values well.

4.8.4 Modeling DO Oversaturation

Model sensitivity analyses indicated that increasing the algal growth rate in the water column would significantly raise the algal biomass (chlorophyll *a*) level and exhaust the inorganic nutrients (Lung and Zou 2000). The high algal biomass levels (significantly exceeding the observed chlorophyll *a* levels) however, do not yield a

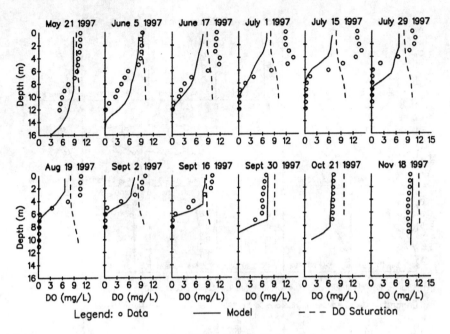

Fig. 4.43 Model results of dissolved oxygen in Triadelphia Reservoir

companion oversaturation of dissolved oxygen due to losing oxygen to the air via reaeration. At the same time, depleted inorganic nutrients cannot sustain the algal growth. There should be an internal mechanism responsible for the excessive oxygen production and such an oxygen production process would need to be at an extremely rapid rate to overcome the oxygen release across the air–water interface. A reasonable hypothesis is that sufficient nutrients must be recycled at high rates to sustain the phytoplankton growth and produce oxygen at a fast pace, supported by another route of nutrient recycling, which could come from zooplankton in the water column. The model was therefore modified to include zooplankton as a predator of phytoplankton in the water column. Zooplankton death would then recycle the needed nutrients to support the excessive phytoplankton growth. A new state variable, zooplankton, was introduced in the model to simulate zooplankton biomass in mg carbon /L. The grazing term in the phytoplankton kinetics is expressed as:

$$k_{gz} = \frac{a}{k_{sa} + a} C_{gz} z \theta_{gz}^{(T-20)} \tag{4.3}$$

where C_{gz} = zooplankton grazing rate (m^3 g C^{-1} day^{-1}).

θ_{gz} = temperature correction factor.

z = zooplankton biomass concentration (mg C/L).

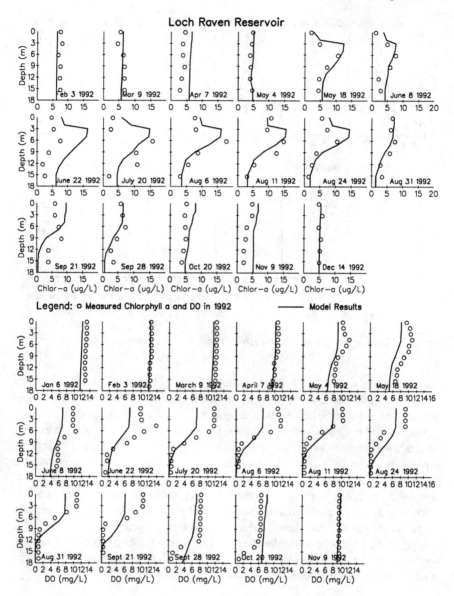

Fig. 4.44 Model results versus data for chlorophyll *a* and DO in Loch Raven Reservoir, 1992

a = phytoplankton biomass (μg Chl *a*/L).

k_{sa} = the half-saturation constant for zooplankton grazing on algae (μg Chl *a*/L).

T = water temperature.

At high levels of phytoplankton biomass, zooplankton grazing levels off according to Eq. (4.3). The zooplankton gains biomass by assimilating phytoplankton and loses biomass by respiration, excretion, and death. These processes can be incorporated into a zooplankton balance (Chapra 1997):

$$\frac{dz}{dt} = a_{ca}\varepsilon \frac{a}{k_{sa} + a} C_{gz}\theta_{gz}^{(T-20)}za - k_{dz}Z \tag{4.4}$$

where

a ratio of carbon to chlorophyll a in the phytoplankton biomass (μg C/mg Chl$a$)

ε a grazing efficiency factor

k_{dz} a first-order loss rate for respiration, excretion, and death (day^{-1})

Figure 4.45 presents model recalibration results vs. 1992 data at Station GUN0142 at the dam (Lung and Zuo 2001). The water quality constituents shown are dissolved oxygen, chlorophyll a, total phosphorus, and ammonia in the surface and bottom layers of the water column. Results from the original model (in dots) and the enhanced model (in solid lines) are displayed for comparison with the data (depth-averaged values and max/min values). Figure 4.45 shows that results from the enhanced model match the surface-layer dissolved oxygen levels much better than the results from the original model in 1992. The chlorophyll a match also improves with the enhanced model and so are the total phosphorus results in both surface and bottom layers.

The most significant change in the model coefficients is the instantaneous (not daily averaged) algal growth rate at about 6 day^{-1} in the enhanced model, resulting in rapid uptake of nutrients in the water column. This high algal growth rate can be sustained by the nutrients recycled from zooplankton. In addition, the algal biomass is quickly consumed by the zooplankton, thereby completing the nutrient cycling

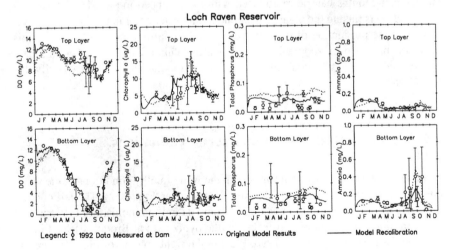

Fig. 4.45 Model recalibration with Zooplankton Grazing in Loch Raven Reservoir, 1992

Fig. 4.46 Zooplankton
biomass and algal primary
productivity rate in Loch
Raven reservoir

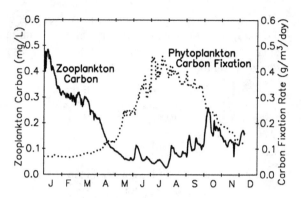

process in the water column. This high algal growth rate yields a much higher photo-synthesis rate to produce dissolved oxygen significantly exceeding the saturation level and maintaining that level after releasing it to the atmosphere via reaeration.

Figure 4.46 presents the zooplankton biomass (in mg Carbon/L) calculated by the model for the surface layer of the model near the dam of Loch Raven Reservoir. While no field data is available for comparison to validate the results, the calculated biomass levels are well within the range of values reported in the literature. Also shown in Fig. 4.46 is the calculated phytoplankton primary production rate (in mg Carbon/L/day) in the water column near the dam in 1992. Again, no field data is available to check this result. As expected, the algal carbon fixation rate is highest in the summer months, thereby producing oxygen at a super high rate. The zooplankton biomass, on the other hand, is the lowest during this period, recycling nutrients at a fast pace to sustain the high algal growth rate in this modified model. The calculated levels of primary production rate are well within the reported range of values. Super-saturation of dissolved oxygen in the Loch Raven Reservoir during the summer months can be reproduced with an enhanced model. The key to this recalibration is raising the algal growth rate in the water column to a sufficient level to sustain the production of dissolved oxygen. The nutrient needed to sustain this growth is provided by zooplankton in the water column. However, additional field data on zooplankton biomass and primary production rates are needed to verify this hypothesis.

4.9 Sediment Systems in Long-Term Simulations

Sediment systems play a significant role in the water quality conditions in the water column as can be seen from the Patuxent Estuary (see Fig. 4.12). The left column plots in Fig. 4.47 present the model results and data of ammonia, ortho-P, and DO concentrations in the bottom layer (below the pynocline) at Broomes Island in the Patuxent Estuary from 1997 to 1999. Plots in the right column of Fig. 4.47 are the corresponding nutrient and SOD fluxes during the same period. It is clear that the high

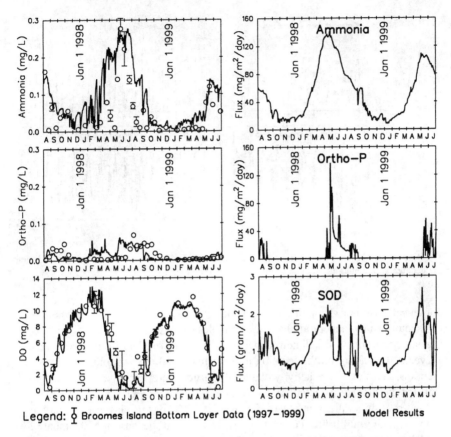

Legend: ⚲ Broomes Island Bottom Layer Data (1997–1999) ——— Model Results

Fig. 4.47 Nutrient and DO Levels below Pynocline, Sediment Nutrient Fluxes, and SOD at Broomes Island (1997–1999)

nutrient levels correspond with low DO levels in the summer months. In addition, the peak ammonia and ortho-P levels are approximately two months behind the peak sediment ammonia and phosphorus fluxes, respectively. Similarly, the anoxic condition in the water column is also about two months following the peak SOD flux!

Due to the nature of the sediment flux models, initial conditions were required for POP, particulate phosphate (PPO$_4$), PON, and POC for long term simulations (Nice 2006; Lung and Nice 2007). To establish initial conditions in the sediment for these variables, long-term simulations of the model were conducted in a fashion as suggested by Di Toro (2001). The long-term simulations were performed by executing a series of back-to-back, two-year simulations. Concentration values for POP, PPO$_4$, PON, and POC in the sediment, including different values for two reactivity class concentrations considered for POP, PON, and POC, and two layers of concentrations for PPO$_4$ were output to file at the end of each two-year simulation. The following two-year simulation in the series utilized the previous simulation

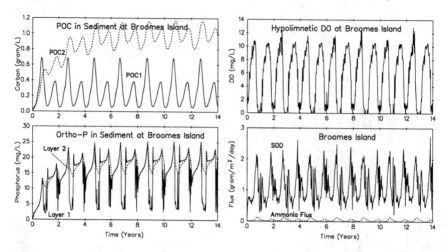

Fig. 4.48 Long-term particulate carbon and phosphorus in sediment, DO in hypolimnion, sediment oxygen demand, and ammonia flux in Patuxent Estuary

output values as initial conditions. The series of two-year simulations were continued until equilibrium concentrations of the variables were reached in the sediment. Equilibrium conditions for sediment concentrations were reached in approximately 8– 12 years, depending on the location in the estuary and the variable considered. As mentioned earlier, two reactivity classes of particulate organic matter were considered for the diagenesis calculation (G_1 and G_2) for POP, PON, and POC. Because of the lower mineralization rate of the G_2 class, POC of this class required a longer time to reach equilibrium concentrations. In addition, because the phosphate flux model is a true mass-balance model which tracks concentrations of phosphate in the sediment, equilibration time for phosphate was slightly longer. Concentrations for POC, Ortho-P, along with hypolimnetic dissolved oxygen, SOD, and ammonia flux at Broomes Island (where persistent hypoxia is observed) are shown in Fig. 4.48.

Interestingly, while nutrients in the sediment required long-term simulations to equilibrate, SOD within the model equilibrated fairly quickly (within the two-year simulation period). Consequently, in the lower estuary, model results indicated that levels of dissolved oxygen were not impacted greatly by long-term simulations. Model results also indicated that water column kinetics and downstream boundary conditions for nutrients and CBOD appear to be more controlling of dissolved oxygen in the lower estuary than long-term buildup of nutrients in the sediment. For these reasons, a 10-year simulation period was utilized to establish initial condition concentrations in the sediment for POP, PON, POC, and PO_4 for subsequent model simulations. In addition, since long-term simulations had only a slight impact on dissolved oxygen levels in the lower estuary using the developed model, long-term simulations were not implemented for model predictions under hypothetical loading scenarios.

Recalling the response time of large lakes (see Fig. 4.32), we see some similarity with the response time of sediment systems. As seen in the above discussions, a key factor is whether the sediment system is in equilibrium with the overlying water serving as a nutrient supplier. Since the time scale associated with the sediment system is much longer than the overlying water, the response time of a sediment system is even longer. Rarely do we see a sediment system that is in equilibrium with the nutrient loads from the overlying water. Such physical insights into the sediment–water interactions are important for a modeler to possess prior to launching the projection modeling analysis for nutrient controls. As modelers, we must keep in mind the response time in receiving water and sediment systems for large lakes and estuaries and convey this concept to regulators in formulating nutrient control strategies. Again, data holds the key to this modeling analysis.

4.10 Algae and Dissolved Oxygen

The Santa Fe River (see map in Fig. 2.5), from the Santa Fe WWTP to Cochiti Pueblo, was on New Mexico's Clean Water Act Section 303(d) list for TMDL development. Upstream of the Santa Fe WWTP, the Santa Fe River is generally a dry arroyo with upstream flow during some snowmelt periods in the spring and after some storm events. Surface water quality monitoring stations were established to characterize the water quality of the Santa Fe River and to evaluate the impact of the plant. Significant diurnal fluctuations of DO and pH observed at two locations (Stations 1 and 2 in Fig. 2.5) downstream from the Santa Fe WWTP in 1999 are presented in Fig. 2.5.

Measured hydraulic geometry of the Santa Fe River in the study area such as average velocity and water column depth was used to develop the model segmentation. On-site measurements of key environmental parameters like water temperature are incorporated into the model. Since observed solar radiation data were not available for the City of Santa Fe, hourly historical data measured at Albuquerque, NM were used instead.

The QUAL2E model was initially selected for this modeling study. A careful review of the modeling framework indicated that while the nutrients in the water column are transported in the downstream direction, the attached algae remain stationary in the river bed. In addition, the attached algae growth rate depends on the velocity in the water column, which is not accounted for in the QUAL2E model. Maintaining the attached algae as stationary and incorporating a velocity dependent attached algae growth rate would therefore require a significant effort to modify the model code. Also, the light intensity level for the QUAL2E model is applied to the entire water column depth. In this environment, the attached algae residing at the river bed receives only attenuated light levels. The depth-averaged light reading in the QUAL2E model would not be accurate for attached algae even through the water column is shallow. On the other hand, the WASP/EUTRO5 model is much easier to adapt to address the above technical difficulties. First, the WASP model allows the turning off of mass transport, i.e., advective and dispersive flows, for any given system variables (water quality constituents). This option would render the attached algae stationary but maintain mass transport for nutrients. The correct light intensity level for the attached algae can be quantified by configuring a second, thin layer in the model segmentation for the attached algae. As such, light levels for the attached algae would be quantified accurately via attenuation through the first layer. [Note that the QUAL2E model does not allow multiple vertical layers in the water column.]

A 3-mile reach of the Santa Fe River from the plant discharge to about 1¼ miles below Preserve is divided into 12 segments, each of which is ¼ miles long. Because the attached algae reside at the bottom of the water column, another 12 segments are configured to accommodate them as a second layer below the water column. This second layer is very thin with a depth of 0.324 ft (0.1 m). A total of 24 segments are therefore configured for the Santa Fe River. Since the attached algae do not transport with the river flow, the mass transport for the algae in this thin layer was turned off. The water column depth ranges from 0.67 to 0.77 ft and the advective velocity associated with the river flow (essentially 100% wastewater flow, about 6.5 cfs during the period from June 7 to June 10, 1999) ranges from 0.42 to 0.55 ft/s along the study area. The water column is shallow and moves at a moderate speed. A vertical dispersion coefficient of 2.1×10^{-5} ft^2/s (or 0.02 cm^2/s) is used between the two water column layers. This value is relatively low, yet large enough to generate sufficient mixing between the two layers, making the thin, second layer a virtual layer.

Since the attached algae were in the bottom of the river channel, they were not simulated in the surface layer of the water column. Thus, the algal concentration and growth rate in the surface layer were set to zero throughout the simulation. Further, no attached algae will be transported from the thin, second layer to the surface layer. Other system variables such as nutrients are freely transported between the two layers via the vertical dispersion coefficient, thereby making nutrients from the water column readily available for the attached algae in the second layer. Since the attached algae reside at the river bottom, their settling velocity is set to zero.

The key feature of this modeling analysis is diurnal simulations of DO and pH. To simulate them requires small time-step calculations of algal growth. To mimic instantaneous algal growth, the light reduction on algal growth should not integrate over time and depth. Instead, the following light reduction equation is used:

$$r = \frac{I}{I_s} e^{\left[-\frac{I}{I_s} + 1 \right]}$$

(4.5)

where

r light reduction factor for algal growth rate (dimensionless number)
I instantaneous light intensity (langley/day) at the surface of the second layer, and
I_s saturated light intensity for optimum algal growth (set at 525 langley/day)

Light attenuation through the water column, i.e., the surface layer, is incorporated to transmit the proper light level to the attached algae in the second layer. Since the attached algae is located at the bottom of the water column with no suspended algae in the water column, the self-shading effect of algae must be turned off.

Attached algal growth is also limited by the availability of nutrients via the Michaelis-Menton kinetics (see Sect. 4.1). The biomass level at which half the maximum growth rate occurs is 5.0 gm-C/m^2 (Warwick et al. 1997). This mechanism allows maximum rates of primary productivity at low levels of biomass with decreasing rates of primary productivity as the benthic community expands.

Using hourly data for Albuquerque, the maximum solar radiation for each hour of each day of the year is determined by searching through the 1961–1990 period. An assumption is made that given the 30-year period, clear-sky conditions would have occurred at least once each hour of each day during that period. The clear-sky solar radiation is taken as the maximum value found in the first step. The clear-sky solar radiation values are then adjusted using the hourly cloud cover conditions recorded for Santa Fe in the climatology data set downloaded from NOAA/NCDC. The equation for adjusting the solar radiation due to cloud cover is:

$$SR_{adj} = \left(1.0 - 0.65CC^2\right)SR_{clear} \tag{4.6}$$

where

SR_{adj} = adjusted solar radiation

CC = cloud cover fraction (0.0 = clear; 1.0 = overcast)

SR_{clear} = clear-sky solar radiation.

The SR values are then corrected to photosynthetically active radiation (PAR) by multiplying the adjusted solar radiation by 0.43 for model input.

The model was run for 16 days until the computation reached a dynamic steady state. Figure 4.49 shows the results for the 4-day period of June 7–June 10, 1999 for attached algal biomass (in chlorophyll a) and dissolved oxygen at locations immediately below the Santa Fe WWTP and Preserve, which is 1.6 miles downstream from the plant. The model calculated dissolved oxygen fluctuation (i.e., phase) follows the observed pattern closely. The diurnal trends of DO between these two locations are quite similar. While the model results match the observed trend of dissolved oxygen at both locations, there are differences between the calculated maximum/minimum values and the observed data. The rising benthic algal biomass results in the dissolved oxygen maximum in the afternoon. The subsequent decline of the algal biomass in the later part of the day produces a minimum dissolved oxygen by mid-night and early morning of the next day.

While the model results match the observed trend of dissolved oxygen and pH at both locations, there are differences between the calculated maximum/minimum values and the observed data. Because the model produces equilibrium values while the Santa Fe River may not reach a steady-state condition, the graph shows some day to day variation in dissolved oxygen and pH.

One of the key model parameters in algal-DO kinetics is reaeration coefficient, k_a. High reaeration coefficients tend to produce the instantaneous DO curve not diverging

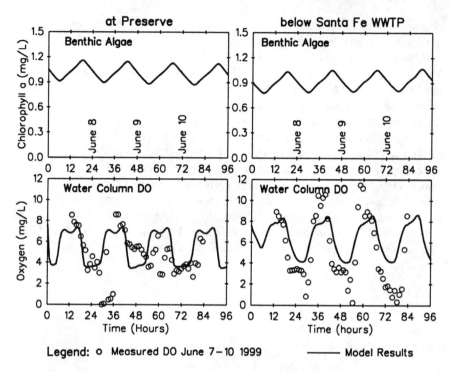

Fig. 4.49 Diurnal benthic algae and DO in Santa Fe River

much from the saturation level. In addition, the shape of the curve is similar to that of the forcing function (in this case, the time-variable water temperature and light intensity levels), with the maximum oxygen concentration (minimum deficit) occurring shortly after solar noon (Chapra 1997). However, as k_a decreases, the dissolved oxygen level moves farther away from saturation. In addition, the shape departs from that of the forcing function and becomes more sinusoidal. The maximum DO concentration occurs later in the afternoon, and a clearly defined DO minimum occurs just after dawn. The small k_a retards the rate of reaeration transfer, allowing the effect of the previous day to carry over into the present day. This condition accounts for the minimum dissolved oxygen concentration (maximum deficit) occurring slightly after dawn. The reaeration equation by Moog and Jirka (1998) was found most appropriate for the study site (with a stream channel slope > 0.0004):

$$k_a = 1740V^{0.46}S^{0.79}H^{0.74} \tag{4.7}$$

An examination of the topographical map of the study area reveals that the average channel slope is approximately 0.005. With an average velocity of 0.152 m/s and a depth of 0.213 m for the study area, one can calculate an average $k_a = 3.44$ day^{-1} at 20°C. Subsequent model calibration analyses result in a value of 2.50 day^{-1} (see Fig. 4.50). A model sensitivity run shows that varying k_a between 2.5 day^{-1} and

Fig. 4.50 Reaeration and Diurnal DO in the Santa Fe River

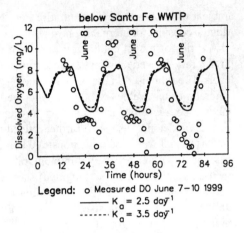

Legend: o Measured DO June 7–10 1999
———— K_a = 2.5 day^{-1}
------ K_a = 3.5 day^{-1}

3.5 day^{-1} has a small impact on the diurnal range of DO but does not affect the peak DO level in the Santa Fe River. This reaeration coefficient value is also used for the exchange of CO_2 across the air–water interface in modeling the diurnal pH fluctuation (see pH modeling in Chap. 5).

Another key model parameter is the saturated dissolved oxygen level, which is calculated as a function of temperature as follows (Lung 2001):

$$C_s = \frac{468}{31.6 + T} \qquad (4.8)$$

where C_s is saturated DO concentration in mg/L and T is temperature in °C. Values generated by this equation must be multiplied by a factor of 0.78 to account for the thin air at the high elevation of the study area, 6200 ft above the sea level. At 20°C, the saturated dissolved oxygen level in the Santa Fe River is only 7.07 mg/L.

Since the wastewater flow is the total river flow during the field survey, the measured plant effluent concentrations are used as upstream boundary conditions for the model input. No other point or nonpoint source loads are included in the model.

A critical point to calibrate the model is the sharp rising and declining rates of dissolved oxygen levels in the water column, with a maximum increase rate at over 240 mg O_2/L/day in the Santa Re River. These rates are supported by the dissolved oxygen gradients measured in the field at sampling sites, thereby further substantiating the validity of the model results. Two factors contribute to this sharp daily swing of dissolved oxygen concentrations: the high benthic algae biomass and the carbon to chlorophyll *a* ratio in the algal biomass. The high algal biomass calculated by the model is about 1200 μg/L, a level that has been reported for the benthic algae in the east branch of Brandywine Creek in Pennsylvania (Knorr and Fairchild 1987). Their growth is primarily limited by the inorganic nitrogen concentrations in the water column and to an extent by the CO_2 concentrations. The ambient orthophosphate concentrations are high, much higher than the Michaelis-Menton constant for inorganic phosphorus, and therefore not a factor limiting growth.

A series of model calibration runs have revealed that a high carbon to chlorophyll a ratio is needed to match the significant daily fluctuation of dissolved oxygen levels. The ratio used in the model is $266 \frac{\mu g\, Carbon}{\mu g\, Chlorophyll\, a}$ following a series of model calibration runs. Values of other kinetic coefficients related to benthic algal growth and nutrient recycling are taken from a recent modeling study of Carson River, NV by Warwick et al. (1997) and further refined in model calibration.

This model was used to demonstrate that it was not possible to meet the current water quality criterion (daily minimum: 6.0 mg/L) with any technically viable treatment alternative (e.g., plant upgrade) short of no discharge. Based on subsequent model validation, the New Mexico Water Quality Control Commission (August 8, 2000) adopted a new site-specific water quality criterion (daily minimum: 4.0 mg/L, minimum 24-h average: 5.0 mg/L).

4.11 Eutrophication Modeling Recap

With respect to eutrophication modeling: the aquatic plant/nutrient problems are the most difficult models with which we have worked with because of the complexity of the plant biology, the nonlinear interactions between nutrients and aquatic plants, and the interactions with the sediment (Thomann 1987). Case studies presented in this chapter further reaffirm his assessment. Comparing with BOD/DO modeling in Chap. 3, eutrophication modeling requires many more skills from the modelers and significant amounts of data to configure, calibrate, and verify models. While BOD/DO models are, in general, linear in formulation, many kinetics relationships in eutrophication models are nonlinear, thereby presenting additional degrees of difficulties in the modeling analysis.

The modeling work by Di Toro et al. (1971) on the Sacramento-San Joaquin Delta, California launched eutrophication modeling, benefiting from work of algal growth kinetics in natural waters at that time. They consolidated estuarine studies into model development and configuration, setting the stage for eutrophication modeling. As pointed by Thomann (1987), sediment–water interactions hold another key component of eutrophication modeling. Di Toro (1986) made a big leap by presenting his sediment flux work. Another leap in eutrophication modeling includes the Potomac Estuary blue-green algal blooms of 1983, revealing aerobic release of sediment phosphorus under high pH conditions (see Chap. 1). Eutrophication models with these continuing improvements substantially enhance and elevate our modeling capabilities.

Increasing model sophistication obviously demands more data to support the model configuration, calibration, and verification analyses, which is simply not available. For example, while the modeling frameworks have become more and more complex, there is not a commensurate spatial and temporal expansion of the Bay database in the past two decades. There is no strong urge to develop new modeling frameworks as there is no shortage of them. Instead, how to enhance data collection to support eutrophication modeling is the key.

References

Ambrose RB et al (1993) The water quality analysis simulation program, WASP5 part a: model documentation. US EPA Environmental Research Lab, Athens, GA

Ariathurai R, Krone RB (1976) Mathematical modeling of sediment transport in estuaries. In Estuarine Processes Vol II circulation, sediments, and transfer of material in the estuary. Academic Press, pp 98–106

Boynton WR et al (1990) Long-term characteristics and trends of benthic oxygen nutrient fluxes in the Maryland portion of Chesapeake Bay. Chesapeake Bay Conference, Baltimore, MD

Boynton WR et al (2008) Nutrient budgets and management actions in the Patuxent River Estuary, Maryland. Estuaries Coasts 31(4):623–651

Breitburg DL et al (2003) The pattern and influence of low dissolved oxygen in the Patuxent River, a seasonally hypoxic estuary. Estuaries 26(2A):280–297

Brockenbrough A (2016) Personal communications. Virginia Department of Environmental Quality

Bryhn AC, Håkanson L (2009) Eutrophication: model before acting. Science 324(5928):723

Cerco CF (1995) Response of chesapeake bay to nutrient load reductions. J Environ Eng 121(8):549–557

Chapra SC (1997) Surface water-quality modeling. McGraw-Hill, New York

Chapra SC, Canale RP (1991) Long-term phenomenological model of phosphorus and oxygen in stratified lakes. Water Res 25(6):707–715

Ciou SK et al (2012) Optimization model for BMP placement in a reservoir watershed. J Irrig Drain Eng 138(8):736–747

Cole TM, Wells SA (1999) CE-QUAL-W2: a two-dimensional, laterally average, hydrodynamic and water quality model. U.S. Army Corps of Engineers, Washington, DC

Conley DJ et al (2009) Ecology. controlling eutrophication: nitrogen and phosphorus. Science 323(5917):1014–1015. https://doi.org/10.1126/science.1167755

Cvitanic I et al (2002) Water quality modeling of the Vrhovo Reservoir with CE-QUAL-W2. NATO/CCMS project modeling nutrient loads and response in river and Estuary systems

Di Toro DM (1986) A diagenetic oxygen equivalents model of sediment oxygen demand. In Hatcher K (ed) Sediment oxygen demand. University of Georgia, Athens

Di Toro DM (2001) Sediment flux modeling. Wiley-Interscience, New York

Di Toro DM et al (1971) A dynamic model of phytoplankton population in the Sacramento-San Joaquin Delta. In Gould RF (ed) Nonequilibrium systems in natural water chemistry. American Chemical Society, Washington, DC

Di Toro DM et al (1990) Sediment oxygen demand model: methane and ammonia oxidation. J Environ Eng 116(5):945–986

Geider RJ (1987) Light and temperature dependence of carbon to chlorophyll ratio in microalgae and cyanobacteria: implications for physiology and growth of phytoplankton. New Phytol 106(1):1–34

Hagy JD et al (2000) Estimation of net physical transport and hydraulic residence times for a coastal plain estuary using box model. Estuaries 23(3):328–340

Hagy JD et al (2004) Hypoxia in Chesapeake Bay, 1950–2001: long-term change in relation to nutrient loading and river flow. Estuaries 27(4):634–658

Hunt GA (2005) Harmful algal bloom modeling framework for the Pocomoke River Estuary. Ph.D. dissertation, University of Virginia

HydroQual (1981) Water quality model analysis of the Patuxent. Report prepared for the Maryland Office of the Environment Programs

Jakobsen HH, Markager S (2016) Carbon-to-chlorophyll ratio for phytoplankton in temperate coastal waters: seasonal patterns and relationship to nutrients. Limnol Oceanogr 61:1853–1868

Kemp WM, Boynton WR (1992) Benthic-pelagic interactions: nutrient and oxygen dynamics. In: Smith DS et al (eds) Oxygen dynamics in the Chesapeake Bay. Maryland and Virginia Chesapeake Bay Programs, College Park, Maryland, pp 149–221

Knorr DE, Fairchild GW (1987) Periphyton, benthic invertebrate, and fishes as biological indicators for water quality in each branch of Brandywine Creek. Proc PA Acad Sci 61(1):62–66

Kuo AY, Neilson BJ (1987) Hypoxia and salinity in Virginia estuaries. Estuaries 10(4):277–283

Kuo JT et al (2006) Eutrophication modelling of reservoirs in Taiwan. Environ Modelling Softw 21:829–844

Lung WS (1986) Assessing phosphorus control in the James River Basin. J Environ Eng 112(1):44–60

Lung WS (1992) A water quality model for the Patuxent Estuary. Water resources engineering report No.8, Department of Civil Engineering, University of Virginia

Lung WS (1996) Fate and transport modeling using a numerical tracer. Water Resour Res 32(1):171–178

Lung WS (2001) Water quality modeling for wasteload allocations and TMDLs. Wiley, New York

Lung WS (2017) Preliminary modeling report of the Tranlin manufacturing plant on the James River. Report submitted to Tranlin Corporation

Lung WS, Bai S (2003) A water quality model for the Patuxent Estuary: current conditions and predictions under changing land-use scenarios. Estuaries 26(2A):267–279

Lung WS, Larson CE (1995) Water quality modeling of upper Mississippi River and Lake Pepin. J Environ Eng 121(10):691–699

Lung WS, Nice AJ (2007) A eutrophication model for the Patuxent Estuary: advances in predictive capabilities. J Environ Eng 133(9):917–930

Lung WS, O'Connor DJ (1984) Two-dimensional mass transport in estuaries. J Hyd Eng 110(10):1340–1357

Lung WS, Testerman N (1989) Modeling fate and transport of nutrients in James Estuary. J Environ Eng 115(5):978–991

Lung WS, Zou R (2000) Water quality modeling of the Loch Raven and Pretty Boy Reservoirs. University of Virginia

Lung WS et al (1993) Eutrophication analysis of embayments in Prince William Sound, Alaska. J Environ Eng 119(5):811–824

Martin JL, McCutcheon SC (1999) Hydrodynamics and transport for water quality modeling. Lewis Publishers, Boca Raton

Moog DB, Jirka GH (1998) Analysis of reaeration equations using mean multiplicative error. J Environ Eng 124(20):104–110

Nice AJ (2006) Developing a fate and transport model for arsenic in estuaries. Ph.D. dissertation, University of Virginia

O'Connor DJ, Lung WS (1981) Suspended solids analysis of estuarine systems. J Environ Eng 107(1):101–120

Odd NUM, Owen NW (1972) A two-layer model of mud transport in the Thames Estuary. Proceedings of Institution of Civil Engineers, London, England, pp 175–205

Schindler DW et al (2008) Eutrophication of lakes cannot be controlled by reducing nitrogen input: results of a 37-year whole-ecosystem experiment. PNAS 105(32):105–112

Testa JM et al (2008) Long-term changes in water quality and productivity in the Patuxent River Estuary: 1985–2003. Estuaries Coasts 31:1021–1037

Thomann RV (1977) Comparison of lake phytoplankton models and loading plots. Limno Oceanogr 22(2):370–373

Thomann RV (1987) System analysis in water quality management—a 25 year retrospect. In Beck MB (ed) System analysis in water quality management. Pergamon Press, Tarrytown, NY, pp 1–14

Thomann RV, Mueller JA (1987) Principles of surface water quality modeling and control. Harper & Row, New York

Thomann RV et al (1976) Mathematical modeling of phytoplankton in Lake Ontario, part 2: simulations using Lake 1 model. US EPA 600/3-76-065

Thompson PA et al (1992) Effects of variation in temperature .1. on the biochemical composition of 8 species of marine phytoplankton. J Phycol 28(4):481–488

U.S. EPA (2010) Ambient water quality criteria for dissolved oxygen, water clarity, and chlorophyll-a for the Chesapeake Bay and its tidal tributaries: 2010 technical support for criteria assessment protocols addendum. 903-R-10-002, Annapolis, MD

U.S. EPA (2003) Setting and allocating the Chesapeake Bay basin nutrient and sediment loads: the collaborative process, technical tools and innovative approaches. EPA 903-03-007. Annapolis, MD

Vollenweider RA (1968) Water management research. Scientific fundamentals of the eutrophication of lakes and flowing waters with particular reference to nitrogen and phosphorus as factors in eutrophication. OECD, Paris

Walker WW (1998) Analysis of monitoring data from Platte Lake, Michigan. Report prepared for Michigan Department of Natural Resources

Warwick JJ et al (1997) Estimating nonpoint source loads and associated water quality impacts. J Water Resour 123(5):302–310

Weller DE et al (2003) Effects of land-use change on nutrient discharges from the Patuxent River Watershed. Estuaries 26(2A):244–266

Chapter 5
Modeling Analysis for pH

One of the most fundamental and significant chemical characteristics of any water is pH, defined as the concentration of hydrogen ion (more properly, hydrated proton). Nearly every reaction of interest in natural waters, in water treatment, and in wastewater treatment involves an effect on pH, and conversely, is affected by pH (Weber 2000). [The diurnal pH fluctuation due to significant benthic algal growth in Chap. 4 is a good example]. Acidification and eutrophication are the two extremes in a wide spectrum of water quality conditions in ambient systems. Compared with eutrophication, acidification has received much less attention to date despite being equally important to water quality. Lake acidification research during the global energy crisis in 1970–1980 resulted in many scientific studies (Eshleman and Hemond 1985; Gherini et al. 1984; Henriksen 1979, 1980; Johannes et al. 1981; Kramer and Tessier 1982; Schindler et al. 1980; Schofield et al. 1985; Small and Sutton 1986), gathering significant amounts of water quality data. It is no surprise that acidification has been closely associated with energy production, resulting in the water-energy nexus (e.g., pH and sulfate modeling of acid mine drainage). Interestingly, the pH-enhanced sediment phosphorus release that led to the 1983 blue-green algal blooms in the Potomac Estuary reported in Chap. 1 was associated with reverse acidification, the key to that successful investigation. Separately, measures such as liming an acidic water system would also require modeling studies to evaluate the mitigation action. For natural waters, quantifying pH is needed in many situations for water quality management and planning.

5.1 Carbonate Equilibrium and Mass Balance

The key water quality substance related to acidification is alkalinity. Lung (1984) summarized a number of simplified methods to evaluate acidification of freshwater lakes using available data from Scandinavia by comparing long-term decreases in alkalinity, resulting from increases in calcium and magnesium (metals) and decreases

© The Author(s), under exclusive license to Springer Nature Switzerland AG 2022
W.-S. Lung, *Water Quality Modeling That Works*,
https://doi.org/10.1007/978-3-030-90483-8_5

in ligands (e.g. sulfate). A simplified underlined ionic balance due to electroneutrality in natural waters (Di Toro 1976) can be written as alkalinity = M (Metals) – L (Ligands), which can then be simplified to the following:

$$[\text{Alk}] = \left[C_a^{2+}\right] + \left[M_g^{2+}\right] - \left[SO_4^=\right]$$

where [Alk] is alkalinity in equivalent weight concentration. Note that $\left[SO_4^=\right]$ equals zero under pre-acidification conditions. It is clear that a simple outcome of acidification is the increase of $\left[SO_4^=\right]$ resulting in the reduction of alkalinity (Lung 1984). A plot of $\left[C_a^{2+}\right] + \left[M_g^{2+}\right]$ versus $\left[SO_4^=\right]$ based on data (in micro-equivalent per liter) in Norwegian freshwater lakes by Henriksen (1979) is presented in Fig. 5.1, showing zones of bicarbonate lakes, acidic lakes, and intermediate lakes. Data on Panther Lake and Woods Lake in the Adirondacks, New York and Bickford Reservoir in Massachusetts (Chen et al. 1984; Gherini et al. 1984; Lung 1987) lie in these three separate zones with different pH levels. Two straight lines with pH 5.28 and 4.69 are plotted in this figure. Lakes located in the bicarbonate zone would have pH above 5.28 while lakes with pH below 4.69 are categorized as acidic. While these three waters have similar pH levels, their alkalinity as reflected from the above equation, put them into three different categories in Fig. 5.1.

The governing equations describing the carbonate equilibria system serve as the basis for the modeling of pH in natural waters. A classic description of this system is available in Stumm and Morgan (1996). The major species considered in this equilibrium are dissolved carbon dioxide [CO_2], bicarbonate [HCO_3^-], carbonate

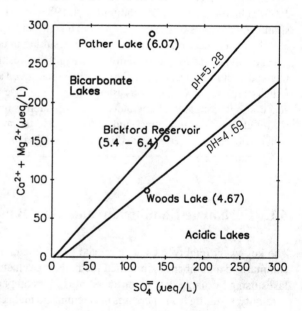

Fig. 5.1 Status of acidification in freshwater lakes

$[CO_3^=]$, hydrogen $[H^+]$, and hydroxyl $[OH^-]$ ions. Since these species undergo rapid reversible ionization reactions with respect to the time scales of the physical processes of mixing and gas transfer, it is mathematically infeasible to formulate direct mass balance equations for each of the these species (Di Toro 1976). In other words, these species reach equilibrium almost instantly and the kinetics of these reversible reactions are extremely fast. Therefore, at a point where mixing of several waters takes place, it is reasonably appropriate to use total inorganic carbon, [TIC] to perform mass balance calculations (Di Toro 1976). [TIC] is defined as:

$$[TIC] = [CO_2] + \left[HCO_3^-\right] + [CO_3^=] \tag{5.1}$$

The reversible ionization reactions among the carbonate species will not affect the amount of [TIC] in the natural water system and therefore it can be treated as a conservative constituent. Next, the metals and ligands balance leads to a second conservative substance, [Alk] as:

$$[Alk] = \left[HCO_3^-\right] + 2\left[CO_3^=\right] + \left[OH^-\right] - \left[H^+\right] \tag{5.2}$$

Since the metal ions and ligands are unaffected by the ionization reactions, alkalinity must also be conservative with respect to these reactions. Combining Eqs. 5.1 and 5.2 yields:

$$\left[CO_{2-acy}\right] = [CO_2] - \left[CO_3^=\right] - \left[OH^-\right] + \left[H^+\right] = [TIC] - [Alk] \tag{5.3}$$

where $\left[CO_{2-acy}\right]$ is CO_2 acidity. Therefore, $\left[CO_{2-acy}\right]$ must also be conservative at a point of mixing, since it is the difference between two conservative parameters (Di Toro 1976). Also note that [Alk] and $\left[CO_{2-acy}\right]$ can be either negative or positive, depending on the relative concentrations of the various ions. In contrast, total inorganic carbon is non-negative. The following analysis demonstrates the carbonate equilibrium in natural waters. Winter and summer 1970 data of temperature, pH, and alkalinity from Horwer Bucht (Lake Lucerne), Switzerland are shown in Fig. 5.2.

The carbonate equilibrium equations are as follows (Snoeyink and Jenkins 1980):

$$CO_{2(g)} \leftrightarrow CO_{2(aq)} \quad K = K_H = 10^{-1.5} \quad \text{Henry's Law} \tag{5.4}$$

$$CO_{2(aq)} + H_2O \leftrightarrow H_2CO_3 \quad K_m = 10^{-2.8} \tag{5.5}$$

$$H_2CO_3 \leftrightarrow H^+ + HCO_3^- \quad K_1' = 10^{-3.5} \tag{5.6}$$

$$H_2CO_3^* \leftrightarrow H^+ + HCO_3^- \quad K_1 = 10^{-6.3} \tag{5.7}$$

$$HCO_3^- \leftrightarrow H^+ + CO_3^= \quad K_2 = 10^{-10.3} \tag{5.8}$$

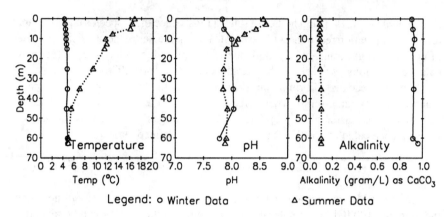

Fig. 5.2 Water temperature, pH, and Alkalinity in Horwer Bucht

$$H_2O \leftrightarrow H^+ + OH^- \quad K_w = 10^{-14} \tag{5.9}$$

From Eq. 5.5, $K_m = 10^{-2.8} = \frac{[H_2CO_3]}{[CO_{2(aq)}]} = 1.6 \times 10^{-3}$.

Thus, $[H_2CO_3]$ is only 0.16% of $[CO_{2(aq)}]$. $[CO_{2(aq)}] \gg [H_2CO_3]$. Because it is difficult to distinguish between $CO_{2(aq)}$ and H_2CO_3 by analytical procedures such as acid–base nitration, a hypothetical species $(H_2CO_3^*)$ is used to represent H_2CO_3 plus $CO_{2(aq)}$.

Thus, $[H_2CO_3] + [CO_{2(aq)}] = [H_2CO_3^*]$, where $[H_2CO_3^*]$ is a hypothetical species only. It should be pointed out that the equilibrium constants in Eqs. 5.4 to 5.9 are temperature dependent and strictly speaking, minor adjustments to their values are needed for the winter and summer conditions.

From the winter data (Fig. 5.2), alkalinity in winter = 0.9 g/L or 900 mg/L as C_aCO_3 and at pH = 7.8 in the first 10-m depth of the lake, $[OH^-] \cong [H^+]$. Thus, Eq. 5.2 yields

$[Alk] = [HCO_3^-] + 2[CO_3^=] = 900$ mg/L as C_aCO_3 and from Eq. 5.8, $\frac{[H^+][CO_3^=]}{[HCO_3^-]} = 10^{-10.3}$

Solving these two equations gives $[HCO_3^-] = 17.888 \times 10^{-3}$M and $[CO_3^=] = 0.056 \times 10^{-3}$M

The majority of the carbonate species is $[HCO_3^-]$.

Also from Eq. 5.7, $\frac{[H^+][HCO_3^-]}{[H_2CO_3^*]} = 10^{-6.3}$ and $[H_2CO_3^*] = 0.566 \times 10^{-3}$M

and $[CO_{2(aq)}] \cong [H_2CO_3^*] = 0.566 \times 10^{-3}$M = 24.90 mg/L.

Results of the carbonate equilibrium calculation for the top 10 m depth of in winter of 1970 are:

$$[Alk] = 18 \times 10^{-3}M = 900 \text{ mg/L as } C_aCO_3 \text{(from data)}$$
$$[TIC] = [CO_2] + \left[HCO_3^-\right] + [CO_3^=]$$
$$= [0.566 + 17.888 + 0.056]10^{-3}$$
$$= 18.51 \times 10^{-3}M = 10^{-1.733}M$$
$$\left[CO_{2-acy}\right] = [CO_2] - [CO_3^=] - [OH^-] + [H^+]$$
$$= [0.566 - 0.056]10^{-3} = 0.510 \times 10^{-3}M$$

At pH of 7.8 in the wintertime, $\left[HCO_3^-\right]$ dominates [TIC]. A plot of the concentrations of different species pC, as $-Log(C)$ versus pH using the graphical procedures (Snoeyink and Jenkins 1980; Stumm and Morgan 1996) to present the above calculation results is shown in Fig. 5.3. [TIC] of $10^{-1.733}$M converted to pC as 1.733, envisioned as the horizontal line across the pH scale from left to right. At different pH range, the major species of [TIC] are shown as $\left[H_2CO_3^*\right]$, $\left[HCO_3^-\right]$, and $\left[CO_3^=\right]$, respectively. In the normal pH range from 7 to 9, the major species of [TIC] belongs to $\left[HCO_3^-\right]$, which is up to two orders of magnitude higher than $\left[H_2CO_3^*\right]$ and $[CO_3^=]$ and provides defense against acidic input. The above calculations demonstrate that the carbonate equilibrium can be solved from knowing any two of the three conservative quantities: pH, [Alk], and $\left[CO_{2-acy}\right]$.

At 4.2°C the solubility of CO_2 at is 0.964 mg/L is much lower than 24.9 mg/L, thereby losing CO_2 to the atmosphere. Similar calculations for the summer conditions at 16.5°C in the top 10 m depth yield the following results: $\left[HCO_3^-\right] = 1.6 \times 10^{-3}$M; $\left[CO_{2(aq)}\right] \cong \left[H_2CO_3^*\right] = 0.008 \times 10^{-3}$M; and$\left[CO_3^=\right] = 0.025 \times 10^{-3}$M. The low $CO_{2(aq)}$ level in the surface of the water column would generate a CO_2 influx from the atmosphere. The dominant species of alkalinity in winter and summer is $\left[HCO_3^-\right]$ and $[CO_2]$ is insignificant. The CO_2 flux will not change [Alk] nor $\left[HCO_3^-\right]$.

Combining Eqs. 5.1, 5.2, 5.7, 5.8, and 5.9 leads to the following:

Fig. 5.3 Graphical solution of carbonate equilibrium

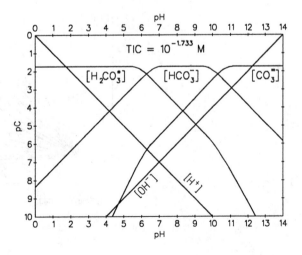

$$[Alk] = \alpha[TIC] + \gamma[H^+] \tag{5.10}$$

where

$\alpha = \dfrac{K_1[H^+]+2K_1K_2}{[H^+]^2+K_1[H^+]+K_1K_2}$ and $\gamma = \dfrac{K_w}{[H^+]} - [H^+].$

The numerical solution of Eq. 5.10 can be displayed in a plot of CO_2 acidity versus alkalinity in Fig. 5.4 (Di Toro 1976). For example, for a CO_2 acidity of 140 mg/L and an alkalinity of 20 mg/L, the pH is 5.5. For pH values less than approximately 4.5, the pH value is essentially determined by the negative alkalinity (mineral acidity) alone; for a given pH, CO_2 acidity can vary widely but the negative alkalinity remains essentially constant. For example, for a pH of 2.8, the alkalinity is approximately $-$ 79 mg/L and the CO_2 acidity can be any value greater than 79 mg/L.

Alkalinity unit conversions:

1 mol $CaCO_3$ = 100 g
1 mol $CaCO_3$. = 1 mol $[Ca^{++}]$ = 2 equivalents
2 equivalents = 1 mol $[Ca^{++}]$ = 1 mol $[Mg^{++}]$ = $\frac{2}{3}[Fe^{+++}]$ = etc.
1×10^{-3} mol Fe^{+++} = 3 milliequivalents
2×10^{-3} mol Ca^{++} = 4 milliequivalents
1.5×10^{-3} mol Mg^{++} = 3 milliequivalents.

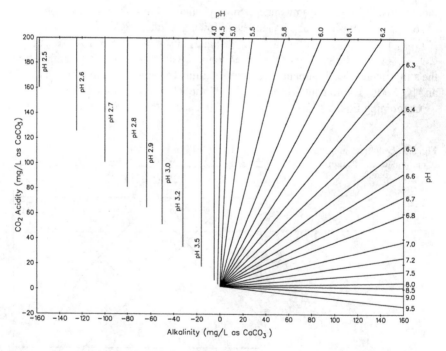

Fig. 5.4 CO_2 acidity versus alkalinity for pH values

For example, 10 milliequivalents alkalinity $= 500$ mg as $CaCO_3$.

Another application of the carbonate equilibrium calculation is presented in the following example. A chemical plant produces a mineral acid wastewater at the rate of $100 \, ft^3/m$ for a 24 h production schedule once each week. The pH of the wastewater is 2.0. The plant is located on a river with a flow rate of $3500 \, ft^3/m$, a pH of 8.4, and an ambient alkalinity of 200 mg/L as C_aCO_3. To meet the state water quality standards of pH 7.5 in the receiving river, what is the allowable acid wastewater flow rate to be specified in the NPDES permit?

First, in the river upstream of the waste discharge, pH $= 8.4$ and alkalinity $= 200$ mg/L as C_aCO_3 yields total inorganic carbon $= 200$ mg/L as C_aCO_3. Therefore, $\left[CO_{2-acy} \right] = 0$. Next, the waste discharge: from the equation: $[Alk] = \left[HCO_3^- \right] + 2 \left[CO_3^= \right] + \left[OH^- \right] - \left[H^+ \right]$. At pH $= 2$, $[Alk] < 0$ and $[Alk] = -\left[H^+ \right] = -10^{-2}$M. As a result, $[Alk] = -500$ mg/L as C_aCO_3. From the equation:

$$\left[CO_{2-acy} \right] = [CO_2] - \left[CO_3^= \right] - \left[OH^- \right] + \left[H^+ \right],$$

where $\left[CO_3^= \right]$ and $\left[OH^- \right]$ are insignificant compared with $\left[H^+ \right]$. Therefore, $\left[CO_{2-acy} \right] = [CO_2] + \left[H^+ \right]$. Assuming that $[CO_2]$ is slightly smaller than $\left[H^+ \right]$, or to be $10^{-2.7}$ M or 99.76 mg/L as C_aCO_3, resulting in $\left[CO_{2-acy} \right] = 99.76 + 500 = 599.76$ mg/L as C_aCO_3, and $[TIC] = \left[CO_{2-acy} \right] + [Alk] = 599.76 - 500 = 99.76$ mg/L as C_aCO_3. Would a waste flow rate of $100 \, ft^3/m$ meet the water quality standards in the receiving water?

The mixture:

$$[Alk] = \frac{(200)(3500) + (-500)(100)}{3600} = 180.55$$

and

$$[TIC] = \frac{(200)(3500) + (99.76)(100)}{3600} = 197.22$$

and $\left[CO_{2-acy} \right] = [TIC] - [Alk] = 16.66$ mg/L as C_aCO_3. From the chart in Fig. 5.4, we get pH $= 7.35$ (slightly below 7.5), not meeting the water quality standard. Next, assuming a lower flow rate of $60 \, ft^3/m$ of the waste, the mass balance for alkalinity is:

$$[Alk] = \frac{(200)(3500) + (-500)(60)}{3,560} = 188.20$$

and

$$[TIC] = \frac{(200)(3500) + (99.76)(60)}{3560} = 198.31$$

and $\left[CO_{2-acy}\right] = [TIC] - [Alk] = 10.11$ mg/L as C_aCO_3. From the chart in Fig. 5.4, we get pH $= 7.5$, meeting the standard. Therefore, the allowable wastewater discharge rate is 60 ft^3/m. Such a trial-and-error solution can be improved with the concept of buffer capacity.

Weber and Stumm (1963) derived the buffer capacity (β, in equivalent/unit pH) in the water with respect to strong acids and strong bases:

$$\beta = \frac{d[Alk]}{d(pH)} = \frac{d\left[CO_{2-acy}\right]}{d(pH)}$$

Therefore,

$$\beta = 2.3\left[[TIC]\frac{\alpha_1^2}{K_1}\left([H^+] + \frac{K_1 K_2}{[H^+]} + 4K_2\right) + [H^+] + [OH^-]\right]$$

Or, in terms of alkalinity,

$$\beta = 2.3\left[\frac{\alpha_1\left([Alk] - [OH^-] + [H^+]\right)\left([H^+] + \frac{K_1 K_2}{[H^+]} + 4K_2\right)}{K_1\left(1 + \frac{2K_2}{[H^+]}\right)} + [H^+] + [OH^-]\right]$$

where

$$\alpha_1 = \frac{[HCO_3^-]}{[TIC]} = \left(\frac{K_1}{K_1 + [H^+] + \frac{K_1 K_2}{[H^+]}}\right)$$

where $\left[HCO_3^-\right]$ in the main fighting force against acid or base input and α_1 represents the percentage of [TIC] in bicarbonate. The physical meaning of β represents the consumption of alkalinity (mainly in the form of $\left[HCO_3^-\right]$) to fight against the change per unit pH. In natural waters with pH in the range between 6 and 9, the majority of [TIC] is $\left[HCO_3^-\right]$, which offers buffer capacity against pH changes. Although the above equations for β appears rather formidable, its use for calculation of buffer capacity is facilitated by the fact that several of its terms can be neglected for a given set of conditions.

Then $\alpha_1 = \frac{10^{-6.4}}{1.1 \times 10^{-6.4}} = 0.91$ (after dropping $10^{-9.4}$), indicating 91% of [TIC] is $\left[HCO_3^-\right]$.

$$\left([Alk] - [OH^-] + [H^+]\right) = \left(2 \times 10^{-3} - 10^{-6.6} + 10^{-7.4}\right) = 2 \times 10^{-3},$$

$$\left([H^+] + \frac{K_1 K_2}{[H^+]} + 4K_2\right) = \left(10^{-7.4} + 10^{-9.4} + 4 \times 10^{-10.4}\right) = 10^{-7.4},$$

$$K_1\left(1 + \frac{2K_2}{[H^+]}\right) = (10^{-6.4} + 2 \times 10^{-9.4}) = 10^{-6.4},$$

$$\text{and } \beta = 2.3\left(\left[\frac{(0.91)(2 \times 10^{-3})(10^{-7.4})}{10^{-6.4}}\right] + 10^{-7.4} + 10^{-6.6}\right)$$

$$= 2.3(0.91)(2 \times 10^{-4})$$

$$= 4.2 \times 10^{-4} \text{ Equivalent/pH unit} = 0.42 \text{ meq/pH unit.}$$

The above sample calculation is given in more detail than would be necessary for most calculations for β, simply to illustrate the manner in which many of the terms in the equation may be readily neglected to facilitate calculation.

In this problem, the river upstream of the waste discharge has a pH $= 8.4$ and [Alk] $= 200$ mg/L as $CaCO_3$ (Note: 100 mg/L $CaCO_3 = 2 \times 10^{-3}$ equivalents/L). Thus [Alk] $= 4 \times 10^{-3}$ equivalents/L. For pH $= 8.4$ in the river water, $[H^+] = 10^{-8.4}$, pOH $= 14 - 8.4 = 5.6$, and $[OH^-] = 10^{-5.6}$, $K_1 = 10^{-6.4}$, and $K_2 = 10^{-10.4}$.

$$\alpha_1 = \left(\frac{10^{-6.4}}{10^{-6.4} + 10^{-8.4} + \frac{10^{-6.4}10^{-10.4}}{10^{-8.4}}}\right) = 0.980, \text{ indicating 98\% of } C_T \text{ is}[HCO_3^-]\text{and}$$

$$\beta = 2.3\left[\frac{0.98[(10^{-8.4})(4 \times 10^{-3})(2 \times 10^{-8.4})]}{(10^{-8.4})(10^{-6.4})} + 10^{-5.6} + 10^{-8.4}\right]$$

$$= 1.826 \times 10^{-4} \text{ Equivalent/unit pH}$$

Now, calculate the permissible acid discharge:

Existing pH in river $= 8.4$
Permissible pH in river $= 7.5$
Permissible pH drop, n pH $= 0.9$

The increment of acid (or base) required to incur a pH change, n pH, in one liter of river water is, $nC = \beta n$ pH $= 1.826 \times 10^{-4}(0.9) = 1.643 \times 10^{-4}$ equiv. This means that the maximum amount of mineral acid that can be added to one liter of river water (if the pH is not to drop below 7.5) is 1.643×10^{-4} equivalents. Next, calculate the permissible rate of discharge of acid waste. The river can accept 1.643×10^{-4} equivalents of acid per liter. The waste contains 10^{-2} equivalents of acid per litter (pH $= 2$). Therefore, the maximum permissible ratio of acid waste to the river water is $\frac{1.643 \times 10^{-4}}{10^{-2}} = 1.643 \times 10^{-2}$. Hence, the maximum rate of wastewater discharge is 3500 cfm $\times 1.643 \times 10^{-2} = 57.5$ cfm, slightly lower than the previous result of 60 cfm obtained from a trial and error run. Note that alkalinity is treated as a conservative substance for mass balance calculations throughout this exercise. This extremely important feature will be used in the remaining part of this chapter for more complicated pH modeling of streams, rivers, lakes, and reservoirs.

5.2 Carbon Dioxide Exchange Across the Air–Water Interface

Recalling the carbonate equilibrium calculations for Horwer Bucht in the wintertime at water temperature of 4.2°C, the surface layer CO_2 is 24.9 mg/L, which is much higher than the saturated concentration of 0.964 mg/L based on the following equation:

$$CO_{2(s)} = 1.1088 - 0.03894T + 0.000693T^2 - 0.00000495T^3 \qquad (5.11)$$

where T is water temperature in °C. During the summer time, the surface layer CO_2 level is 0.353 mg/L, which is lower than the saturation concentration of 0.633 mg/L. Therefore CO_2 would be transported from air to the water column. Since the dominant species of alkalinity in winter and summer is $\left[HCO_3^-\right]$ and $[CO_2]$ is insignificant, the CO_2 transfer will not change the total alkalinity, nor $\left[HCO_3^-\right]$.

However, for waters affected by acid discharges, carbon dioxide is often greatly supersaturated. Since there is exchange of CO_2 with the atmosphere, the CO_2 acidity and the TIC vary with time as the water moves downstream in the river. The higher the reaeration rate, the more rapidly the TIC changes. As will be shown, for pH values greater than approximately 4.5, exchange of CO_2 with the atmosphere can cause significant changes in pH. For pH below approximately 4.2, however, loss of CO_2 to the atmosphere will not have a significant effect on the equilibrium pH. When pH is below 4.2, alkalinity is not affected by changes in CO_2 concentration and thus remains constant in river segments where no alkalinity is added or precipitated.

The change in dissolved carbon dioxide with time due only to air–water exchange in a stream is given by Di Toro (1976):

$$\frac{d[CO_2]}{dt} = K_a\left(\left[CO_{2(s)}\right] - [CO_2]\right) \qquad (5.12)$$

where K_a is CO_2 reaeration coefficient (day^{-1}) across the air–water interface. Solving Eq. 5.12 yields

$$[CO_2] \text{ at time } t = [CO_2]_0 e^{-K_a t} + \left[CO_{2(s)}\right]\left(1 - e^{-K_a t}\right)$$

where $[CO_2]_0 =$ initial carbon dioxide concentration at $t = 0$ in the water column. For negative alkalinity, [TIC] may be substituted for CO_2 such that

$$[TIC]_t = [TIC]_0 e^{-K_a t} + \left[CO_{2(s)}\right]\left(1 - e^{-K_a t}\right)\frac{50}{44}$$

For positive alkalinity and for pH values below 7.5, CO_2 may be replaced by CO_{2-Acy}

$$[CO_{2-Acy}]_t = [CO_{2-acy}]_0 e^{-K_a t} + [CO_{2(s)}](1 - e^{-K_a t})\frac{50}{44}$$

The factor $\frac{50}{44}$ in the above equations is the ratio of the molar equivalent, [TIC] or $[CO_{2-acy}]$ as $CaCO_3$, to the molar equivalent of CO_2. The substitutions of [TIC] or $[CO_{2-acy}]$ for $[CO_2]$ under the conditions given above are approximations which lead to a straightforward and accurate method of calculating the pH change caused by the loss of CO_2 to the atmosphere. Any inaccuracy involved in substituting [TIC] or $[CO_{2-acy}]$ for $[CO_2]$ in the above equations is minimal and causes essentially no error in the calculation of pH values (Di Toto 1976). The above equations can also be used to quantify the mass flux rate of CO_2 across the air–water interface. Usually the mass transfer rate, K_L (m/day) is given. Dividing K_L by the water depth yields the reaeration rate, K_a in day^{-1}.

5.3 Recovery from Acid Mine Drainage

Acid mine drainage was a problem causing lowering the pH levels in the North Branch Potomac River from river miles 350 to 280 (Fig. 5.5) in the 1950 and 1960's. The Interstate Commission on the Potomac River Basin (ICPRB) conducted a study to assess the recovery from low pH levels in this portion of the Potomac River in 1976. Regulatory staff from ICPRB, the Maryland Water Resources Administration, the Virginia Water Control Board, the West Virginia Division of Water Resources, and the US Environmental Protection Agency Region III provided technical assistance to a study by Hydroscience (1976). Alkalinity and pH data from the US Army Corps of Engineers (1975) are presented in the left column of Fig. 5.5 for the river miles 350 to 280. The upstream portion of the river (miles 350 to 340) shows low pH values at 3.5 while the acidity levels remained high above 80 mg/L as $CaCO_3$. Downstream from this river mile 340 where the Westvaco discharge and UPRC WWTP enter, pH shows a sharp rise resulting from the loss of acidity. Such an improvement continues until pH rises to 9 along with the reduction of acidity to below 10 mg/L as $CaCO_3$.

This part of the river was observed with recovery from low pH primarily because of the alkalinity discharges at the Upper Potomac River Commission (UPRC) WWTP and Westvaco. Although the data show considerable variability, the rapid increase in pH from an average of 3.5 to 5.0 is clearly seen downstream of Westvaco and UPRC discharges near the city of Luke at river mile 339. The pH continues to increase below this point until values averaging 7.5 are attained immediately upstream of the confluence with the South Branch. Average acidity values decrease from values exceeding 75 mg/L (as $CaCO_3$) upstream of Luke to values approaching zero near the South Branch of the Potomac River. Upstream of the Westvaco discharge, maximum acidity $[CO_{2-acy}]$ values exceeding 200 mg/L of $CaCO_3$ and pH values below 3.0 were common.

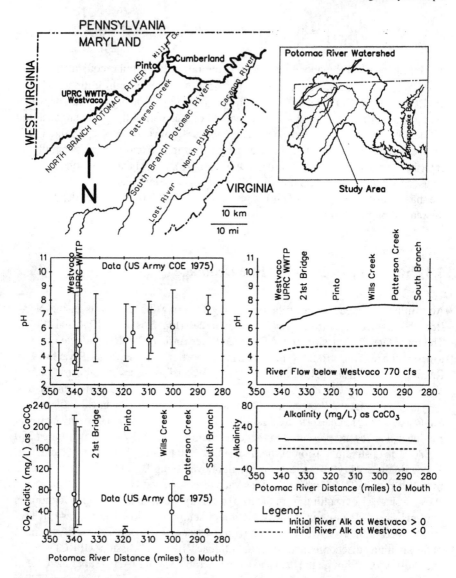

Fig. 5.5 Modeling pH recovery from acid mine drainage in the North Branch of the Potomac River

The carbonate equilibrium concepts presented in Sect. 5.1 can be used to analyze the pH recovery of the North Branch of the Potomac River between the West-vaco/UPRC WWTP and the confluence with the South Branch of the Potomac River with mass balance renewed wherever discharges enter the river. Such computations can be implemented with a simple model of pH supported by mass balance of alkalinity and acidity along the way. The continuing decrease of acidity due to the release of CO_2 from the water column to the atmosphere would result in the rise of pH, which can be substantiated by a modeling framework such as QUAL2kw. The right-hand panels of Fig. 5.5 show the model results for the cases of positive and negative initial (below Westvaco discharge) alkalinity. Note that the same pH scales in the data plot and the model result plot are used. In addition, the acidity and alkalinity scales are also the same for clear comparison. When the initial alkalinity immediately below the UPRC WWTP and Westvaco is positive (i.e. around 15 or 16 mg/L as $CaCO_3$), pH recovery is expected to continue rising to 7.6 at river mile 285 where the South Branch of the Potomac River joins. However, if the initial alkalinity is negative (i.e. -1 mg/L as $CaCO_3$), pH rise is not expected and the alkalinity levels remain negative throughout the river stretch. Such a result is not surprising as Fig. 5.4 shows that pH would not rise above 4.5 when alkalinity is negative! In either case, the alkalinity levels in the North Branch of the Potomac River would remain relatively constant throughout the river portion with or without pH recovery. The simple conclusion is that releasing CO_2 from the water column to the atmosphere is the main driving force behind the pH recovery. Positive alkalinity levels at the initial upstream point are needed to start the recovery process. Higher initial alkalinity levels (i.e. 50 or 100 mg/L as $CaCO_3$ at river mile 340) would produce even better recoveries (achieving pH above 8.5) in the downstream direction at river mile 285.

5.4 pH Affected by Microbial Processes

A number of microbial processes, which commonly occur in the water affect pH changes in the water column:

Algal photosynthesis to increase pH by consuming CO_2 and releasing O_2:

$$6CO_2 + 6H_2O \leftrightarrow C_6H_{12}O_6 + 6O_2 \tag{5.13}$$

Algal respiration to decrease pH by producing CO_2:

$$C_6H_{12}O_6 + 6O_2 \leftrightarrow 6CO_2 + 6H_2O \tag{5.14}$$

Methane fermentation to decrease pH by releasing CO_2:

$$C_6H_{12}O_6 + 3CO_2 \leftrightarrow 3CH_4 + 6CO_2 \tag{5.15}$$

Nitrification to decrease pH by producing NO_3^-:

$$NH_3 + 2O_2 \leftrightarrow NO_3^- + H_3O^+ \tag{5.16}$$

Denitrification to increase pH by consuming NO_3^-:

$$5C_6H_{12}O_6 + 24NO_3^- + 24H_3O^+ \leftrightarrow 30CO_2 + 12N_2 + 66H_2O \tag{5.17}$$

Sulfide oxidation to decrease pH by producing sulfate:

$$HS^- + 2O_2 + H_2O \leftrightarrow SO_4^= + H_3O^+ \tag{5.18}$$

Sulfate reduction to increase pH by consuming sulfate:

$$C_6H_{12}O_6 + 3SO_4^= + 3H_3O^+ \leftrightarrow 6CO_2 + 3HS^- + 9H_2O \tag{5.19}$$

It has been established in studies that microbial processes in anaerobic waters and sediments can supply substantial amounts of alkalinity to acidified waters. Such alkalinity can develop in water bodies acidified to a moderate degree, as by acid precipitation (Kelly et al. 1995; Schindler et al. 1980; Schindler 1986; Cook et al. 1986), and it can also develop in strongly acidified waters, such as those acidified by acid mine drainage (Herlihy et al. 1987). Internal buffering studies have concentrated largely on sulfate reduction (SR) as the prime mechanism of alkalinity production, although several groups have recognized the potential importance of similar processes such as denitrification and iron reduction (Hemond and Eshleman 1984; Bell et al. 1987; Hemond and Fechner 2015). Given the high levels of sulfate (1–20 mM) in the water column of Lake Anna, Virginia (Fig. 5.6), the relative magnitude of other alkalinity producing reactions appears minimal. A portion of Lake Anna (Fig. 5.6) receives acid mine drainage from Contrary Creek which drains an 1820 ha watershed with abandoned pyrite mines and provides 35% of the water entering that part of the lake. Contrary Creek has very acidic water (pH 2.5–3.9) and high concentrations of sulfate (1–20 mM) and Fe (150–450 μM) (Herlihy et al. 1987). Lake Anna also receives drainage from Freshwater Creek (Fig. 5.6) with a 2288 ha watershed and non-acidic water. Its average pH at Point 3 is about 5, almost two orders of magnitude higher

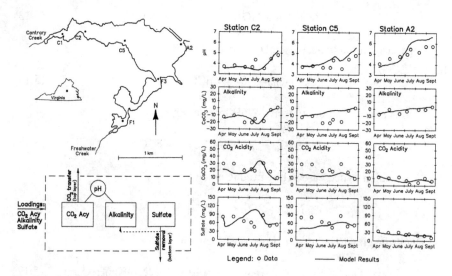

Fig. 5.6 Contrary creek of Lake Anna, Virginia and Model Kinetics

than the acidic inflow from Contrary Creek. A mass balance budget indicated that 48% of the annual sulfate input was retained in the contaminated arm of the lake (between sampling points C1 and C5), and the SR measurements showed that the process accounted for the sulfate retention measured in the budget study (Herlihy et al. 1987). Furthermore, based on the sulfate retention figures, alkalinity generation was more than adequate to account for the increases in pH observed between the mouth of the acid mine stream and the Freshwater Creek outflow, 2 km down the lake (Lung et al. 1988).

The three major constituents in the model are $[CO_{2-acy}]$, [Alk], and sulfate. Sulfate removal via microbial processes would increase alkalinity. Once $[CO_{2-acy}]$ and [Alk] are calculated, pH can be obtained from the following equations (Kemp 1971) depending on the alkalinity level in the water column.

$$[H^+] = \frac{K_{a,1}\left([CO_{2-acy}] - K_{a,2} - \frac{K_w}{K_{a,1}}\right)}{[Alk] + K_{a,1}} \text{ for } [Alk] > 0 \qquad (5.20)$$

$$[H^+] = \frac{-K_{a,1}}{2} + 0.5\left(K_{a,1}^2 + 4K_w + 4K_{a,1}[CO_{2-acy}]\right)^{1/2} \text{ for } [Alk] = 0 \quad (5.21)$$

$$[H^+] = -[Alk] \text{ for } [Alk] < 0 \qquad (5.22)$$

where K_1 and K_2 are from Eqs. 5.7 and 5.8 and K_w is the dissociation constant of water (temperature dependent) in Eq. 5.9.

5.5 Stream Diurnal pH Variations

Recall that in the Santa Fe River in New Mexico (see Fig. 2.6 in Sect. 2.2.2), significant diurnal pH variations due to algal photosynthesis and respiration were observed. Algal photosynthesis can be represented by the following chemical equilibrium: $6CO_2 + 6H_2O \leftrightarrow C_6H_{12}O_6 + 6O_2$ (Eq. 5.13), which consumes CO_2 while algal respiration in the form of $C_6H_{12}O_6 + 6O_2 \leftrightarrow 6CO_2 + 6H_2O$ (Eq. 5.14) produces CO_2. Both processes affect the pH level but have no effect on alkalinity. Another key process in the CO_2 budget in the water column is the mass transfer across the air–water interface, as presented in Sect. 5.2. Another source of CO_{2-acy} is the recycled carbon from dead attached algae via $CBOD$, which must be included in the calculation. Once the $[CO_{2-acy}]$ and [Alk] concentrations are determined, pH can then be calculated using the chart in Fig. 5.4.

One of the key coefficients in the calculation is reaeration coefficient for CO_2, K_a. Quantifying the reaeration coefficient for small streams at low flows is extremely difficult. Commonly used empirical equations are not suitable for the Santa Fe River because they tend to overestimate the reaeration rates for extremely low flows and shallow waters found in the study area (see Sect. 4.6). A more reasonable equation by Moog and Jirka (1998) were used in this study:

$$K_a = 1.74 V^{0.46} S^{0.79} H^{0.76} \qquad (5.23)$$

for stream channel slope, $S > 0.0004$ and

$$K_a = 5.59 S^{0.16} H^{0.73} \qquad (5.24)$$

for $S < 0.0004$. An examination of the study area topographical map reveals that the average channel slope is approximately 0.005. Using the above equation with an average velocity of 0.128 m/s and a depth of 0.235 m for the study area, an average $K_a = 3.52$ day^{-1} at 20°C is calculated. Subsequent model calibration analyses result in a value of 2.50 day^{-1} for DO and CO_2. Equation 5.12 is used to determine the saturated CO_2 concentrations. A factor of 0.78 was used to scale down the saturation level due to the high altitude (5000 ft above sea level) of the study area.

An important gage to calibrate the model is the sharp rising and declining rates of dissolved oxygen levels in the water column, with a maximum increase rate at over 240 mg O/L/day in the Santa Re River. Two factors contribute to this sharp daily swing of dissolved oxygen concentrations: the high benthic algae biomass and the carbon to chlorophyll a ratio in the algal biomass. For example, the high algal biomass calculated by the model is about 1000 µg/L, a level that has been reported for the benthic algae in the east branch of Brandywine Creek in Pennsylvania (Knorr and Fairchild 1987). A series of model calibration runs have revealed that a high carbon to chlorophyll a ratio is needed to match the significant daily fluctuation of dissolved oxygen levels. The ratio used in the model is 266 µg C/µg chlorophyll a. Values of other kinetic coefficients related to benthic algal growth and nutrient recycling are taken from a modeling study of Carson River, NV by Warwick et al. (1997) and further refined in model calibration.

Figure 5.7 shows the model results for the 4-day period of June 7–June 10, 1999 in attached algal biomass (in chlorophyll a), $[CO_{2-acy}]$, dissolved oxygen, and pH at two locations: at Preserve and immediately below the Santa Fe WWTP. The model has reached a steady-state condition with repeated daily patterns for these water quality constituents. As stated earlier, the calculated chlorophyll a level of 1,000 µg/L is considered reasonable for the benthic algae, as similar levels have been observed elsewhere. While the benthic algae biomass is rising in the middle of the day, $[CO_{2-acy}]$ is declining. The rising benthic algal biomass results in the dissolved oxygen maximum in the afternoon. The subsequent decline of the algal biomass in the later part of the day produces a minimum dissolved oxygen by mid-night and early morning of the next day. The model calculated dissolved oxygen fluctuation (i.e., phase) follows the observed pattern closely. The calculated pH pattern also follows the temporal trend of $[CO_{2-acy}]$ concentrations, with lower pH levels matching higher $[CO_{2-acy}]$ concentrations.

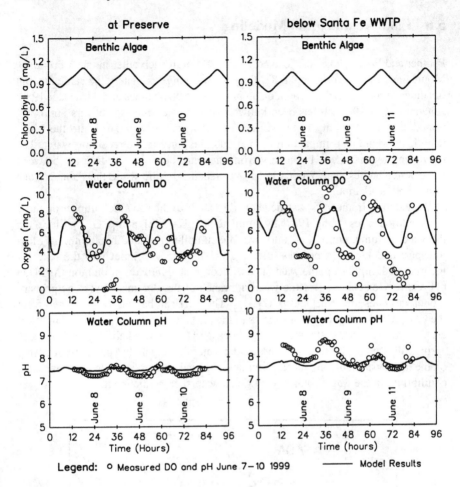

Fig. 5.7 Model results versus data in Santa Fe River

Model results are also compared with the data observed at Preserve (Fig. 5.7). The temporal trend of $[CO_{2-acy}]$, dissolved oxygen, and pH is very similar to that at the location immediately below the Santa Fe WWTP. The significant daily fluctuation of dissolved oxygen is also reproduced. Note the timing of minimum CO_2 concentrations matches the time of the day for maximum dissolved oxygen levels. Compared with the results below the WWTP, the slightly less pronounced daily fluctuation of $[CO_{2-acy}]$ dissolved oxygen, and pH at Preserve is due to the reduced algal growth at this downstream (Preserve) location (see Fig. 2.5 for map).

While the model results match the observed trend of dissolved oxygen and pH at both locations, there are differences between the calculated maximum/minimum values and the observed data. Because the model produces equilibrium values while the Santa Fe River may not reach a steady-state condition, the graph shows some day to day variation in dissolved oxygen and pH when compared to the collected data.

5.6 Lake Acidification Modeling

Panther and Woods lakes are located within 30 km of each other in the Adirondack Mountains of New York (Fig. 5.8). They received approximately the same amount of acidic deposition per unit area on an annual basis (Johannes and Altwicker 1980; Johannes et al. 1981). It has been known that snowmelt release delivers unproportioned amount of acidity in the surface runoff from the watershed into the lakes, resulting a significant impact on the pH and alkalinity in the receiving water. The interesting fact is: the pH recovery from the acid input is very different between these water bodies. The Woods Lake Watershed is acidic while the Panther Lake Watershed is alkaline.

A modeling framework was developed to analyze the snowmelt impact on these two acidified lakes (Lung 1987) by simulate the pH, [Alk], and $\left[CO_{2-acy}\right]$ levels in the water column, configured into two layers: epilimnion and hypolimnion. Mass transport and kinetic processes affecting the fate of acid input into the lake are incorporated. Input from the watershed is incorporated into the epilimnion. As in the pH recovery calculation for the western branch of the Potomac River, the primary model variables are [Alk] and $\left[CO_{2-acy}\right]$. Then pH can be readily calculated from the $CO_2/ HCO_3^-/ CO_3^=$ equilibrium as in the pH calculations for streams. The lakes are viewed as a water system receiving shock acid (negative alkalinity) loads in early spring followed by dilution from less acidic water flowing in the spring. By focusing on the lake system, the model can simulate seasonal dynamics of the lake carbonate equilibrium in the water column with given watershed acidic input.

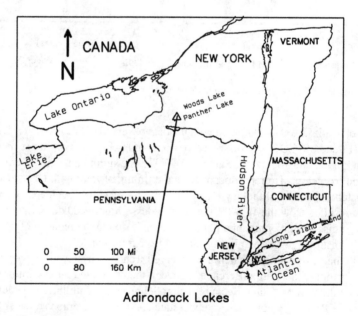

Fig. 5.8 Adirondack lakes: woods and panther

Key mass transport includes watershed input to the lake, deposition on the lake surface, and snowpack releases during the spring months. Output through lake outlets is also included and is only from the surface layer, i.e. epilimnion. In many lakes, groundwater seepage or infiltration may be significant, and the model allows such external interchanges. Mass transport within the water column is characterized by advection and dispersion in the vertical direction. Another key mass transport process is CO_2 transfer across the air–water interface and is characterized by the differential between the saturated and actual CO_2 levels in the epilimnion (see Eq. 5.12). The CO_2 level may be approximated by $[CO_{2-acy}] - [H^+]$ in the epilimnion (Di Toro 1976).

The $CO_2 / HCO_3^- / CO_3^=$ equilibrium is not the only chemical equilibrium operative in many lakes and impoundments affecting the [Alk] and $[CO_{2-acy}]$ levels in the water column. Other processes, primarily of a biochemical nature, may also affect the lake alkalinity and $[CO_{2-acy}]$ levels. For example, hypolimnetic sulfate reduction (see Eq. 5.19) during a period of anoxia tends to increase the alkalinity levels by decreasing the total concentration of nonprotolytic anions (i.e. $SO_4^=$). Schindler (1980) reported that in Lake 223 of the Experimental Lakes Area in Canada, over 60% of the acid neutralization was accounted for by sulfate reduction and denitrification. In some lakes, denitrification (Eq. 5.17) is a source of alkalinity because it generates N_2O and N_2, which are volatile gases and could be lost from the water system, thereby increasing alkalinity. In addition, ammonification (i.e., ammonia release from sediments) and nitrate uptake by algae generate alkalinity. Conversely, sulfide oxidation (Eq. 5.18), nitrification (Eq. 5.16), and ammonia uptake by algae decrease alkalinity. Finally, algal photosynthesis (Eq. 5.13) reduces CO_2 levels and increases pH levels while algal respiration (Eq. 5.14) increases CO_2 levels and decreases pH levels. Figure 5.9 summarizes the interactive kinetics processes between alkalinity, $[CO_{2-acy}]$, and pH in the water column (Lung 1987).

A significant amount of data on the watershed and the lake water column were generated from the Integrated Lake-Watershed Acidification Study (ILWAS) sponsored by the Electric Power Research Institute (EPRI) (Gherini et al. 1984; Schofield et al. 1985). Chen et al. (1984) conducted a hydrologic analysis of these two lakes under ILWAS. The snowmelt runoff rate was determined from the changes in the snow storage volume during February, March, and April using the data from Johannes et al. (1981). Independent measurements of flow rates at the lake outlets supported the construction of hydraulic budgets for the two lakes during snowmelt period (Tetra Tech Inc. 1983). The quantity of snowmelt runoff was significant for both lakes during this period, although the groundwater contribution was more significant to Panther Lake than Woods Lake. The acid loads (in negative alkalinity) to the lakes were derived from the actual loss of major ions $SO_4^=$ and NO_3^- in the snow storage (Johannes et al. 1981). When the lakes were under ice and snow cover from December to April, no CO_2 exchange across the lake surface was allowed in the model, accomplished by turning off the mass transfer process across the water surface. Data from these studies were used to develop, configure, and calibrate a snowmelt response model for these two lakes.

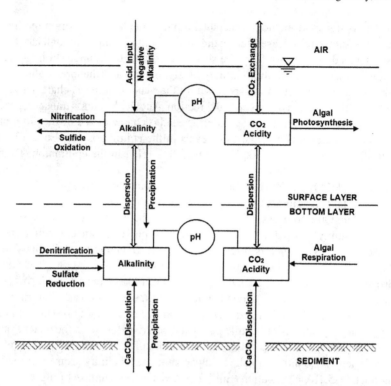

Fig. 5.9 Alkalinity, CO_2 acidity, and pH in water column

Results from the applications to Panther Lake and Woods Lake are summarized in Fig. 5.10. Model calculated pH, [Alk], and $\left[CO_{2-acy}\right]$ levels versus measured data of Panther Lake and Woods Lake are shown in the left and right columns of Fig. 5.10, respectively. Data on alkalinity in the epilimnion and hypolimnion are available for comparison with the model results. Only pH measurements in the epilimnion are available and no CO_2 acidity data are available from both lakes. Note that same scales for plotting are used between these two lakes for a good perspective. Alkalinity and CO_2 acidity concentrations are expressed in mg/L as $CaCO_3$. In general, model results match the measured alkalinity and pH levels reasonably well for both lakes. During the period of ice and snow cover, biological activities affecting the change in alkalinity are probably at a minimal level, making alkalinity a more or less "conservative" tracer with respect to the lake system (all the kinetics coefficients $= 0$). That is, the alkalinity dynamics in the lake simply become an input–output balance for this constituent. The lake acts as a system receiving shock acid (negative alkalinity) loads during the snowmelt period followed by a recovery due to dilution from less acid input.

The impact of the snowmelt on Panther Lake is clearly shown. The strong acidic input drives the alkalinity levels in the epilimnion to negative levels, reaching a season low of -2 mg/L as $CaCO_3$. In the same period, pH drops from 7 to 4.6, quite a substantial reduction. In the meantime, pH and alkalinity in the hypolimnion stay

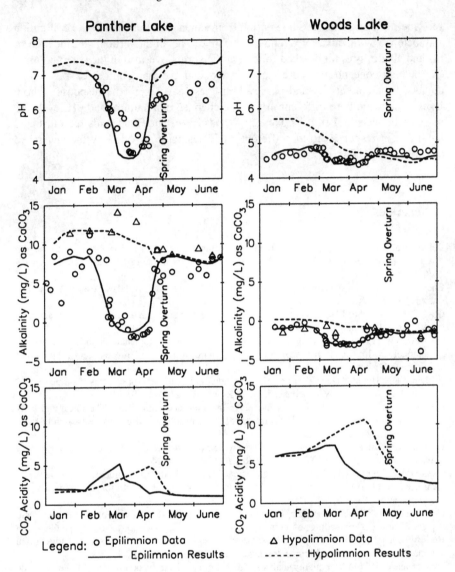

Fig. 5.10 Acidification modeling of panther lake and woods lake

high, showing no impact from the strong acidic input from snowmelt, mostly assisted by the winter reversed thermal stratification. By April, the snowmelt effect is easing to raise the alkalinity and pH levels in the epilimnion, then further assisted by the spring overturn to mix the two layers in the water column. The recovery is provided by the alkaline watershed runoff.

In the meantime alkalinity levels in both layers of Woods Lake stayed negative as the pH levels in the epilimnion remained below 5, taking the acidic punch from the snowmelt. Throughout the same period from January to June, pH in the epilimnion stay below 5 while alkalinity remains negative. The slight vertical stratification of pH and alkalinity is diminished following the spring overturn in the water column. CO_2 acidity concentrations are higher in Woods Lake than Panther Lake, strongly dictated by the alkalinity and pH levels in these two lakes. The response of both lakes to snowmelt are different in that the changes in alkalinity and pH levels are more pronounced in Panther Lake than Woods Lake. The EPRI study has attributed the different responses to the different buffer capacity provided by the watershed (Gherini et al. 1984).

References

Bell PE et al (1987) Biogeochemical conditions favoring magnetite formation during anaerobic iron reduction. Appl Environ Microbiol 53(11):2610–2616. https://doi.org/10.1128/aem.53.11.2610-2616.1987

Chen CW et al (1984) Hydrologic analyses of acidic and alkaline lakes. Water Resour Res 20(12):1975–1882

Cook RB et al (1986) Mechanisms of hydrogen ion neutralization in an experimentally acidified lake. Limnol Oceanogr 31:134–148

Di Toro DM (1976) Combining chemical equilibrium and phytoplankton models—a general methodology. In: Canale RP (ed) Modeling biochemical processes in ecosystems. Ann Arbor Science Publishers, pp 233–255

Eshleman KN, Hemond HF (1985) The role of organic acids in the acid-base status of surface waters at Bickford Watershed, Massachusetts. Water Resour Res 21(10):1503–1510

Gherini SA et al (1984) The ILWAS model: formulation and application. The integrated lake-watershed acidification study vol 4: summary of major results. Electric Power Research Institute report EA-3221, Palo Alto, California

Hemond HF, Eshleman KN (1984) Neutralization of acid deposition by nitrate retention at Bickford Watershed. Massachusetts. Water Resour Res 20(11):1718–1724

Hemond HF, Fechner EJ (2015) Chemical fate and transport in the environment. Academic Press

Henriksen A (1979) Acidification of freshwaters: a simple approach for identification and quantification. Nat 278:542–545

Henriksen A (1980) Acidification of freshwaters—a large scale titration. Ecological impact of acid precipitation. In: Proceedings of an international conference, Sandefjord, Norway, pp 68–74

Herlihy AT et al (1987) The importance of sediment sulfate reduction to the sulfate budget of an impoundment receiving acid mine drainage. Water Resour Res 23(2):287–292

Hydroscience (1976) Water quality analysis of the Potomac River. Report prepared for the Interstate Commission on the Potomac River Basin (ICPRB), p 211

Johannes AH, Altwicker ER (1980) Atmospheric input into three Adirondack lake watershed. Ecological impact of acid precipitation. In: Proceedings of an international conference, Sandefjord, Norway, p 256

Johannes AH et al (1981) Snow pack storage and ion release. ILWAS, Electric Power Research Institute Report EA-1825, Palo Alto, California

Kelly CA et al (1995) Disruption of sulfur cycling and acid neutralization in lakes at low pH. Biogeochemistry 28:115–130

Kemp PH (1971) Chemistry of natural waters-I. fundamental relationships. Water Res 5:297–311

Knorr DF, Fairchild GW (1987) Periphyton, benthic invertebrates, and fishes as biological indicators of water quality in the East Branch Brandywine Creek. Proc PA Acad Sci 61(1):61–66

Kramer JR, Tessier A (1982) Acidification of aquatic systems: a critique of chemical approaches. Environ Sci Technol 16(11):606A-615A

Lung WS (1984) Understanding simplified lake acidification models. J Environ Eng 110(5):997–1002

Lung WS (1987) Lake acidification model: practical tool. J Environ Eng 113(4): 900–9. Lung WS et al (1988) Modeling fate and transport of sulfate and alkalinity in an acidified lake. Water Air Soil Pollut 37:157–170

Moog DB, Jirka GH (1998) Analysis of reaeration equations using mean multiplicative error. J Environ Eng 124(20):104–110

Schindler DW (1980) Experimental acidification of a whole lake: a test of the oligotrophication hypothesis. In: Proceedings of an international conference on the ecological impact of acid precipitation, Norway

Schindler DW (1986) The significance of in-lake production of alkalinity. Water Air Soil Pollut 30:931–944

Schindler DW et al (1980) Experimental acidification of Lake 223, experimental lake area: background data and first three years of acidification. Canadian J Fishery Aquatic Sci 37(3):342–354

Schofield CL et al (1985) Surface water chemistry in the ILWAS basins. Water Air Soil Pollut 26:403–423

Small MJ, Sutton MC (1986) A regional pH-alkalinity relationship. Water Res 20(3):335–343

Snoeyink VL, Jenkins D (1980) Water chemistry. Wiley, New York

Stumm W, Morgan JJ (1996) Aquatic chemistry, chemical equilibria and rates in natural waters. Wiley, New York

Tetra Tech, Inc (1983) Workshop notes: ILWAS aquatic system workshop. Lafayette, California

US Army COE (1975) North branch Potomac River mine drainage study

Warwick JJ et al (1997) Estimating nonpoint source loads and associated water quality impacts. J Water Resour Planning Manag 123(5):302–310

Weber WJ (2000) Environmental systems and processes: principles, modeling, and design. Wiley, New York

Weber WJ, Stumm W (1963) Mechanism of hydrogen ion buffering in natural waters. J AWWA 55(12):1553–1578. https://doi.org/10.1002/j.1551-8833.1963.tb01178.x

Chapter 6
Coliforms, Pathogens, and Viruses

Waterborne disease and pathogens have been a concern of environmental engineers for more than a century. Detecting and understanding the behavior of pathogens in water is very critical to protecting the public health. While directly detecting pathogens provides the most accurate information, there are various types of pathogens existing in water, making it very difficult and expensive to detect all pathogens. Therefore, indicator microorganisms are used in monitoring the water quality. Fecal coliform, a group of bacteria from the intestine of humans and other warm-blooded animals, is an indicator microorganism widely used due to its easy detection and strong correlation with pathogens (Thomann and Muller 1987). Fecal coliform bacteria include *Escherichia, Klebsiella,* and *Citrobacter,* among which more than 90% of fecal coliform are *Escherichia Coli* (*E. coli*). While a high density of fecal coliform does not necessarily cause waterborne diseases, it does indicate the possible existence of pathogens. Although the reliability of using fecal coliform as an indicator microorganism has been questioned in recent years and new techniques such as DNA identification have been proposed to replace indicator microorganisms, fecal coliform is still the most widely used in the U.S. and around the world. In the meantime, fecal coliform continues to be water quality issue in many parts of the world.

6.1 Coliforms

Figure 6.1 shows the fecal and total coliforms levels measured in the Yamuna River in India in the stretch from Delhi to about 450 km from entering the Ganges River based on data from India's CPCB. The coliform levels are extremely high, well above the criteria, particularly in the 25 km section in Delhi.

Fecal coliform in water comes from both point and nonpoint sources. In urban areas, sewer systems and combined sewer overflow (CSO) systems carry large amounts of fecal coliform; in agriculture and residential areas, animal manure and

© The Author(s), under exclusive license to Springer Nature Switzerland AG 2022   203
W.-S. Lung, *Water Quality Modeling That Works*,
https://doi.org/10.1007/978-3-030-90483-8_6

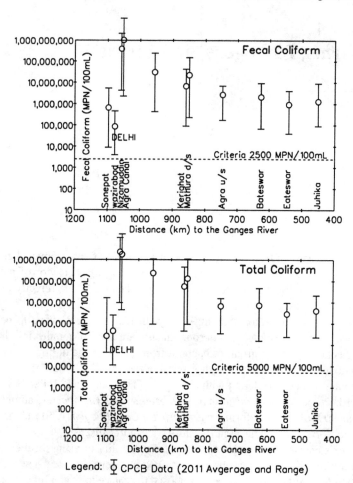

Fig. 6.1 Fecal and total coliform levels in the Yamuna River

failure of septic system are the major sources (Psaris and Hendricks 1982). In some areas, wild animals contribute a significant amount of fecal coliform. Although the secondary treatment level of wastewater is applied at municipal plants throughout the United States and the effluent meets the standard for fecal coliform, violations of the water quality standard for fecal coliform have been reported from coastal salt waters to inland fresh waters. These facts imply that other sources such as combined sewer overflow (CSO) may play a key role in fecal coliform contamination. In the United States, high levels of fecal coliform may impair natural water, causing the failure of shellfish farms and causing the closure of common swimming areas (Brock et al. 1985; Connolly et al. 1999; Crabill et al. 1999). In developing countries, a large percent of industrial and agricultural wastewater is not treated and directly discharge into rivers and lakes. In addition, residential wastewater is often directly

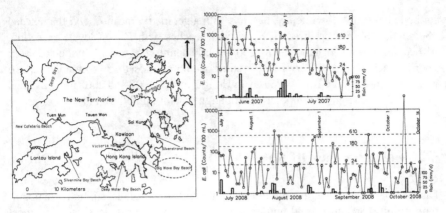

Fig. 6.2 Hong Kong beaches and *E. coli* in the Big Wave Bay beach in 2007 and 2008

discharged into surface waters. Organic wastewater reuse for irrigation practices is widely applied in many developing countries (Saqqar and Pescod 1991). All these sources cause very high fecal coliform population density in surface waters and affect the quality of drinking water, since drinking water treatment and distribution system are not fully developed in these countries (Athayde et al. 2000; Muller et al. 2001).

Colony forming unit (CFU) and most probable number (MPN) are two methods used to enumerate microorganisms in samples. The main difference between CFU and MPN is that CFU is calculated from the bacterial and fungal colonies growing on a solid agar plate while MPN is calculated from viable bacteria growing in a liquid medium. The fecal coliform density is recorded as the number of organisms per 100 ml. Sometimes the unit of colony producing units per 100 ml of water (CPU/100 mL) is used; this is equal to the number of organisms per 100 mL.

One of the common pathogen problems is *E. coli* contamination in natural waters. As a key indicator of water pollution, various standards have been used for water quality management ranging from 180 to 250 counts/100 mL to protect the safety of water uses. Hong Kong (Fig. 6.2) is a coastal city with abundant beaches for recreation use. Millions visit these beaches during the summer months to enjoy the waters. However, swimming and surfing in fecal-contaminated water may result in gastrointestinal and skin illnesses. To protect public safety, the Hong Kong Environmental Protection Department has been monitoring the water quality of bathing beaches since 1986. To establish scientific criteria for assessing beach water quality, epidemiological studies were conducted in late 1980s on bathers at these beaches. *E. coli* as found to be a good indicator of fecal pollution and its concentration showed a strong correlation with the incidence rate of swimming associated illness such as skin and gastrointestinal illness. In particular, it is found that *E. coli* levels of 24, 180, and 610 counts/100 mL are associated with illness rates of approximately 0, 10 and 15 cases per 1000 swimmers, respectively (Thoe 2010). Beaches with an *E. coli* level below 24 counts/100 mL are considered 'good', or suitable for swimming and bathing. Readings between 24 and 180 counts/100 mL are graded as suitable for body contact. Subsequently, beaches with a geometric mean *E. coli* that exceed 180 counts/100 mL are graded 'poor', and beaches with *E. coli* levels exceeding 610 counts/100 mL or last reading above 1,600 counts/100 mL are graded as 'very poor'. Beaches with a 'very poor' grading would be closed to swimmers. *E. coli* levels measured at the Big Wave Bay in Hong Kong in 2007 and 2008 are presented in Fig. 6.2, showing the temporal pattern aligned with rainfall events.

6.2   Coliform Die-Off Rate

Factors that determine the die-off rate of coliforms include sunlight, temperature, salinity, predation, and nutrient deficiencies, toxic substances, etc. The most commonly used is the first order decay kinetics, presented in the literature as die-off rate, or the time required for a population reduction of 90% (Mancini 1978). Figure 6.3 summaries the derivation of the 1st order die-off rates using monitored data from Rock Creek (a short 10 mile section) in Pennsylvania, a forty-mile stretch

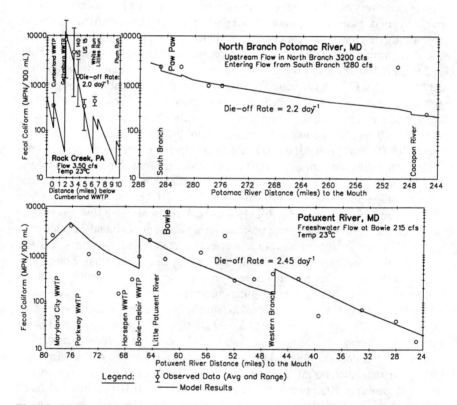

Fig. 6.3 Coliform die-off rates in Rock Creek, north branch Potomac River, and the Patuxent River and Estuary

of the North Branch Potomac River from a point below Cumberland to a point upstream of Hancock, and the Patuxent River covering the riverine section upstream of Bowie and the estuarine section to a point 24 miles above the mouth. Simple 1-D stream models were used to back calculate the die-off rates in each case. Such an approach collapses various factors as listed above into a single kinetic coefficient, a parameterization.

The same coliform and distance scales are used for these three study sites. The derived die-off rates are quite close, ranging between 2.0 and 2.45 day^{-1}. Rock Creek is a very small stream with the Gettysburg WWTP wastewater loads providing the main coliform input. Because of the slow water movement associated with the low flow, the die-off rate of fecal coliforms progressed swiftly within a very short distance. On the other hand, the significant river flow in the North Branch Potomac River moved the coliforms fast in the downstream direction, thereby taking a much longer distance to lower the coliform levels. The coliform levels rise immediately following the input of the WWTPs in the upper Patuxent River. The spatial attenuation of fecal coliforms in the Patuxent River is somewhat between Rock Creek and the Potomac. These physical insights reveal the importance of travel time with respect to these rivers. Note that the derivation of the CBOD deoxygenation rates, k_d, in the upper Patuxent and upper Mississippi Rivers is presented in Fig. 2.26, showing plots of CBOD loads versus time of travel, a similar idea.

While the above approach of deriving coliform die-off rates using coliform data is relatively straightforward, it lacks predictive capability as the environmental condition changes. A more physically based approach to modeling the fate and transport of coliforms would require the quantification of many environmental factors affecting the die-off rate. Fecal coliform enters surface waters along with discharge or storm runoff in two states. Fecal coliform bacteria are either in a free-moving (i.e. free-floating, free-swimming) state flowing with water or in an attached (particle-associated) state moving with sediment particles. After fecal coliform enters surface water, the free bacteria move along with the advective movement of water. The free fecal coliform bacteria can also disperse in horizontal and vertical directions in receiving waters. The attached bacteria will settle to the bottom along with sediment particles in addition to advective and dispersive movement with sediment particles. As a member of *Enterobacteriaceae* family, fecal coliform bacteria will die more quickly in surface waters than in the intestine of humans and animals. The residual fecal coliform will adapt to the living conditions and may continue moving to the water bottom and even re-grow after adaptation to the changed living conditions (Burn 1987; Camper et al. 1996). When bottom shear stress is high enough to cause sediment resuspension, fecal coliform may re-enter the water column (U.S. EPA 1985; Thomann and Mueller 1987; Chapra 1997). Hence, the fate and transport processes for fecal coliform bacteria include die-off, advection, dispersion, the movement of free-floating bacteria, the settling of attached bacteria, and resuspension. The growth of fecal coliform is minimal although it is possible after adaptation to the living conditions.

Various factors, such as solar radiation, dissolved oxygen, temperature, pH, salinity, and toxins contribute to the die-off of fecal coliform. Previously, it was believed that ultraviolet (UV) radiation was the major component of solar radiation that killed most fecal coliform in surface waters. However, research in the past two decades found that the photochemical reaction associated with visible light plays a more important role for the die-off of fecal coliform because UV is very limited after penetrating the water column (Joyce et al. 1996). The die-off of fecal coliform caused by solar radiation is also influenced by other environmental conditions in the water column. DO was found to be a possible catalyst in the photochemical reaction (Curtis et al. 1992). When the DO concentration is low, the die-off rate of fecal coliform is low even though the solar radiation is intensive, and when the DO concentration is high, even weak light intensity causes the die-off of large amounts of fecal coliform.

Temperature is another factor that controls the rate of growth and die-off of microorganisms by influencing the activity of certain enzyme. For fecal coliform, only the die-off rate is considered to be temperature related. The die-off rate of fecal coliform has been assumed to have a positive correlation with temperature, which means that the die-off rate increases as temperature increases (Thomann and Mueller 1987; Chapra 1997). Salinity is another important factor resulting in different die-off rates for coliforms in freshwater and sea water. In a landmark investigation, Mancini (1978) suggested the following equation to independently derive the die-off rate by factoring in water temperature, salinity, and available light:

$$k_T = [0.8 + 0.006(Sal)]1.07^{(T-20)} + \frac{I_a}{k_e H}\left[1 - e^{-k_e H}\right] \tag{6.1}$$

where

k_T = the die-off rate at water temperature, T (°C),
Sal = salinity (ppt),
I_a = daily average water surface solar radiation (langleys/day),
k_e = light extinction coefficient (m^{-1}),
H = completely mixed water depth (m).

In a study of sun bathing beaches in Hong Kong, Thoe (2010) developed a model to forecast the beach conditions in terms of *E. coli* levels following rainfalls. Since the recreational beaches are shallow, the depth-specific die-off rates of *E. coli* are calculated as follows:

$$k(z) = [k_b + k_s S]\theta_T^{(T-20)} + k_I I e^{-k_e z} \tag{6.2}$$

z = depth (m)
k_b = basic die-off rate (day^{-1})
k_s = salinity dependent die-off rate (day^{-1})
k_I = solar radiation dependent decay rate
θ_T = temperature correction factor
S = salinity (ppt)
T = water temperature (°C)

I = solar radiation (W/m^2 = 0.0864 MJ/m^2)

k_e = light extinction coefficient (m^{-1}).

Note that Eq. 6.1 is for depth averaged die-off rate while Eq. 6.2 is for depth-specific die-off rate.

In addition to the factors contributing to the die-off of fecal coliform mentioned above, pH is usually known as a key factor controlling the microbial population (Presser et al. 1997, 1998). Fecal coliform can survive in a relatively wide range of pH values in surface waters when the pH values are below 9. However, when the pH is higher than this value, the fecal coliform population will significantly decrease which implies the increase of the die-off rate (Finch and Smith 1986). Studies in wastewater stabilization ponds found that the die-off rate of fecal coliform will dramatically increase if the pH is close to or greater than 9.0 (Curtis et al. 1992; Fernandez et al. 1992; Mayo 1995). Furthermore, pH and solar radiation together will cause greater die-off of fecal coliform than simply the sum of the die-off rates caused by these factors when considered separately (Curtis et al. 1992; Awuah et al. 2002).

Unlike indigenous bacteria, nutrients such as nitrogen, phosphorus, and dissolved organic carbon (DOC) or assimilable organic carbon (AOC) are found to have little impact on the fecal coliform population in most natural water bodies. Some researchers found that DOC might play a role for the fecal coliform population in sediment, while others obtained the result that the fecal coliform population has a weak correlation with DOC (Bastviken et al. 2001). Although nutrients do not influence the fecal coliform population in natural waters, it is highly possible for them to become the limiting factors of fecal coliform growth in drinking water distribution system, which implies that nutrients could be limiting factors in pristine waters (Bois 1997; Escobar et al. 2001; Heldal et al. 1996; Lechevallier et al. 1996; Volk and Lechevallier 1999). As allochthonous bacteria, fecal coliform cannot compete with indigenous bacteria in obtaining nutrients. When fecal coliform is in a starvation situation, a protection mechanism is activated causing the bacteria to be less susceptible to death than when nutrients are sufficient (Varnam and Evans 2000).

For the fecal coliform population in freshwaters, like inland rivers, the salinity is very low and has no impact on the die-off of fecal coliform. However, for estuaries and coastal areas where the salinity is high and varies both in time and in space, salinity becomes an important factor contributing to the die-off of fecal coliform (Davies et al. 1995; Martinez et al. 1989). Correlation of the fecal coliform population and the salinity in the water column has been reported in the literature, and the die-off rate of fecal coliform increases when salinity increases (Brock et al. 1985; Mancini 1978).

The factors mentioned above affect the die-off rate of fecal coliform and also directly or indirectly change the physiology of fecal coliform. Sediment particles or suspended solids play a special role in affecting both the die-off rate and the settling rate of fecal coliform in the water column. Sediment particles affect the population of fecal coliform in several ways. First, suspended particles greatly attenuate the solar radiation intensity in the water column and protect the fecal coliform from being killed (Davies-Colley et al. 1994; Emerick et al. 2000). Second, organic particles usually release certain nutrients, and fecal coliform bacteria have a tendency to swim toward the high nutrient environment and easily reach the particles (Jackson 1987, 1989; Milne 1986). The attached fecal coliform then settles together with the particles and disappears from the water column (Baudart et al. 2000; Gannon et al. 1983; Howell et al. 1996).

After fecal coliform is deposited to the sediment bed the living conditions become more ideal than those in the water column. For example, fecal coliform can avoid the harmful solar radiation and high DO concentration in the water column. In addition, there are sufficient nutrients available in the sediment and predators are limited (Epstein and Shiari 1992). Fecal coliform in sediment can survive for a long time and can even reproduce (Burton et al. 1987; Floderun et al. 1999). Many studies found that the fecal coliform level in sediment is 10–10,000 times higher than that in the water column (Buckley et al. 1998; Crabill et al. 1999; Doyle et al. 1992; Irvine and Pettibone 1993). Davies et al. (1995) found that sediment bed fecal coliform in both freshwater and the marine environment can survive much longer than in the water column. When sediment particles are resuspended from the bottom during storms, fecal coliform will be transported back to the water column and cause high fecal coliform population density.

6.3   Fecal Coliform Modeling of the Tidal Basin and Washington Ship Channel

Over half of the 50,000 the TMDLs completed in the U.S. from the late 1990's to 2010 are bacteria TMDLs. The Tidal Basin and Washington Ship Channel (TBWC) on the Potomac River in Washington, District of Columbia were on the 1996 303(d) list because of excessive counts of fecal coliform bacteria that exceeded the District's water quality standards (WQS). A bacteria TMDL study was therefore required. The Potomac River Watershed covers 14,679 square miles in four states and the District of Columbia. The river is more than 380 miles long from its start in West Virginia to Point Lookout on the Chesapeake Bay. The Potomac River provides 75% of the Washington metropolitan area's drinking water and all of the District's drinking water. The river also receives discharges from wastewater treatment plants, including the District's Blue Plains WWTP serving Arlington and Alexandria located just upstream of the DC/MD line in Virginia. There are no drinking water intakes from the river downstream of the District. The TBWC is located nearby the confluence of the Anacostia River and Potomac River (Fig. 6.4) and is the key destination for visitors during the spring Cherry Blossom Festival. The Tidal Basin was constructed in the late nineteenth century by the Army Corps of Engineers as a part of the comprehensive management of the Potomac River and land development of the District.

The main function of Tidal Basin is to flush the Washington Ship Channel with the freshwater from the Potomac River. Two floodgates exist in the system, one linking

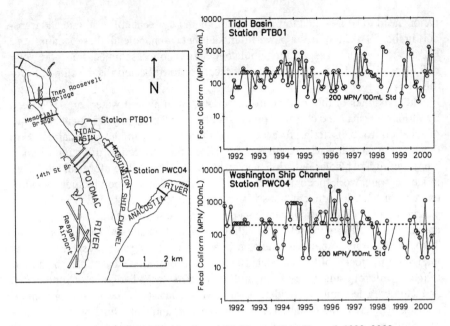

Fig. 6.4 Fecal voliforms in the Tidal Basin and Washington Ship Channel, 1992–2000

the Tidal Basin and the Potomac River, and the other linking the Tidal Basin and Washington Ship Channel. Freshwater flows into the Tidal Basin through the flap gate when the tidal elevation changes and the elevation in the Potomac River are higher than that in the Tidal Basin. In the same way, the freshwater flushes into the Washington Ship Channel as the water surface elevation become higher in the Tidal Basin. The direction of water flow is unidirectional from the Potomac River to the Tidal Basin then to the Washington Ship Channel. The other end of the Channel is connected to the Anacostia River. The Tidal Basin is shallow with an average depth of around 2 m and a surface area of about 0.4 km^2. The Channel is about 122 m wide and the depth varies from 1 to 8 m. The observed fecal coliform concentrations in both of the TBWC from 1992 to 1998 are shown in Fig. 6.4 using the data from EPA STORET. Consistent violation of the 200 MPN/100 mL fecal coliform standard for Class A (primary contact recreation) is seen in the TBWC during this period. The TBWC was on District of Columbia's official list of polluted water, therefore requiring the development of a TMDL for fecal coliform bacteria. As such, a fate and transport model of these contaminants is needed for the TMDLs.

The flow field in the TBWC is governed by the tidal fluctuation, floodgate operation, and wind. Therefore, a hydrodynamics model is necessary to provide the transport information to fecal coliform modeling. In addition to advection and diffusion of fecal coliform, sediment transport is believed to be important for the transport of fecal coliform. Hence, a sediment transport model is also required. The die-off of fecal coliform is influenced by various environmental factors such as solar radiation, pH, salinity, and temperature. To determine the die-off rate of fecal coliform, these factors need to be modeled too. As a result, modeling fecal coliform fate and transport in the TBWC requires a powerful model that can handle all these factors. As a result, a model that is capable to simulate hydrodynamics, sediment transport, solar radiation, pH, and fate and transport of fecal coliform bacteria is developed to for the TBWC.

The dynamics of the fecal coliform population in natural waters are influenced by various physical and chemical processes of the water and physiological processes of fecal coliform bacteria. To account for these factors, modeling fecal coliform dynamics in natural waters requires information about advection, diffusion, temperature, salinity, pH, solar radiation, and sediment transport. Therefore, an integrated modeling framework for complete fecal coliform fate and transport includes a hydrodynamic model, a sediment transport model, a water quality model, and a fecal coliform model. The hydrodynamic model drives the sediment transport sub-model and the fecal coliform model by providing the advection and diffusion information. The sediment transport model generates the sediment settling and resuspension information for the fecal coliform model. The fecal coliform model calculates the die-off of fecal coliform using a die-off rate influenced by solar radiation, temperature, pH, and salinity. The complete modeling framework consists of the following components: a hydrodynamic model; a sediment transport and water quality model; and a fecal coliform model. The transport of salt and heat is included in the hydrodynamic model because both of these factors affect the hydrodynamics by changing water density. Solar radiation and pH can be obtained by running a water quality model. The hydrodynamic model drives the sediment transport and water quality model. The hydrodynamic, sediment transport, and water quality models drive the fecal coliform model.

The pH-alkalinity calculations in the water quality model is based on the carbon dioxide-bicarbonate-carbonate equilibrium in natural surface waters reported in Chap. 5. The model variables include alkalinity and carbon dioxide acidity. The mass balance of alkalinity includes advection, diffusion, and alkalinity decrease due to ammonia uptake, nitrification, and sulfide oxidation, alkalinity increase due to ammonification, denitrification, and sulfate reduction. The mass balance of carbon dioxide (CO_2) acidity includes advection, diffusion, mass transfer across the air–water interface, acidity decrease due to photosynthesis, and acidity increase due to respiration.

6.3.1 Coliform Bacteria Die off

Die-off is a very important process for the decrease of fecal coliform population in surface water. Current models for natural surface waters usually use very simple methods to determine the first-order die-off rate. The growth and die-off of microorganisms in natural environments is a complex process that is affected by the interactions among the same population, among diverse populations, between microbial population, plants, and animals, and interactions with the environment. For the practical purpose of modeling fecal coliform bacteria in surface waters, the interactions among the microbial populations, plants, and animals are neglected since it requires large numbers of experiments and the supporting data is difficult to obtain. In this study, only the abiotic factors are considered. The factors considered in this study include solar radiation, pH, salinity, and temperature.

Solar radiation is a major factor that causes the die-off of fecal coliform in the water column. Short wave radiation below 500 nm, especially UVB, was believed to contribute most to the die-off of fecal coliform by direct damaging the bacteria DNA (Mancini 1978). The relationship between the solar radiation intensity and fecal coliform die-off rate is usually expressed as $k_I = \alpha I$, where α is the solar radiation proportionality constant; I is the solar radiation intensity in the water column; and k_I is the die-off rate caused by solar radiation. This relationship is easy to apply in models and is widely used in modeling surface water fecal coliform die-off (Mancini 1978).

The solar radiation intensity is calculated in the water quality module of the EFDC model (Bai and Lung 2006). Meteorological data including cloud cover, solar radiation and water column conditions such as the light extinction coefficient, suspended solid concentration or turbidity are used to calculate the light intensity in the upper boundary of each water column layer in EFDC. The solar radiation intensity is then averaged over each water column layer and saved for the calculation of the die-off rate caused by solar radiation. For free-swimming fecal coliform bacteria, the above formula is used first. For attached bacteria, an adjustment must be included because the attached bacteria are protected by sediment particles and receive less solar radiation. Emerick et al. (2000) derived a function to account for the protection from the sediment particles as below $C_p(t) = \frac{C_p(0)}{k_I t}\left(1 - e^{-k_I t}\right)$, where $C_p(0)$ is the number of attached fecal coliform at the beginning of exposure to solar radiation; $C_p(t)$ is the number of survived attached fecal coliform after exposure to solar radiation to time t; and k_I is the die-off rate of fecal coliform caused by solar radiation and enhanced by the pH value.

The pH value in the water column affects the die-off rate of fecal coliform in two ways. First, if the pH is close to or higher than 9, the die-off rate of the fecal coliform will increase even in dark conditions. Second, if solar radiation exists, pH either decreases the resistance of fecal coliform to the light or increases toxic forms of oxygen (Curtis et al. 1992), resulting in more fecal coliform deaths by solar radiation. Hence, in this study, the effect of pH will be considered in two situations. When the model simulates the dark conditions, the die-off rate is a function of the pH value. When the model computes fecal coliform dynamics in daytime, the influence of pH will be incorporated into the die-off rate caused by solar radiation. The increase of the die-off rate by the high pH value was also derived based on experiments by Curtis et al. (1992). As they point out, the combination of solar radiation and high pH value causes more death of bacteria than the sum of the death caused during high pH conditions without solar radiation and the death caused by solar radiation under neutral pH conditions.

The die-off rates were calculated based on the data from the paper by Curtis et al. (1992). The die-off rate caused by solar radiation and pH was compared with the die-off rates caused only by pH, and it was assumed that the increase of the die-off rate caused by pH has a linear relationship with the pH value. The pH values reported are between 7.5 and 9.0. The resultant formula for the die-off rate with the effect of pH is $\Delta k = \gamma pH + \lambda$, where γ and λ are two regression constants. In this study, it is assumed that this relationship is valid for pH above 7. For the conditions where the pH value is below 7, the effect of pH will be omitted under the assumption that fecal coliform either adapts to this pH range or has been inhibited, meaning neither growth nor death.

In dark condition, the pH was found to increase the fecal coliform die-off rate only when pH is high. Curtis et al. (1992) suggested 9.3 and others reported 9.0 as the threshold value. Hence, in dark condition, the die-off caused by pH is considered only when pH is higher than 9.0 as $k_{pH} = c(pH)$ for $pH > 9$. Because different studies were based on different waters and the experiments were conducted in different ways, the threshold value is site-specific.

In addition to pH, salinity must be considered when simulating fecal coliform dynamics in estuaries and coastal areas. Synergism between sunlight and salinity was observed (Davies-Colley et al. 1994). However, the observed data is not adequate to quantify this combined effect. Consequently, a linear relationship between salinity and the die-off rate of fecal coliform as reported in the literature (Mancini 1978; Brock et al. 1985) was adopted for the model. When Mancini derived the linear regression model for the die-off rate and the salinity level, the die-off rate includes a constant term, which can be considered as a natural die-off rate or a base die-off rate even without salinity. In addition, the salt level in the formula is expressed as percent of seawater instead of salinity originally at 20 °C as $k_s = 0.8 + 0.006P_s$, where k_s is the die-off rate of fecal coliform in salt water; and P_s is the percent of seawater. Chapra (1997) modified the formula to use salinity instead of percent of seawater with the assumption that the seawater salinity is between 30 and 35 ppt. The formula is then written as $k_s = 0.8 + 0.02S$, where S is salinity in ppt. This formula implies that the die-off rate in fresh water without solar radiation and pH effect is approximately equal to 0.8 day^{-1} at 20°C.

The growth of microorganisms is strictly governed by a temperature range according to Shelford's Law of Tolerance (Atlas and Bartha 1998). When the temperature is out of the range, microorganism population density stops increasing. Either the growth rate or die-off rate are decreased to zero when temperature is lower than the lowest temperature for growth or growth rate is decreased and die-off rate is greatly increased when temperature is higher than the highest temperature for growth. While the growth rate of microorganisms has an optimal temperature range and usually exhibits a parabolic curve, the die-off rate increases as temperature increases. The most widely used relationship is the Arrhenius equation, which is represented as $k_T = k_{20}\theta^{(T-20)}$, where k_T (day^{-1}) is the die-off rate in dark affected by salinity at temperature; k_{20} is the die-off rate (day^{-1}) in dark affected by salinity at temperature 20°C; θ is a dimensionless constant; and T is the water temperature (°C). Linear regressions are also reported in the study concerned with the relationship between temperature and fecal coliform die-off rate. In this model, only the Arrhenius type formula is considered since only one constant θ needs to be assigned.

Temperature was found to affect the die-off caused by solar radiation in some studies. Temperature is also believed to have impact on the photo oxidation reaction. However, others found that temperature's influence is negligible. The different conclusions are probably due to the different experiment designs. In this model, temperature is assumed to have an impact on the die-off rate caused by the die-off rate and salinity. Using Mancini's approach, the die-off rate caused by salinity and adjusted by temperature is $k_T = (0.8 + 0.02S)\theta^{(T-20)}$, where k_T is the die-off rate caused by salinity and adjusted by temperature; S is salinity; θ is the temperature adjustment coefficient; and T is water temperature. The original k_{20} is replaced by a natural mortality and salinity affected die-off rate in this formula.

After these environmental factors are calculated, the die-off rates of fecal coliform caused by these factors are then updated for free-swimming bacteria, attached bacteria in the water column, and for the total bacteria in the sediment bottom. In addition, the model checks the solar radiation intensity and determines whether the die-off rate is calculated in light conditions or dark conditions. For free-swimming fecal coliform bacteria in light conditions, the die-off rate caused by solar radiation is first calculated with the combined effect of DO. This rate is further adjusted for the enhancement of a high pH value. The die-off caused by salinity is then calculated and adjusted using temperature. The total die-off rate is the sum of these individual die-off rates based on the assumption that these individual processes are independent from each other. Hence, the total die-off rate for free-swimming fecal coliform bacteria during light conditions is

$$k_T = (0.8 + 0.02S)\theta^{(T-20)} + \alpha \bullet I + \gamma \bullet pH + \lambda$$

For the attached fecal coliform bacteria in light condition, a two-step approach is applied. The die-off rate caused by solar radiation is calculated in the first step.

$$k_I = \alpha \bullet I + \gamma \bullet pH + \lambda$$

The density is then updated as

$$C_P^* = \frac{C_P^n}{k_I \Delta t}\left(1 - e^{-k_I \Delta t}\right)$$

In the next step, the die-off rate related to salinity and temperature is calculated as

$$k_T = (0.8 + 0.02S)\theta^{(T-20)}$$

When the water is in dark condition, both the free-swimming fecal coliform and attached fecal coliform are assumed to have the same die-off rates. The factors considered for calculating the fecal coliform die-off rates are salinity, pH, and temperature.

Therefore, the total fecal coliform die-off rate during the dark conditions is

$$k_T = (0.8 + 0.02S)\theta^{(T-20)} + c \bullet pH$$

Fecal coliform bacteria can survive in the sediment bed significantly longer than in the water column for various reasons, and some researchers even found fecal coliform growth in the sediment bed (Davies et al. 1995). Hence, the die-off rate calculated from the water column cannot be applied directly to the fecal coliform in sediment bed. Unfortunately, no research has been conducted to link the environmental factors in the sediment bed to the fecal coliform die-off/growth rate. In addition, the salinity, pH values are not calculated for sediment layers. Therefore, a temperature dependent first-order die-off rate is set as the total die-off rate.

$$k_T = k_{20}\theta^{(T-20)}$$

6.3.2 Sediment–Water Partition of Fecal Coliforms

Suspended solids have significant influences on the fate and transport for fecal coliform bacteria in surface waters. Sediment particles in the water column indirectly protect all the bacteria from solar radiation by decreasing the solar radiation intensity. Sediment particles also directly protect the attached bacteria by providing a shade area for them. Attached fecal coliform move with the sediment particles to the bottom bed and disappear from the water column. Whenever resuspension of settled sediment particles occurs, the bottom bacteria can re-enter the water column. If the water column bacteria density is extremely high, bacteria enhance the flocculation process of the cohesive sediments. Since the fecal coliform level in surface water is usually lower than the density for enhancing flocculation, this interactive process is not modeled in this study.

To include the impact of sediment transport processes to the fate and transport of fecal coliform, the free-swimming bacteria and the attached bacteria must be considered separately. The first step is to determine the ratio between the free bacteria and the attached bacteria. The ratio is calculated by assuming an equilibrium adsorption process in this study. Equilibrium adsorption is usually used to describe the partition of contaminants to sediment particles. Langmuir isotherm, Freundlich isotherm, and BET isotherm are three widely used models of adsorption isotherms. When the contaminant concentration is low, a linear isotherm can be assumed (Chapra 1997).

In the water column, let C_F and C_P represent free and particle attached fecal coliform bacteria with the unit MPN/L. The total fecal coliform concentration is

$$C = C_F + C_P$$

If K_P is the partition coefficient (L/mg) and m is the sediment concentration in the water column (mg/L), the attached coliform bacteria is

$$C_P = K_P m C_F$$

with the linear adsorption assumption.

Let F_F be the fraction of the fecal coliform bacteria that is in free-swimming state. F_F can be calculated as

$$F_F = \frac{1}{1 + K_P m}$$

and the fraction for the attached bacteria F_P is

$$F_P = \frac{K_P m}{1 + K_P m}$$

Hence, the free bacteria C_F can be calculated as

$$C_F = F_F C$$

and the attached bacteria C_P can be calculated as

$$C_P = F_P C$$

In the model, the fractions F_F and F_P are calculated after the sediment concentration in the water column is updated.

In the sediment bed, the bulk density of fecal coliform is C (MPN/L)

$$C = C_F + C_P$$

where C_F (MPN/L) is the bulk density of fecal coliform in the free moving state in porous water and C_P (MPN/L) is the bulk density or concentration of fecal coliform in the attached state.

According to the linear adsorption assumption,

$$C_P = \frac{K_P \beta_B}{\varepsilon} C_F$$

where K_P (L/mg) is the partition coefficient; β_B (mg/L) is the bulk density of the sediment; ε is the porosity of the sediment. The fraction of free bacteria and attached bacteria F_F and F_P are then calculated as

$$F_F = \frac{1}{1 + \frac{K_P \beta_B}{\varepsilon}}$$

and

$$F_P = \frac{\frac{K_P \beta_B}{\varepsilon}}{1 + \frac{K_P \beta_B}{\varepsilon}}$$

The free bacteria and attached bacteria are thus calculated with

$$C_F = F_F C$$

and

$$C_P = F_P C$$

The density of fecal coliform in the water column is expressed as a volumetric density with the unit MPN/L. In the sediment bed, the volumetric density is converted into an areal density with the unit MPN/m^2. The conversion is conducted during the calculation of the processes associated with sediment bed.

6.3.3 Settling of Attached Fecal Coliform in Water Column

The direct physical impact of sediment transport processes to the attached bacteria is the settling effect that brings these bacteria to the sediment bed. The settling velocity of attached bacteria is the same as the particle settling velocity, which is dependent on the properties of the particles such as the size and composition. For cohesive sediment, it is even more complicated because of flocculation, which is a very complex process related to the cohesive sediment materials, concentration, water ionic characteristics, shear stress and turbulence intensity of the flow. The settling velocity of flocs does not have a simple linear relationship with the sizes of flocs. Empirical parameters have to be set to calculate the settling velocity. In the EFDC model, several empirical formulas are provided for calculating the settling velocity for the cohesive sediment particles and flocs.

Unlike the disappearance of fecal coliform from die-off in the water column, the destination of fecal coliform settling is the bottom sediment bed. Hence, the bottom fecal coliform density must be updated after calculating the settling loss. Using the areal concentration, the increase of the bottom fecal coliform density is equal to the density decrease multiplied by the water layer thickness from the overlaying water column. Once fecal coliform bacteria settle to the bottom, it is assumed to evenly distribute among the receiving sediment layer. The movement of fecal coliform among the sediment layers is modeled as a diffusive process for free moving bacteria and will be discussed in next section.

6.3.4 Resuspension of Fecal Coliform in Sediment Bed

When the bottom shear stress is stronger than critical shear stress of erosion or sediment bed shear strength, the bottom sediment can be resuspended into the overlaying water column. Correspondingly, all the fecal coliform bacteria living in the resuspended sediment enter the water column. The sediment concentration in the water column is much lower than that in the bottom bed layers. Therefore, part of the attached fecal coliform will change to the free-swimming state since the adsorption process is reversible. Depending on the relationship between the shear stress and the critical shear stress and bed strength, two types of resuspension, mass erosion and surface erosion, may occur. When the bottom shear stress is higher than the bed shear strength, mass erosion will occur and the whole layer of sediment will be resuspended. If the bottom shear stress is lower than the bed shear strength and higher than the critical shear stress for resuspension, surface erosion will occur and only the top sediments will be resuspended. The bed density of fecal coliform is updated simply by using the ratio of the eroded sediment to the total sediment, which is equal to the decreased depth over the original depth before erosion. Since the lost fecal coliform from the sediment bed re-enters the water column, the water column density of fecal coliform is updated using that information.

6.3.5 Fecal Coliform Transport in Sediment and Across Sediment–Water Interface

The transport of bacteria in porous media has been widely studied in groundwater remediation. The full governing equation for total bacteria fate and transport in porous media is an advection/diffusion/adsorption/chemotactic/reaction equation. In this study, the fecal coliform movement in the sediment bed due to chemotactic response is not included since the knowledge about the preference of fecal coliform is very limited. The reaction term is solved in the kinetic portion of the model. Since the horizontal scale for modeling surface water is much greater than the vertical scale in the sediment bed and the bacteria transport in porous media is slow, only the vertical direction is considered in the model development. In addition, it is assumed that the attached bacteria in the sediment bed do not move at all. Hence, the resultant transport equation for fecal coliform is an advection–diffusion-adsorption equation in the vertical dimension for free-moving fecal coliform. In addition to the transport inside the sediment layers, the diffusive movement of bacteria across the interface between the sediment layer and the water column is also considered in the model.

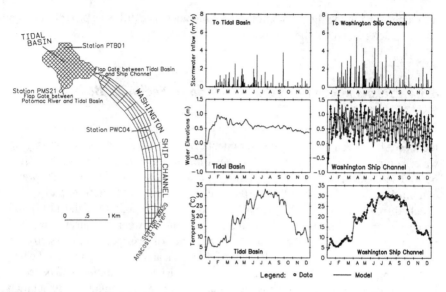

Fig. 6.5 Computational Grid for Tidal Basin and Washington Ship Channel and Hydrodynamic Results

Since adsorption is calculated in the step of calculating sediment–water interaction, only diffusion is evaluated in this step.

6.3.6   Hydrodynamic Computations

First, the two-dimensional segmentation of the TBWC is shown in Fig. 6.5. The water column is divided into two layers to form a three-dimensional computation grid for a total of 530 cells. The three-dimensional configuration has the advantage of eliminating the assignment of the horizontal dispersion in the system albeit the water is shallow and the horizontal mixing is relatively weak. In addition, the vertical segmentation would accommodate the solar radiation effect, a major factor in producing much higher die-off rates in the surface layer than the bottom layer, as presented in a later section.

The meteorological data including the air pressure, wet bulb air temperature, dry bulb air temperature, cloud cover, and precipitation are directly obtained from the Reagan National Airport. Solar radiation was calculated using the short-wave solar radiation algorithm based on the time and the cloud cover. Wind speed and direction data are also from the Reagan National Airport. Tidal elevation determines the flow rate entering the Tidal Basin from the Potomac River and the flow field in the Washington Channel. Hence, the tidal elevation at the boundary between the Potomac River and the Tidal Basin and the boundary between the Anacostia River and the Washington Ship Channel were specified to the model. The tidal elevation data are obtained from the NOAA tidal station in the Washington Ship Channel. Since the area of the modeling domain is very small, the differences between the two boundaries are minimal. For modeling the freshwater flushing of the Washington Ship Channel, the flood flap gate operation tables relating the water elevation differences and the flow rates are very important. Unfortunately, no information regarding the gate operation is available to date. Therefore, the gate operation tables were assumed and adjusted to generate reasonable water elevation in the Tidal Basin, which is to have the flushing process and not to have extremely high or low surface water elevation.

In addition to the driving force from tidal variation, storm water can affect the hydrodynamics calculation. There are six storm sewers discharging into the Tidal Basin and nine storm sewers draining into the Washington Ship Channel. Unfortunately, neither monitored nor modeled storm water runoff data are available. Hence, an approximation of the impact of storm water was made and included in the model. It was assumed that the stormwater empties into the TBWC mainly from two sewers, one to the Tidal Basin, and the other to the Washington Ship Channel. A time series of daily storm water flow rate was calculated by assuming two infiltration rates for the two basins that drain the TBWC, respectively (Bai and Lung 2006). The storm water runoff was then estimated by multiplying the precipitation rate, infiltration loss percentage, and the drainage area. The stormwater runoff discharging into the Washington Ship Channel and Tidal Basin for 1998 are shown in Fig. 6.5.

The model calculated surface elevation in the Washington Ship Channel match the observed tidal elevations at station PWC04 very well. No observed water surface elevation in the Tidal Basin was available for comparison with the model results. However, the calculated water surface elevations show significant flushing with water from the Tidal Basin into the Washington Ship Channel when the water surface elevation in the Tidal Basin is higher than the Washington Ship Channel, consistent with the design purpose of the Tidal Basin.

The modeled water temperature results are shown in Fig. 6.5 for stations PTB01 and PWC04, respectively. The time-variable water temperature in the Washington Ship Channel is reproduced by the model. Such a match is critical as temperature plays an important role in determining the die-off rate of fecal coliform. Again, no observed water temperature was available.

The layer-averaged TSS results from the sediment transport model are shown in Fig. 6.6 for the TBWC. The observed data are also displayed on these two figures to represent the real magnitude of the suspended solid concentration. The modeled suspended solids showed that the spatial variation is high in TBWC.

The fecal coliform density was modeled after the hydrodynamics and the sediment transport. The model used the same time step as for execution of the hydrodynamics and sediment transport models. In addition to the advective and diffusive transport, which was calculated using the same algorithm for temperature and suspended solids, the adsorption of fecal coliform to sediment particles and die-off of fecal coliform were computed. Equilibrium adsorption was applied to calculate the partition of attached and free-swimming fecal coliform. Since very limited data are available on partition of the free swimming and particle associated fecal coliform, the partition coefficient was set to 0.1 L/mg to obtain a high adsorption rate of fecal coliform to

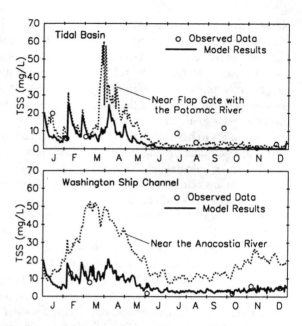

Fig. 6.6 Model calculated TSS results versus data for Tidal Basin and Washington Ship Channel

sediment particles as reported by Steets and Holden (2003). The ratio of adsorbed fecal coliform varies over a wide range because of the wide range of sediment concentration. The coefficient that relates the solar radiation intensity to die-off rate is a critical parameter. In this study, the coefficient for the die-off rate caused by solar radiation was set to 0.02 m^2/W d^{-1} similar to the value adopted by Canale (1993). The base die-off rate was set to 0.8 day^{-1} at 20°C as suggested by Mancini (1978). For the sediment bed fecal coliform, the die-off rate was set to 0.2 d^{-1} at 20°C based on the study of Davies et al. (1995). The temperature adjustment parameter was set to 1.024. Since data are very limited to calibrate the model results, values for these parameters were chosen to obtain reasonable die-off rates. Figure 6.7 clearly shows the die-off rates for the surface and bottom layers for a period of 6 days from July 17,

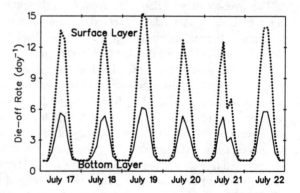

Fig. 6.7 Model calculated die-off rates (July 17–July 22, 1998)

Fig. 6.8 Model calculated
die-off rates in surface and
bottom layers, 1998

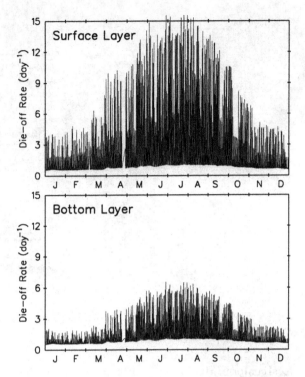

1998 to July 22, 1998. The bottom layer die-off rate curve shows the same diurnal trend as the surface layer die-off rate curve with peak rates in mid-day. The calculated time-variable die-off rates for the surface and bottom layers for the whole year are shown in Fig. 6.8. Clearly, the die-off rate is much lower for the bottom layer due to the low solar radiation intensity.

Model results show that the Tidal Basin is under the impact of both stormwater and the Potomac River. Figure 6.8 shows the fecal coliform results on February 5, 1998 following a rainfall event. The stormwater impact is usually limited to the grid cells in the area near the outfall, while the fecal coliform input from the Potomac River is able to reach a much larger area depending on the tidal elevation. Fecal coliform from the Potomac River reaches approximately half of the Tidal Basin (Fig. 6.9). Fecal coliform in the Washington Ship Channel show a different spatial pattern, displaying minimum stormwater impact. The larger water volume and stronger mixing in the Washington Ship Channel versus the Tidal Basin provided rapid dilution in the system. The spatial trend in Fig. 6.9 shows the peak fecal coliform levels at the boundary with the Anacostia River and the fecal coliform gradually decrease toward the upper portion of the Washington Ship Channel. The fecal coliform results at stations PTB01 and PWC04 are compared with field data in Fig. 6.9. The fecal coliform concentrations in the Tidal Basin vary dramatically in time by mimicking the temporal pattern of rainfall, suggesting that the fecal coliform concentration in the Tidal Basin is significantly influenced by the stormwater runoff. Since the modeled

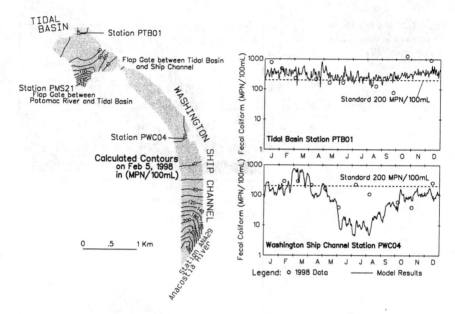

Fig. 6.9 Spatial and temporal patterns of fecal coliform model results and data in Tidal Basin and Washington Ship Channel, 1998

fecal coliform in the Washington Ship Channel are closely controlled by the boundary condition, the station in the system is not able to capture the highest fecal coliform. TMDLs based on this station may be biased to meet the fecal coliform standard for the entire water system.

Although the model was calibrated with very limited data, the model results still provide insights into the fecal coliform fate and transport pattern in the TBWC. The modeled results show significant spatial variation of the fecal coliform distribution. In the Tidal Basin, the highest fecal coliform density is near the storm water outlet (Station PMS21) as expected. The fecal coliform densities in the segments close to the flood gate connecting to the Potomac River are highly related to the fecal coliform density in the water of the Potomac River. In the Washington Ship Channel, the fecal coliform densities in the segments near the boundary are strongly affected by the tidal influence of the Anacostia River. The fecal coliform density near the storm water outlet (Station ANA29) shows a pattern similar to that in the Tidal Basin. However, the magnitude is much lower than that in the Tidal Basin although the storm water load is similar, implying that the dilution in the Washington Ship Channel is much stronger than in the Tidal Basin.

Another potentially important issue in the management of fecal coliform is the sediment bed, which plays an important role in storing fecal coliform as stated in the literature review. There is no measured sediment bed fecal coliform density for the TBWC. Hence, the results here are only useful to demonstrate the possibility of sediment bed storage of fecal coliform. In the model, the initial sediment bed fecal coliform was assumed to be 0 MPN/100 mL. The sediment bed receives the fecal coliform from the settling of particle associated fecal coliform. Based on the studies related to the survival of fecal coliform in freshwater and marine sediment beds, the die-off rate for the sediment bed fecal coliform was set to 0.2 day^{-1} for both the free-swimming and particle associated states (Davies et al. 1995). The modeled sediment bed fecal coliform density is several magnitudes higher than that in the water column as shown in Fig. 6.10, which is within the range of values reported in the literature.

Following the submission of the TMDL study report, in which the above modeling results (Lung and Bai 2003) were included, the U.S. EPA approved the District's TMDL on the Tidal Basin and Washington Ship Channel on December 15, 2004. Subsequently, the original TMDL was revised to incorporate a new water quality standard for *Escherichia coli* (*E. coli*) that the District promulgated in October 2005 after the approval of the original TMDLs. The allocations specified in the original TMDL are still in effect; the revision provides a translation of those loads to *E. coli*, the parameter on which the existing standard is based. The translation was performed using a translator equation developed from analysis of paired fecal coliform/*E. coli* sampling data concurrently collected from waters in the District. When the original

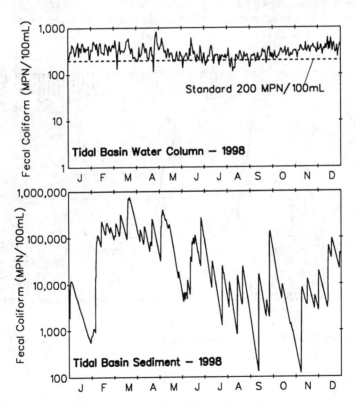

Fig. 6.10 Fecal coliform in water column and sediment of Tidal Basin

2004 fecal coliform bacteria TMDL was developed for the TBWC, the standard for Class A (primary contact recreation) waters was a maximum 30 day geometric mean of 200 MPN/100 ml. Effective January 1, 2008, the District bacteriological water quality standard changed from fecal coliform to *E. coli*. The current Class A water standard is a geometric mean of 126 MPN/100 ml.

6.4   *E. Coli* Modeling of Beaches

Using a different approach, Thoe (2010) successfully developed statistical and data-driven predictive models for the Hong Kong marine beaches. One of the models is the Multiple Linear Regression (MLR) method, which relates the natural log of *E. coli* counts, in a linear combination, to a number of select hydro-environment variables, including rainfall, wind speed and direction, global solar radiation, tidal level, water temperature, and salinity. Figure 6.11 shows the model results and data for *E. coli* at Big Wave Bay in 2007. The *E. coli* data from Fig. 6.2 is used in this model calibration. The MLR model can track the changes in water quality and follows quite closely to the ever-changing observed *E. coli* level, achieving very good correlation at 0.75. The methodology was later applied to forecast the faecal coliform levels in California beaches (Thoe et al. 2014).

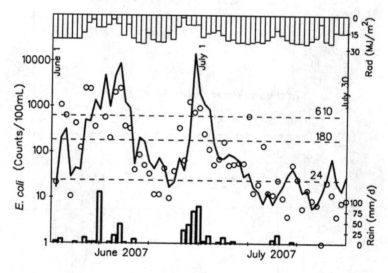

Fig. 6.11   Model tesults versus *E. coli* data for Big Wave Bay, Hong Kong (Thoe 2000)

References

Athayde et al (2000) faecal coliform die-off in wastewater storage and treatment reservoirs. Water Sci Technol 42:139–147

Atlas RM, Bartha R (1998) Microbial ecology fundamentals and applications. Benjamin/Cummings Publishing Company, Inc, CA

Auer MT, Niehaus SL (1993) Modeling fecal coliform bacteria-I: field and laboratory determination of loss kinetics. WR 27(4):693–701

Awuah E et al (2002) The effect of pH on enterococci removal in pistia-duckweed- and algae-based stabilization ponds for domestic wastewater treatment. Water Sci Technol 45(1):67–74

Bai S (2004) Developing a fate and transport model of fecal coliform bacteria for surface waters. PhD dissertation, University of Virginia

Bai S, Lung WS (2006) Three-dimensional modeling of fecal coliform in the Tidal Basin and Washington Channel, Washington, DC. J Environ Sci Health Part A 1327–1346.https://doi.org/10.1080/10934520600656984

Bai S, Lung WS (2005) Modeling sediment impact on the transport of fecal bacteria. WR 39:5232–5240

Bastviken D et al (2001) Similar bacterial growth on dissolved organic matter in anoxic and oxic lake water. Aquat Microb Ecol 24:41–49

Baudart J et al (2000) Salmonella spp. and fecal coliform loads in coastal waters from a point versus nonpoint source of pollution. J Environ Qual 29(1):241–250

Blumberg AF, Mellor GL (1987) A description of a three-dimensional coastal ocean circulation model. In: Heaps NS (ed) Three-dimensional coastal ocean models. Coastal Estuarine Sci 4:1–19

Bois FY et al (1997) Dynamic modeling of bacteria in a pilot drinking-water distribution system. WR 31(12):3146–3156

Borst M, Selvakumar A (2003) Particle-associated microorganisms in storm water runoff. WR 37:215–223

Brock RL et al (1985) Relationship of rainfall, river flow, and salinity to faecal coliform levels in a mussel fishery. NZ J Mar Freshwater Res 9(4):485–494

Buckley R et al (1998) Coliform bacteria in streambed sediments in a subtropical rainforest conservation reserve. WR 32(6):1852–1856

Burn DH (1987) Modeling coliform bacteria subject to chlorination. J Environ Eng 113(3):585–594

Burton GA et al (1987) Survival of pathogenic bacteria in various freshwater sediments. Appl Environ Microbiol 53(4):633–638

Camper AK et al (1996) Effect of growth conditions and substratum composition on the persistence of coliforms in mixed-population biofilms. Appl Environ Microbiol 62(11):4014–4018

Canale RP et al (1993) Modeling fecal coliform bacteria-II: model development and application. WR 27(4):703–714

Cerco C et al (2003) Eutrophication and pathogen abatement in the San Juan Bay Estuary. J Environ Eng 129(4):318–327

Chapra S (1997) Surface water quality modeling. McGraw Hill, NY

Connolly JP et al (1999) Modeling fate of pathogenic organisms in coastal waters of Oahu, Hawaii. J Environ Eng 125(5):398–406

Crabill C et al (1999) The impact of sediment fecal coliform reservoirs on seasonal water quality in Oak Creek, Arizona. WR 33(9):2163–2171

Curtis TP et al (1992) The effect of sunlight on faecal coliforms in ponds: implications for research and design. W Sci Technol 26(7–8):1729–1738

Curtis TP et al (1992) Influence of pH, oxygen, and humic substances on ability of sunlight to damage fecal coliforms in waste stabilization pond water. Appl Environ Microbiol 58(4):1335–1343

Davies CM et al (1995) Survival of fecal microorganisms in marine and freshwater sediments. Appl Environ Microbiol 61(5):1888–1896

Davies-Colley RJ et al (1994) Sunlight inactivation of enterococci and fecal coliforms in sewage effluent diluted in seawater. Appl Environ Microbiol 60(6):2049–2058

Doyle JD et al (1992) Instability of fecal coliform populations in waters and bottom sediments at recreational beaches in Arizona. WR 26(7):979–988

Emerick RW et al (2000) Modeling the inactivation of particle-associated coliform bacteria. Water Environ Res 72(4):432–438

EPA (1985) Rates, constants, and kinetics formulations in surface water quality modeling

Epstein SS, Shiari MP (1992) Rates of microbenthic and meiobenthic bacterivory in a temperate muddy tidal flat community. Appl Environ Microbiol 58(8):2426–2431

Escobar IC et al (2001) Bacterial growth in distribution systems: effect of assimilable organic carbon and biodegradable dissolved organic carbon. Environ Sci Technol 35:3442–3447

Feeney C (1998) Transport of fecal coliform in Barrington River Estuary. NEWEA J 32(2):170–195

Fernandez A et al (1992) Influence of pH on the elimination of fecal coliform bacteria in waste stabilization ponds. Water Air Soil Pollut 63:317–320

Finch GR, Smith DW (1986) Batch coagulation of a lagoon for fecal coliform reductions. WR 20(1):105–112

Floderun S et al (1999) Particle flux and properties affecting the fate of bacterial productivity in the benthic boundary layer at a mud-bottom site in south-central Gulf of Riga. J Marine Sys 23:233–250

Gannon JJ et al. (1983) Fecal coliform disappearance in a river impoundment. WR 17(11):1595–1601

Garvey E et al (1998) Coliform transport in a pristine reservoir: modeling and field studies. Water Sci Technol 37(2):137–144

Hampton C (1997) Determination of the volume of contaminated sediment in the Anacostia River of the District of Columbia, Technical Report Submitted to Government of the District of Columbia

Heldal M et al (1996) The elemental composition of bacteria: a signature of growth conditions. Mar Pollut Bulletin 33:3–9

Howell JM et al (1996) Effect of sediment particle size and temperature on fecal bacteria mortality rates and the fecal coliform/fecal streptococci ratio. J Environ Qual 25:1216–1220

Irvine KN, Pettibone GW (1993) dynamics of indicator bacteria populations in sediment and river water near a combined sewer outfall. Environ Technol 14:531–542

Jackson GA (1987) Simulating chemosensory responses of marine microorganisms. Limnol Oceanog 32(6):1253–1266

Jackson GA (1989) Simulation of bacterial attraction and adhesion to falling particles in an aquatic environment. Limnol Oceanog 34(3):514–530

Joyce TM et al (1996) Inactivation of fecal bacteria in drinking water by solar heating. Appl Environ Microbiol 62(2):399–402

Lechevallier MW et al (1996) Full-scale studies of factors related to coliform regrowth in drinking water. Appl Environ Microbiol 62(7):2201–2211

Lung WS, Bai S (2003) Fecal coliform and pH-alkalinity modeling of the tidal basin and Washington ship channel. Department of Civil & Environmental Engineering, University of Virginia Report submitted to the Department of Health, Washington, DC

Lung WS (1987) Lake acidification model: practical tool. J Environ Eng 113(4):900–915

Mancini JM (1978) Numerical estimates of coliform mortality rates under various conditions. J Water Poll Control Fed 50(11):2477–2484

Martinez J et al (1989) Estimation of *Escherichia coli* mortality in seawater by the decrease in H-label and electron transport system activity. Microb Ecol 17:219–225

Mayo AW (1995) Modeling coliform mortality in waste stabilization ponds. J Environ Eng 121(2):140–149

Mellor GL, Yamada T (1982) Development of a turbulence closure model for geophysical fluid problems. Rev Geophy Space Phy 20(4):851–875

Menon P et al (2003) Mortality rates of autochthonous and fecal bacteria in natural aquatic ecosystems. WR 37:4151–4158

Mills SW et al (1992) Efficiency of faecal bacterial removal in waste stabilization ponds in Kenya. Water Sci Technol 24(7–8):1739–1748

Milne DP et al (1986) Effects of sedimentation on removal of faecal coliform bacteria from effluents in estuarine water. WR 20(12):1493–1496

Muller EE et al (2001) The occurrence of *E. coli* O157:H7 in south African water sources intended for direct and indirect human consumption. WR 35(13):3085–3088

PBS&J (1999) Bacterial indicator study. Technical Report to Guadalupe Blanco River Authority and Texas Natural Resource Conservation Commission

Presser KA et al (1997) Modelling the growth rate of *E. Coli* as a function of temperature, pH, lactic acid concentration. Appl Environ Microbiol 63(6):2335–2360

Presser KA et al (1998) Modelling the growth limits (growth/no growth interface) of *E. coli* as a function of temperature, pH, lactic acid concentration, and water activity. Appl Environ Microbiol 64(5):1773–1779

Psaris PJ, Hendricks DW (1982) Fecal coliform densities in a western watershed. Water Air Soil Pollut 17(3):253–262

Qin D et al (1991) Bacterial (total coliform) die-off in maturation ponds. Water Sci Technol 23:1525–1534

Saqqar MM, Pescod MB (1991) Microbiological performance of multi-stage stabilization ponds for effluent use in agriculture. Water Sci Technol 23:1517–1524

Saqqar MM, Pescod MB (1992) Modeling coliform reduction in wastewater stabilization ponds. Water Sci Technol 26(7–8):1667–1677

Scarlatos PD (2001) Computer modeling of fecal coliform contamination of an urban estuarine system. Water Sci Technol 44(7):9–16

Schillinger JE, Gannon JJ (1982) Coliform attachment to suspended particles in Stormwater. Technical Report, University of Michigan

Schillinger JE, Gannon JJ (1985) Bacterial adsorption and suspended particles in urban stormwater. J Water Poll Control Fed 57(5):384–389

Smolarkiewicz PK, Grabowski WW (1990) The multidimensional positive definite advection transport algorithm: nonoscillatory option. J Comput Phys 86:355–375

Sperling MV (1999) Performance evaluation and mathematical modelling of coliform die-off in tropical and subtropical waste stabilization ponds. WR 33(6):1435–1448

Sperling MV (2002) Relationship between first-order decay coefficients in ponds, for plug flow, CSTR and dispersed flow regimes. Water Sci Technol 45(1):17–24

Steets BM, Holden PA (2003) A mechanistic model of runoff-associated fecal coliform fate and transport through a coastal lagoon. WR 37:589–608

Tetra Tech Inc. (2000) Theoretical and computational aspects of sediment transport in the EFDC model. Technical Report Prepared for USEPA

Thoe W (2010) A daily forecasting system of marine beach water quality in Hong Kong. PhD dissertation, University of Hong Kong

Thoe W et al (2012) Daily prediction of marine beach water quality in Hong Kong. J Hydro-Environ Res 6(3):164–180

Thoe W, Lee JHW (2013) Daily forecasting of Hong Kong beach water quality by multiple linear regression (MLR) models. J Environ Eng 140(2):04013007. https://doi.org/10.1061/(ASCE)EE.1943-7870.0000800

Thoe W et al. (2014) Predicting water quality at Santa Monica beach: evaluation of five different models for public notification of unsafe swimming conditions. WR 67:105–117

Thomann RV, Mueller JA (1987) Principles of surface water quality modeling and control. Harper & Row, NY

Uchrin CG, Weber WJ (1981) Modeling suspended solids and bacteria in Ford Lake. J Environ Eng 107(5):975–993

US EPA (1985) Rates, constants, and kinetics formulations in surface water quality modeling, 2nd ed Environmental Research Lab, Athens, GA

US EPA (2004) Decision rationale District of Columbia total maximum daily loads Tidal Basin and Washington Ship Channel for fecal coliform bacteria, December 15, 2004

US EPA (2014) Decision rationale 2014 *E. coli* bacteria allocations and daily loads for the Tidal Basin and Washington Ship Channel TMDL Revision, District of Columbia

US EPA (2013) Decision rationale 2014 *E. coli* bacteria allocations and daily loads for the Tidal Basin and Washington Ship Channel TMDL Revision, District of Columbia, Appendix B *E. coli* bacteria allocations and daily loads for the Tidal Basin and Washington Ship Channel, February 2013

Varnam AH, Evans MG (2000) Environ Microbiol. ASM Press, Washington, DC

Velinsky D et al (1994) Tidal river sediments in the Washington, D.C. area. 1. Distribution and sources of trace metals. Estuaries 17:305–320

Volk CJ, Lechevallier MW (1999) Impacts of the reduction of nutrient levels on bacterial water quality in distribution systems. Appl Environ Microbiol 65(11):4957–4966

Wilson JF Jr (1969) Movement of a solute in the Potomac River Estuary at Washington, D.C. at low inflow conditions. Technical Report Submitted to Department of Interior

Wilkinson J et al (1995) Modelling faecal coliform dynamics in streams and rivers. WR 29(3): 847–855

Xu P et al (2002) Non-steady-state modelling of faecal coliform removal in deep tertiary lagoons. WR 36:3074–3082

Chapter 7
Modeling Toxics and Emerging Chemicals

Thomann (1987) pointed out that toxic substances fate models, linear in nature, tend to be less complex than generally believed. This is an encouraging and welcoming statement for modeling toxics, endocrine disrupting chemicals (EDCs), and pharmaceutical and personal care products (PPCPs). However, modeling their fate and transport in the water systems is becoming an urgent task given the sharp increase in the number of new compounds and the wide spread use of these so-called emerging chemicals. The focus is on estrogens, an EDC most commonly found in domestic wastewaters but generally ignored by regulatory agencies to date. A unique characteristic of many EDCs is the sorption kinetics. The relatively slow process of reaching sorption equilibrium of triclosan in ambient waters renders the universal assumption of instantaneous equilibrium invalid in modeling. Significant concentration differences predicted for the dissolved triclosan due to the slow sorption kinetics are predicted for the Patuxent Estuary. Case studies on modeling metals in the Patuxent Estuary in Maryland and estrogens in the South River Watershed in Virginia are presented in this chapter.

7.1 Estrogens and Pharmaceuticals

The progression of water quality problems over the years from conventional pollutants to emerging chemicals started with the BOD/DO modeling of the Ohio river by Streeter and Phelps in 1925 to the receiving water modeling of TSS, pathogens, nutrients/chlorophyll *a* in the 1960s–1970s. While these water quality problems are still with us through the transition of metals and other toxics such as PCBs, additional attention is now on emerging chemicals, which are the collective term for anthropogenic pollutants that eventually find their way into natural waters, threatening both the aquatic life and clean water supplies. These degradation-resistant chemicals, which have likely gone unnoticed in the ecosystems for a long period, are now being detected at alarming levels in surface waters (Kolpin et al. 2002).

© The Author(s), under exclusive license to Springer Nature Switzerland AG 2022   231
W.-S. Lung, *Water Quality Modeling That Works*,
https://doi.org/10.1007/978-3-030-90483-8_7

Most noticeable are endocrine disrupting chemicals (EDCs) and pharmaceutical and personal care products (PPCPs) including consumer products used in everyday life, such as prescription and over-the-counter drugs (Anderson et al. 2004). By mimicking or blocking hormones that regulate bodily function, EDCs and PPCPs disrupt the normal operation of the endocrine system. The effect of these chemicals on aquatic life was reported as early as two decades ago. Their impact on human health is not exactly known, partly due to the length of time needed to establish conclusive evidence. Moreover, detection and lab analytical methods to measure EDCs and PPCPs in natural waters are still being developed. Nonetheless, because trace levels of these chemicals are found in many metropolitan water supplies, there is cause for concern (Caldwell et al. 2012). Because wastewater treatment facilities cannot fully degrade or remove many of these bioactive compounds, the residuals ultimately end up in ambient waters. Dose–response studies of EDCs, the majority of which have been carried out on fish and amphibians, demonstrate the induction of female characteristics in male fish following chronic and acute exposure. The chemicals have been found to affect biochemical processes in natural waters at concentrations on the order of one ng/L (Nano grams per liter) or parts per trillion (ppt).

To date, some of the major sources of EDCs and PPCPs include:

(a) Domestic wastewater treatment plants (even COVID-19 has been detected)
(b) Nonpoint sources such as cattle farms
(c) Concentrated animal feeding operations (CAFOs) such as swine and poultry farms.

They cover a wide range of chemicals: antibiotics and growth stimulants in livestock operations; prescription and over-the-counter drugs; and consumer products; to name a few. They are mostly degradation resistant, not easily removed (e.g. estrogens and their look-a-likes) at wastewater treatment plants.

Most EDCs and PPCPs are long molecules that have low volatility. Thus, they tend to stay in the aqueous phase. They have also been detected at significant levels in sediments (Juergens et al. 1999). Physical, biological, and chemical transformations occur simultaneously within these two environmental media (water and sediment). Adsorption and uptake by aquatic organisms are the major physical processes. Biological pathways occur under aerobic and anaerobic conditions. A study by Sarmah et al. (2007) demonstrated that the degradation of four EDCs (E2, EE2, BPA, and NP) in river sediment and groundwater aquifer material was over 90% over 2–4 days, but significantly lower thereafter. For their study on solar-induced transformations of EDCs, Lin and Reinhard (2005) used xenon lamp irradiation to simulate the photodegradation of EDCs (EE2, estriol, and E2) in river water. Determined half-lives ranged from 28 h for EE2 to 42 h for E2 (Zhao et al 2019).

7.2 Natural Attenuation of Pharmaceuticals

Pharmaceuticals exhibit different behaviors from each other in natural waters (Robinson 2017). The natural attenuation mechanisms of pharmaceuticals include photolysis, hydrolysis, biodegradation/biotransformation, and sorption (Snyder et al. 2007). Volatilization is not an effective elimination process for most pharmaceuticals because most of them are large molecules, thus exhibiting low Henry's constant (Gurr and Reinhard 2006). Generally, pharmaceuticals are resistant to hydrolysis (Nikolaou et al. 2007). Loffler et al. (2005) investigated the environmental fate of 10 select pharmaceuticals to find that abiotic transformation is unlikely for these pharmaceuticals. Tixier et al. (2003) incorporated overall (i.e., lumped) removal rates for diclofenac, naproxen, and ibuprofen in the water column. Their approach is based on certain degrees of parameterization by lumping processes such as sorption, biodegradation, biotransformation, and photodegradation into one overall attenuation rate. Lin et al. (2006) derived overall attenuation rates for a number of pharmaceuticals (Gemfibrozil, ibuprofen, and naproxen) and Rhodamine WT dye by measuring their concentrations and time of travel along the river. The rates range from 3.08 day^{-1} to 9.78 day^{-1} for these three pharmaceuticals compared with 3.78 day^{-1} for Rhodamine WT dye (Fig. 7.1).

In the aquatic environment, photodegradation may occur via two principal processes: direct and indirect photolysis/photosensitization. Direct photolysis

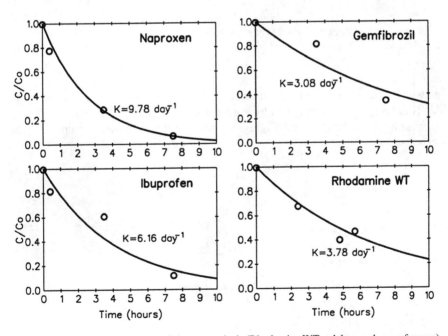

Fig. 7.1 Overall attenuation rates of pharmaceuticals (Rhodamine WT red dye used as a reference)

involves light absorption by the chemical itself, leading to bond cleavage. Indirect photolysis occurs via light absorption by photosensitizers, the most important of which are dissolved organic matter (DOM), nitrate, and nitrite (Hagg and Mill 1989). Excited photosensitizers generate singlet oxygen ($^1O_2$), OH radicals (•OH), DOM-derived peroxy radicals (ROO•), triplet-state DOM ($^3DOM^*$), solvated electrons (e_{aq}^-), and other photoreactants that can react with the compounds of interest. Further, individual processes such as sorption, biodegradation, and photodegradation, to name a few, are not fully understood for independent quantification now.

Independently quantifying the photodegradation rate in the water column, based on up-to-date scientific understanding of this process require incorporating light condition, nutrient levels, dissolved organic matters, etc. Further, lab studies by Lin et al. (2006) have demonstrated how to measure the photodegradation process in the water column. Our next step would be to transform their lab work into a prototype such as a river system to independently quantify the photodegradation rate of antibiotics (Sumpter et al 2006). To compliment this effort, the model will also simulate the environmental conditions referenced above, as well as other key water quality parameters. Subsequent model calculated pharmaceutical concentrations must be checked against the field data, thereby requiring field sampling and lab analysis.

The rate of photodegradation reaction may usually be expressed as a first-order reaction:

$$\frac{dC}{dt} = -k_d C \tag{7.1}$$

The reaction coefficient, k_d may then be expressed as

$$k_d = k_o I_a \tag{7.2}$$

in which k_o is a function of the quantum yield and I_a is the average available light.

The absorption of solar energy by water and its constituents affects the distribution of both light and temperature. Light attenuation in the river water column may be expressed as

$$I = I_o e^{-k_e z} \tag{7.3}$$

in which

I light intensity at depth, z
I_o light intensity at the surface, $z = 0$
k_e light extinction coefficient (m^{-1}).

The average value of light in the river water column is obtained by integrating over and dividing by the depth, H yielding

$$I_a = \frac{I_o}{k_e H}\left[1 - e^{-k_e H}\right] \tag{7.4}$$

If the magnitude of $k_e H \gg 1$, as in relatively deep turbid systems, which is representative of many natural waters, Eq. 7.4 reduces to

$$I_a = \frac{I_o}{k_e H} \tag{7.5}$$

Conversely, if $k_e H \ll 1$, as in clear shallow systems representative of laboratory conditions, Eq. 7.4 becomes

$$I_a = I_o \left[1 - \frac{k_e H}{2} \right] \tag{7.6}$$

and obviously approaches I_o as a limit. Mancini (1978) suggested that at low concentration levels, a negligible portion of the incident solar radiation is absorbed by the photoactive chemicals. Light extinction is closely affected by suspended solids concentrations in the water column. Further, for a wide range of conditions in natural water systems, particularly streams, estuaries and near-shore lakes and oceans, it has been demonstrated that the light extinction coefficient may be dominated by absorption versus scattering (O'Connor 1988).

An independent derivation of the coefficient k_o is derived using laboratory data. Substitution of Eqs. 7.2 and 7.4 into Eq. 7.1 and integration yields

$$C = C_o e^{\frac{-k_o I_o}{k_e H} \left[1 - e^{-k_e H} \right] t} \tag{7.7}$$

The overall photodegradation rate can be derived from Eq. 7.5 as

$$k_d = \frac{k_o I_o}{k_e H} \left[1 - e^{-k_e H} \right] \tag{7.8}$$

Using experimental data of picloram by Hedlund and Youngson (1972), Mancini (1978) developed an analysis framework to quantify appropriate values of k_e and $k_o I_o$ to be 0.85 m^{-1} and 0.229 day^{-1}, respectively. The calculated photodegradation rates for picloram are 0.072 day^{-1}, 0.116 day^{-1}, and 0.211 day^{-1} at solution depths of 3.65 m, 1.82 m, and 0.292 m, respectively (Fig. 7.2). Such a result incorporates the effects of k_e (light extinction), H (water depth), and k_o (quantum yield). For picloram k_o was calculated to be 1×10^{-4}/Langley by Mancini (1978).

Fig. 7.2 Derivation of
photodegradation rates

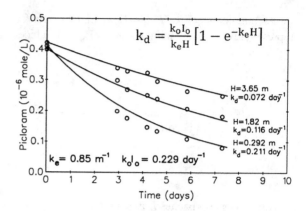

$$k_d = \frac{k_0 I_0}{k_e H}\left[1 - e^{-k_e H}\right]$$

7.3 Estrogens in Watersheds

7.3.1 The South River Watershed

The South River in Virginia has two major tributaries merging eventually to reach
the end of the watershed at USGS 01627500 (Fig. 7.3), making it convenient for
the hydrology calibration. In addition to cattle and poultry farm lands, there are
four municipal wastewater treatment plants (WWTPs): Stuarts Draft, Waynesboro,
Vesper View and Harrisonburg discharging directly into the South River (Fig. 7.6). A

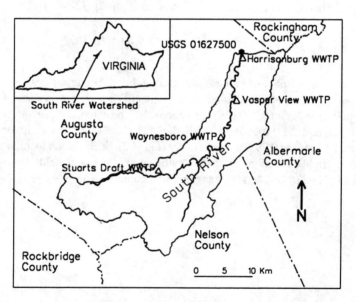

Fig. 7.3 The South River watershed in Virginia

modeling study was initiated to evaluate the fate and transport of a prevalent estrogen, 17β-estradiol (E2), in the South River watershed, Augusta County, Virginia (Zhao 2018). The purpose of this study was to evaluate the impacts of several factors such as E2 loads and flow rates on the E2 in-stream concentrations. The BASINS modeling framework was configured to track the fate and transport of E2.

National Land Cover Database (NLCD) 2011 data was obtained from the Multi-Resolution Land Characteristics Consortium (MRLC) to identify the land use of the study area. The dominant land use categories are forest and agricultural land, which account for 60.6% and 24.4% of the total area, respectively. Hourly precipitation, wind speed, temperature, dew point temperature and cloud cover data at Shenandoah Regional Airport (38°15'50.4" N, 78°53'45.6" W, 111.59 m), Charlottesville-Albemarle Airport (38°08'16.8" N, 78°27'10.8" W, 195.38 m) and Roanoke Regional Airport (37°18'57.6" N, 79°58'26.4" W, 1175 m) from 2010 to 2015 were obtained from the National Oceanic and Atmospheric Administration (NOAA). Hourly solar radiation, evaporation, and evapotranspiration data were calculated and disaggregated by WDMUtil. Cloud cover data was originally reported as clear, scattered, broken or overcast, and was converted to a scale ranging from 1 to 10 using the strategy by Perez et al. (2002). Daily average flow data from 2013 to 2015 at the USGS gage 01627500 was obtained from the USGS database. The cattle and poultry populations in the study area were obtained from Virginia's Animal Feeding Operations (AFO) database. The livestock populations were assumed to remain constant throughout this three-year simulation period. The observed and simulated average daily flow rates at USGS 01627500 from 2013 to 2015 are presented in Fig. 7.4, demonstrating that the hydrologic model is well calibrated by matching the measured flows.

Cattle manure, poultry litter, bio solids, septic systems, and domestic WWTPs are considered as major sources of E2 in this watershed. The fraction of time for cattle spent in feedlots or grazing of the South River watershed are derived from a report by VDEQ (2009). Direct E2 deposition onto pastureland during cattle grazing was quantified by multiplying the daily E2 production of cows by the time fraction that cows spent on pastureland. Direct E2 deposition into streams was estimated by multiplying the daily E2 production of cows by the time fraction that cows spent in streams and the faction of E2 that can be desorbed from cattle manure and released into streams, which was observed to be 18% (Andaluri et al. 2012). E2 loads from manure land application were estimated by multiplying the manure application rate by the E2 content in manure. Annual application rates of dairy cattle manure, beef cattle manure, and poultry litter are 2040, 2700 and 673 g/m^2-year to cropland, and are 1200, 2700 and 673 g/m^2-year to pastureland, with priority given to cropland (VDEQ 2009). Liquid dairy manure receives priority over poultry litter and poultry receives priority over solid cattle manure (VDEQ 2009). It was estimated that liquid dairy manure was applied to 7.01 km^2 of cropland, poultry litter was applied to 12.1 km^2 of cropland and 20.7 km^2 of pastureland, and solid beef manure was applied to 1.57 km^2 of pastureland. The daily E2 excretion from livestock and E2 loads from failed septic system and straight pipes in the study area are estimated by Zhao and Lung (2017). There are approximately 65,000 acres of permitted land application

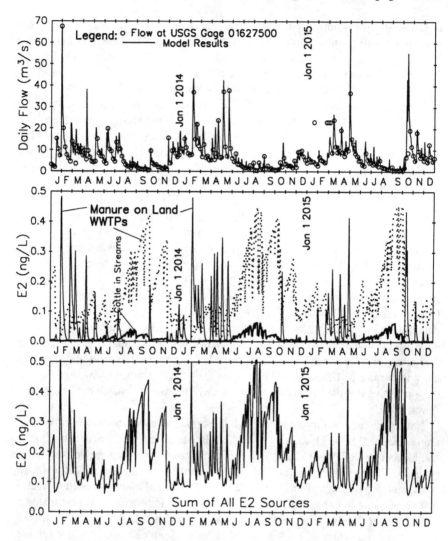

Fig. 7.4 E2 in the South River watershed (2013–2015)

sites for bio solids in Virginia. The identified permitted land area within the study area is 0.675 km^2 in sub-watershed 8 and 0.372 km^2 in sub-watershed 9 based on VDEQ records. The application rate of dry bio solids in Virginia was estimated to be 0.759 dry kg/m^2-year. The content of E2 in bio solids is quite low and the highest content is smaller than 0.48 ng/g dry weight (Yang et al. 2012). In this study, an E2 content of 0.48 ng/g dry wt was used. The daily E2 loads from bio solids were calculated by multiplying the application rate of dry biosolids by the E2 content in dry biosolids.

Table 7.1 E2 loads from WWTPs in the study area

WWTP	Flow (m³/d)	Treatment	Effluent E2 (ng/L)[a]	E2 loads (g/year)
Stuarts draft	15,100	Tertiary	4.6	25.4
Waynesboro	22,700	Tertiary	1.42	11.8
Vesper view	379	Secondary	10.6	1.47
Harrison	379	Secondary	15.2	2.10

[a]Servos et al. (2005)

Table 7.2 Total E2 loads to the South River watershed

Source	E2 loads (g/year)	Percentage
Direct loading to receiving waters		
Cattle	8.56	2.65
Straight pipes	0.162	0.0501
WWTPs	40.7	12.6
Loading to land surfaces		
Cattle	124	38.3
Poultry	149	46.0
Bio solids	0.381	0.118
Failed septic systems	1.11	0.343
Total	135	100

Compared with other sources, domestic WWTPs contribute a significant amount of E2 to the receiving water in the study area. E2 concentrations in WWTP effluents were estimated from the literature based on the treatment level. The WWTP flows in the study area were obtained from the Augusta County Service Authority (ACSA), and the City of Waynesboro government website. The estimated E2 loads from the four WWTPs are summarized in Table 7.1. The majority of the WWTP load is from the Stuarts Draft and Waynesboro plants. Table 7.2 summarizes the total E2 loads to the South River Watershed (Zhao and Lung 2017).

The hydrological calibration results are shown in the top panel of Fig. 7.5. The calculated flow rates at the outlet (USGS 01627500) for the period of 2013–2015 match the data closely, showing the typical temporal pattern of high flows in the spring and low flows during the summer months (Zhao and Lung 2018). The modeling results indicated that flow rate was a major input affecting the E2 levels in the water. When there was little precipitation, E2 accumulated on land where it was readily released to the receiving water. Conversely, during storm events, E2 on the land surface was transported to the rivers by the surface runoff and the E2 released into streams was diluted by the high water flow. Variations of E2 concentrations in the South River depended on the relative magnitudes of the point and nonpoint source loads. Modeling results showed that E2 levels in the South River were below the lowest observable effect level (LOEL) for fish. However, the practices of storing

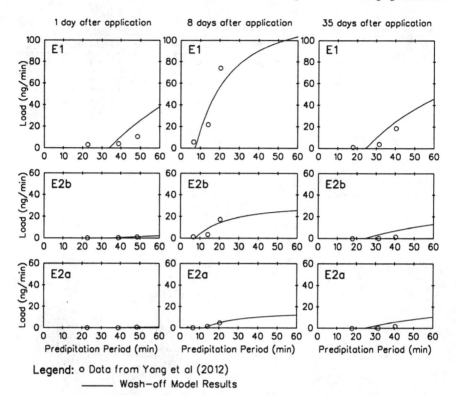

Legend: o Data from Yang et al (2012)
         ——— Wash—off Model Results

Fig. 7.5 The wash-off model results versus data

manure before land application and fencing off rivers to keep cattle out of the water are encouraged to prevent the potential for high E2 levels in streams receiving feedlot runoff.

This study also compared the relative significance of various E2 sources. The calculated E2 concentrations in the South River at the mouth (i.e. USGS 01627500) for the period of 2013–2015 are presented in Fig. 7.4, with the total concentration from all sources displayed in the bottom panel. Results of the component analysis are plotted in the mid panel of Fig. 7.4, showing the E2 concentrations from manure on land, WWTPs, and cattle in streams, respectively. Although Table 7.2 shows significant E2 from cattle and poultry on land, attenuation through the watershed significantly reduces their strength into the South River, yielding to the E2 loads from WWTPs as the leading contributor of E2 concentrations in the receiving water. The calculated E2 concentrations ranging from 0.060 to 0.51 ng/L in the South River are comparable to the measured values of field studies reported in the literature (Soto 2004; Bradley et al. 2009). Also note the temporal pattern of the E2 associated with manure on land follows the hydrologic pattern (in the top panel of Fig. 7.4), suggesting that nonpoint loads are strongly influenced by the runoff. The leading component of the E2 concentration is from the WWTPs as these loads bypass the

Fig. 7.6 Partition
coefficients for copper and
cadmium in the Patuxent
Estuary

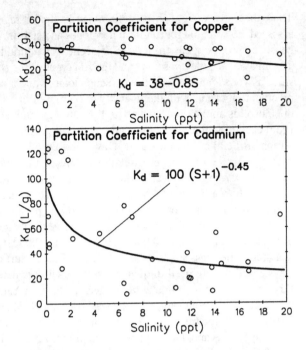

runoff to discharge directly into the South River. Although the WWTP loads are
much lower than those from cattle and poultry, they are the most influential loads to
the South River while loads from cattle and poultry undergo significant attenuation
on land. Cattles in stream follow the temporal pattern of the WWTP loads as their
influence stands out during dry weather conditions.

7.3.2 Tracking Estrogen Wash-Off Following Rainfall Events

Estrogens on cropland and pasture are transported into streams by surface runoff
during storm events (Jenkins et al. 2006; Yang et al. 2012; Zheng et al. 2012; Schoenborn et al. 2015). Estrogens transported by surface runoff is described by Butcher
(2003) and Hossain et al. (2010) as:

$$L_s = C_t\left(1 - e^{-k_w I}\right) \qquad (7.9)$$

where L_s is the amount of E2 transported into streams by runoff; C_t is the E2 mass
on land; k_w is the wash-off coefficient; and I is the surface runoff. Equation 7.9 can
be rewritten as:

$$L_n\left(1 - \frac{L_s}{C_t}\right) = -k_w I \qquad (7.10)$$

Thus, the value of k_w can be determined by linear regression. Yang et al. (2012) assessed the potential for runoff of hormones and sterols, including androgens, estrogens, and progestogens from three adjacent agricultural test plots (Plots 1, 2 and 3). The area of each plot was 6 m^2 and the soil types of the study area were mainly classified as Vona loamy sand and Vona sandy loam. Yang's study modeled four identical precipitation events with the intensity of 65 mm of one-hour duration 5 days before the biosolids application, as well as 1 day, 8 days and 35 days after the application. From each plot, the runoff rates and the hormone mass loads in surface water flow during artificial events on Day 1, Day 8 and Day 35 were measured and recorded. Although these artificial precipitations were of the same intensity, the measured surface flow rate greatly varied due to the variation of antecedent moisture condition (AMC). The variations of AMC on these three days were caused by the interference of the natural storm events and the effects of the previous artificial events. The data reported by Yang et al. (2012) was used to derive the value of the wash-off coefficient of E1, E2β, and E2α to be 0.00015, 0.00021, and 0.00016 min/L, respectively. From that study, the mass loads of E1, E2β, and E2α in surface water flow reported on Days 1 and 8 were used to quantify key model parameters, and the mass loads on Day 35 was used to validate the models. Data from Yang et al. (2012) are used to derive key coefficient values for the wash-off model (Shaw et al. 2009; Hossain et al. 2010) to track the transport of estrogens following rainfall events. The comparison of model results and data is summarized in Fig. 7.5 for E1, E2β, and E2α during one day, 8 days, and 35 days after application of estrogens, respectively. The wash-off model results match the data quite well (Zhao and Lung 2021).

7.4 Fate and Transport Modeling of Copper and Cadmium in the Patuxent Estuary

The role of toxic trace elements in the health of the Chesapeake Bay ecosystem has not been well defined but has always been an environmental concern. While toxicity of certain trace elements is apparent, the impact of contaminants when acting alone or in conjunction with other stressors is not completely understood. This lack of understanding prompted research studies which attempt to define actual concentrations and trends of certain contaminants in estuaries (Riedel et al. 2000) and other studies which investigate possible interaction with other stressors, like nutrients and low dissolved oxygen (Riedel and Sanders 2003; Riedel et al. 2003). One such study has led to the collection of data for a suite of trace elements for the Patuxent River and Estuary, including copper, cadmium, arsenic, nickel, lead, zinc, and mercury for 15 monitoring sites over approximately a two and half year period (Riedel et al. 2000). Such a database offered an opportunity to apply the eutrophication model in the Patuxent Estuary as developed in Chap. 4, with the addition of processes which are key to the fate of copper and cadmium in the estuary. The finely discretized water quality model of the Patuxent developed using the CE-QUAL-W2 modeling framework as presented in Chap. 4 (1993 segments, 2-D laterally averaged) has been expanded to simulate the fate and transport of copper and cadmium and includes mass transport, sorption with suspended solids (dissolved and particulate forms), settling/resuspension of particulates with suspended solids, and flux of dissolved metal from the sediment (Nice 2006).

A common theme among most studies concerning contaminants in rivers and estuaries is the importance of solid–liquid partitioning (considering particulate and dissolved forms) with suspended matter and the sediment (Chen et al. 1996; Ciffroy et al. 2003; Ji et al. 2002; O'Connor and Lung 1981; Pan et al. 1999; Sung 1995; Thouvenin et al. 1997; Turner 1996). Consequently, a sediment transport model was also incorporated into the proposed framework and is utilized in the calculation of solid–liquid partitioning for metals with suspended solids. While highly parameterized, the developed model adequately predicted observed trends in transport of suspended solids, including location of the turbidity maximum, usually occurring in the Patuxent just downstream of Nottingham (see map in Fig. 4.12). The resulting model for copper and cadmium was utilized to explore certain aspects of metal sorption to suspended solids, specifically the relationship to salinity and the so-called "particle concentration effect". Furthermore, the developed model for suspended solids and solid–liquid partitioning were important components incorporated into the arsenic model, as discussed later in this section. The use of a linear isotherm (assuming instantaneous equilibrium between the sorbed and dissolved portions of metals) to approximate partitioning is considered adequate for the low concentrations of copper and cadmium observed in the estuary and is consistent with water quality practice for contaminants (Thomann et al 1993; Chapra 1997; Thomann and Mueller 1987). The formulation using a distribution coefficient, K_d is the same as K_p for coliform bacteria in Chap. 6. Distribution coefficients in the Patuxent Estuary were calculated for both copper and cadmium using data collected for particulate and dissolved concentrations. The calibrated K_d values along the Patuxent Estuary are 22 L/kg–38 L/kg and 23 L/kg–94 L/kg for copper and cadmium, respectively (Fig. 7.6) by Nice (2006).

Spatial results for the copper model are shown in Fig. 7.7 in longitudinal concentration profiles along the Patuxent Estuary on nine dates during the period of 1995–1997. The left column shows model results compared with data in 1995 in terms of dissolved copper and total copper. TSS results are also presented to demonstrate the close relationship between TSS and total copper concentrations along the estuary. The solid line represents model results for dissolved copper concentrations, the dashed line for total copper, and the dotted line for TSS. Likewise, the circles represent data for dissolved copper concentrations and the triangles for total copper. Longitudinal profiles of dissolved copper, total copper, and TSS for 1996 and 1997 are presented in the middle and right columns of Fig. 7.7, respectively.

Vertical profiles of model calculated TSS concentrations at Broomes Island (at 24 km from the mouth of the Patuxent River) match the measured data closely (Nice 2006). More importantly, the turbidity maxima located in the middle portion of the estuary, just downstream of Nottingham (63 km from the mouth of the river) are well reproduced by the model.

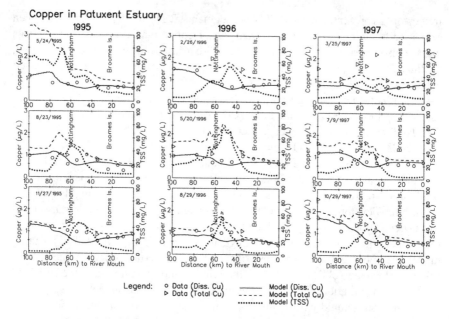

Fig. 7.7 Modeling dissolved copper, total copper, and TSS in the Patuxent Estuary (1995–1997)

As shown in the plots, dissolved concentrations are reproduced very well when compared to observed data. The general trend of slight increases in dissolved concentrations just past mid-estuary is reproduced by the model. In addition, the general trend of slight increases in dissolved copper concentrations just past mid-estuary is reproduced by the model. These figures also show that the solid–liquid partitioning model is meeting intuitive expectations, as TSS concentrations increase, the particulate fraction in the water column increases (total copper less dissolved copper). Temporal results for the copper model for eight monitoring stations within the estuary (excluding stations at upstream and downstream boundaries) compare well with the data (Nice 2006).

The same modeling framework are applied to simulate dissolved and particulate concentrations of cadmium in the Patuxent Estuary (Nice 2006). Concentrations of cadmium in the water column are influenced by mass transport, solid–liquid partitioning, settling/resuspension with solids, and accumulation and flux from the sediment. Like copper modeling, concentration of total cadmium in the sediment is calculated using long-term simulations starting with an initial concentration estimated from sediment core data for the Patuxent. Spatial results for the cadmium model are shown in Fig. 7.8 in the form of longitudinal concentration profiles for nine dates in 1995–1997. These plots are similar to the spatial profiles for copper (Fig. 7.7), however, salinity instead of TSS concentration profiles are shown in dotted lines. Model calculated concentrations of cadmium compare reasonably well with the observed data. Locations of total cadmium concentration peaks and the end of salinity intrusion match very well. Note that a local (about 42 km from the mouth)

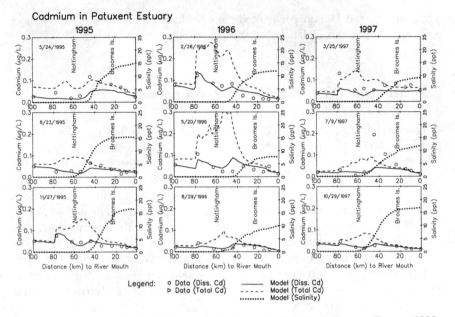

Fig. 7.8 Modeling dissolved cadmium, total cadmium, and salinity in the Patuxent Estuary (1995–1997)

increase in dissolved cadmium concentrations are thought to be due to desorption with the upward gradient of salinity (Nice 2006). One such dramatic increase in observed dissolved cadmium concentrations, which the model did not reproduced, occurred on July 9, 1997 at mid-estuary. This sizable increase either raises questions regarding the validity of the data point or whether some processes influencing cadmium are neglected.

Water quality models for copper and cadmium were developed for the Patuxent Estuary using hydrodynamic and mass transport calculations provided by the CE-QUAL-W2 modeling framework with the addition of new kinetic and equilibrium routines which are key to the fate and transport of the metals. Key processes simulated in the model include solid–liquid partitioning with suspended solids, settling/resuspension of particulates with suspended solids, and flux of dissolved metal from the sediment system. Since suspended solids concentrations are a vital part of the solid–liquid partitioning calculations, a sediment transport model is also incorporated (Nice 2006). Distribution coefficients, K_d derived from dissolved and particulate data from the Patuxent Estuary (Fig. 7.6) were used to quantify solid–liquid partitioning calculations for both copper and cadmium. Figure 7.6 shows that a slight relationship existed between salinity and K_d values for cadmium, while no such relationship could be established for copper. In any case, application of the empirical equation has been proven successful in reducing the level of parameterization required by the model by linking the K_d values to another constituent (salinity).

The resulting model for copper and cadmium produced very good results for copper as compared to observed data, and reasonable results for cadmium.

7.5   Modeling Arsenic in the Patuxent Estuary

Concerns of arsenic in the health of the Chesapeake Bay has prompted studies that attempt to define actual concentrations and trends of contamination in estuaries (Riedel et al. 2000). Figure 2.6 (in Chap. 2) shows the concentrations of arsenic species along the Patuxent Estuary from upstream to downstream in 1995–1997. The first observed characteristic is substantial amounts of methylated forms of arsenic in the lower estuary, believed to be caused by the uptake and methylation of arsenic by algal blooms (Riedel et al. 2000; Sanders and Riedel 1993). The other phenomena is a mid to lower estuary maximum of arsenic occurring in the summer, thought to be due to release of arsenic from the sediments during anoxic conditions in the bottom waters (Riedel et al. 2000). The collection of this substantial dataset for arsenic on the Patuxent provides another opportunity to apply the existing water quality modeling frameworks for copper and cadmium with the addition of key processes which are fundamental to the fate and transport of arsenic. The resulting model for arsenic was utilized to evaluate hypotheses proposed by some researchers regarding two behavioral characteristics of arsenic observed in the Patuxent Estuary. The modeling framework for copper and cadmium was refined to gain insight into these processes and evaluate the validity of theories concerning the behavior of arsenic in aquatic environments (Nice 2006).

Data collection studies have shown that the most common forms of arsenic in coastal and marine waters are arsenate, arsenite, methylarsonate (MMA), and dimethylarsenate (DMA) (Andreae 1978, 1979; Howard et al. 1982; Riedel 1993; Riedel et al. 2000; Sanders and Cibik 1985; Waslenchuk 1978). Arsenate, where As has an oxidation state of $+5$, is the predominant inorganic species found in coastal and marine environments. Arsenite, where As has an oxidation state of $+3$, is the inorganic form typically found in reducing conditions like anoxic bottom waters and interstitial pores of sediments. However, arsenite has also been found in coastal waters during aerobic conditions or conditions which are not thermodynamically appropriate, possibly the product of biological reduction (Francesconi and Edmonds 1994; Sanders et al. 1994). Arsenate has also been observed in hypoxic bottom waters at concentrations which are higher than those predicted by thermodynamic equilibrium (Smith and Butler 1990). Because of potentially slow rates of reduction of arsenate in hypoxic waters and biological oxidation of arsenite in aerobic waters, some researchers conclude that a kinetic approach to describing biogeochemical cycles of metalloids is more appropriate than a thermodynamic equilibrium approach (Cutter 1992). The methylated species of arsenic, MMA and DMA, can both be created biologically by phytoplankton and possibly from other microorganisms, although the exact processes are not currently well understood. The methylated forms are more resilient than arsenite and will degrade at slower rates, but exact rates of transformation to the inorganic arsenate or arsenite species have not been well established (Riedel 1993).

Four arsenic species are configured in the model: arsenate, arsenic, methylarsonate (MMA), and dimethylarsenate (DMA) for the Patuxent Estuary (Nice 2006). Adsorption and desorption of inorganic arsenic, arsenate and arsenite, is quantified based on the partition coefficients (K_d) and suspended solids concentrations. Particulate inorganic arsenic settles and is resuspended along with suspended solids. K_d values for total arsenic were estimated using particulate and dissolved data and are assigned to vary longitudinally along the estuary based on the collected data. Because only total arsenic (not different species of arsenic) was determined for the particulate data, K_d values assigned for arsenate and arsenite are identical. Another simplification, DMA and MMA are assumed to only exist in dissolved form and do not participate in adsorption (or settling/resuspension) with suspended solids. Arsenic loads from the watershed above the fall line and from the Western Branch are evaluated from available data. For inorganic arsenic, As(V) and As(III), total (dissolved and particulate) loading rates are calculated using site specific K_d values and suspended solids concentrations. Kinetic and equilibrium processes simulated by the model are shown in Fig. 7.9 (Nice 2006).

The sediment model for arsenic is based on a sediment flux model for phosphate proposed by Di Toro (2001). Sinks and sources of arsenic in the sediment include settling/resuspension, dispersive flux, particle mixing and burial. Also, in the aerobic layer, arsenite is assumed to be oxidized at a first-order rate to arsenate, while in

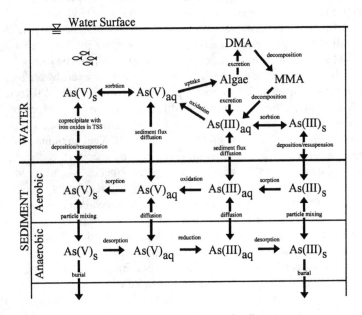

Fig. 7.9 Modeled processes of arsenic in water column and sediment

the anaerobic layer, arsenate is reduced to arsenite based on first-order kinetics. Flux is initiated from the aerobic layer by coupling distribution coefficients in the sediment with the level of dissolved oxygen. As dissolved oxygen falls below a critical level, distribution coefficients decrease, thereby increasing concentrations of dissolved arsenic in the pore water and initiating dispersive flux from the sediment to the overlying water. Again, as simplification, DMA and MMA are assumed not to participate in the sediment calculations. Parameters utilized in the final calibrated model for the water column and sediment component are summarized in Table 7.3.

Rates for oxidation of arsenite and demethylation of MMA and DMA were assigned based on those used by the Hellweger model and those reported in the literature. For model application to laboratory data generated for Lake Biwa, Hellweger et al. (2003) implemented an oxidation rate of 0.8 day^{-1} for arsenite and a demethylation rate of 0.02 day^{-1} for both MMA and DMA. For model application to field data for Lake Biwa, Hellweger used an oxidation rate of 0.2 day^{-1} for arsenite, a demethylation rate of 0.1 day^{-1} for MMA, and a demethylation rate of 0.02 day^{-1} for DMA (Hellweger and Lall 2004). Hellweger notes that the oxidation rates used for arsenite in his model are slightly higher than those reported in literature for abiotic rates (Hellweger et al. 2003). During calibration of the current model application for arsenic, an oxidation rate ($k_{o,AsIII}$) of 0.03 day^{-1} was assigned for arsenite and a demethylation rate ($k_{DM,MMA}$ and $k_{DM,DMA}$) of 0.02 day^{-1} was assigned for both MMA and DMA, which are all close to rates discussed in literature. Furthermore, since oxidation and demethylation could be bacteria mediated, a temperature coefficient (θ) of 1.047 was utilized, which is similar to values used for modeling carbonaceous deoxygenation (Bowie et al. 1985). The limitation constant ($k_{m,PO4}$) for uptake of phosphate by phytoplankton (also the inhibitor constant for uptake of arsenate) was set to 0.005 mg/L or 0.16 μmol/L. This value is well within the range of half-saturation constants for fixed stoichiometry, Michaelis-Menton growth formulations used for various modeling efforts (Bowie et al. 1985). The limitation constant ($k_{m,AsV}$) for uptake of arsenate by phytoplankton (also the inhibitor constant for uptake of phosphate) was also set to 0.16 μmol/L. For application of Hellweger's model to different datasets, the values of limitation/inhibitor constants vary widely, from 0.1 to 140 μmols/L for $k_{m,AsV}$ and from 0.90 to 16 μmols/L for $k_{m,AsV}$ (Hellweger et al. 2003). The maximum uptake rate of arsenate in units of μg As/mg C/day must also be specified. Since the model employs fixed nutrient stoichiometry phytoplankton kinetics, the maximum uptake rate ($V_{max,AsV}$) is linked to phytoplankton growth and represents micrograms of arsenate consumed for every milligram of phytoplankton carbon growth. This value differs from the uptake rate used in Hellweger's model where the uptake is based on phytoplankton carbon biomass rather than carbon growth. Likewise, values for uptake of arsenate derived from mesocosm experiments for the Patuxent Estuary are also reported on a per biomass basis (Sanders and Riedel 1993). Thus, because a value for this parameter cannot be directly borrowed from literature, the maximum uptake rate ($V_{max,AsV}$) is the primary variable adjusted during calibration, where a final value of 2.5 μg As/mg C/day was deemed to produce satisfactory results. While a direct comparison of this value to literature values is not possible, a comparison

Table 7.3 Rates and coefficients for arsenic model (Nice 2006)

Rate or coefficient	Value
Water column	
K_d values for arsenate and arsenite in estuary (L/g)	21–50
K_d value for arsenate and arsenite for tributary input at Western Branch (L/g)	49
K_d value for arsenate and arsenite for tributary input at Patuxent near Bowie (L/g)	21
Oxidation rate for arsenite in the water column, $k_{o,AsIII}$ (day^{-1})	0.03
Demethylation rate for MMA in the water column, $k_{DM,MMA}$ (day^{-1})	0.02
Demethylation rate for DMA in the water column, $k_{DM,DMA}$ (day^{-1})	0.02
Temperature correction coefficient for oxidation and demethylation	1.047
Maximum arsenate uptake rate, $V_{\max,AsV}$ (µg As/mg C/day)	2.5
Limitation constant for uptake of phosphate by phytoplankton, $k_{m,PO4}$ (µmol/L)	0.16
Limitation constant for uptake of As(V) by phytoplankton, $k_{m,AsV}$ (µmol/L)	0.16
Sediment	
Initial concentration of total arsenate in sediment (mg/L)	4.0
Initial concentration of total arsenite in sediment (mg/L)	0.8
Porosity	0.8
Active aerobic layer depth for arsenic flux model (cm) (top layer)	0.1
Active anaerobic layer depth for arsenic flux model (cm) (bottom layer)	9.9
Burial (only from aerobic layer to anaerobic layer) (cm/year)	0.25
Partition coefficient for arsenate and arsenite in anaerobic layer (L/kg)	100
Coefficient for partition coefficient for arsenate and arsenite in aerobic layer in upper estuary	150
Coefficient for partition coefficient for arsenate and arsenite in aerobic layer in lower estuary	300
Critical threshold concentration of dissolved oxygen which determines calculation of partition coefficient for arsenate and arsenite in aerobic layer (mg O_2/L)	2.0
Solids concentration in active sediment layers for arsenic flux model (kg/L)	0.52
Particle mixing DO half-saturation constant for arsenic flux model (mg O_2/L)	4.0
First-order decay rate for accumulated benthic stress for arsenic flux model (day^{-1})	0.03
Diffusion coefficient for particle mixing in arsenic flux model (cm^2/day)	1.2
Temperature correction coefficient for diffusion coefficient for particle mixing	1.117
Diffusion coefficient for dissolved phase mixing in arsenic flux model (cm^2/day)	5.0
Temperature correction coefficient for diffusion coefficient for dissolved phase mixing	1.08
Reference concentration of POC in reactivity class G1 for particle mixing calculation in arsenic flux model (mg C/g)	0.1
Oxidation rate for arsenite in porewater of aerobic sediment layer (day^{-1})	0.03
Reduction rate for arsenate in porewater of anaerobic sediment layer (day^{-1})	0.02

to uptake rates reported by other researchers on a per biomass basis is appropriate. In different applications of his model, which utilizes variable stoichiometry phytoplankton kinetics, Hellweger employed maximum uptake rates ranging from 0.004 to 5.9 μmol As/mg C/day, which converts to 0.3 to 440 μg As/mg C/day (Hellweger et al. 2003). Sanders and Riedel (1993) observed As(V) reduction rates ranging from 50×10^{-18} mol/cell/day to greater than 230×10^{-18} mol/cell/day during a phytoplankton bloom in the Patuxent Estuary (Sanders and Riedel 1993). Assuming that 3.2×10^{10} phytoplankton cells represent one gram of carbon (Hellweger et al. 2003), these transformation rates convert to 0.12 to greater than 0.55 μg As/mg C/day. These transformation rates are slightly lower than the maximum uptake rates presented by Hellweger and might be more representative of actual uptake rates after limitation factors have been applied.

Since sorption characteristics are thought to be similar between phosphate and arsenate, much of the parameters used in the sediment flux model for arsenic are the same as those used for the phosphate flux model presented in Chap. 4. Because of data availability and the role of phosphate in arsenate, arsenate uptake by phytoplankton, some revisions to the eutrophication model (Chap. 4) were implemented to improve model results, including solid–liquid partitioning for phosphate and corresponding linkage to the sediment transport model. Minor differences included changing the diffusion coefficient for particle mixing between layers from 0.6 to 1.2 cm^2/d, which is still within the range of values specified by Di Toro (2001). As with the copper and cadmium models, initial conditions for arsenate and arsenite in the sediment were assigned and long-term simulations were performed until near equilibrium conditions were reached in certain locations.

The arsenic model results are summarized in Fig. 7.10 for comparison with the available data of arsenate, arsenite, DMA and MMA in 1995, 1996, and 1997 respectively. In addition, model calculated TSS concentrations (in dotted lines) along the Patuxent Estuary are also presented. Spatial results of the calibrated model as compared to observed data are shown in Fig. 7.10 in longitudinal concentration profiles for the four forms of arsenic. Note that arsenic concentration scales (in μg/L) are the same for all species in the plots. Model calculated dissolved and total arsenate and arsenite are presented while only the dissolved arsenate and arsenite data are available for comparison with the model results. Turbidity maxima (dotted lines) and the peak total arsenate and arsenite concentrations (dashed lines) are closely related in these plots. Arsenate concentrations are generally higher than arsenite, DMA, and MMA levels in the Patuxent Estuary.

Overall, the model results reproduced concentrations of all four arsenic species reasonably well when compared to observed data. Most importantly, specific trends, which are evident in the observed data, are matched by the model. For instance, a maximum of arsenate in the lower estuary in the summer (August 23, 1995 and August 29, 1996) is shown in both the model results and the observed data. The general (rising) trend of DMA and MMA being present in the lower estuary is also reproduced by the model.

While proven useful for prediction of concentrations of four species of arsenic in the estuary, there is room for improvement. Currently, distribution coefficients for

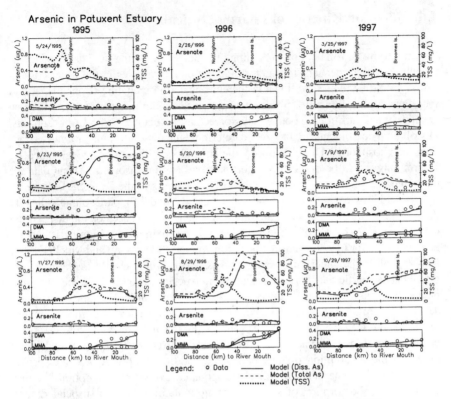

Fig. 7.10 Arsenic modeling results versus data in the Patuxent Estuary (1995–1997)

the water column and sediment are based on total arsenic, not on individual species. When data becomes available for the Patuxent, distribution coefficients should be verified and updated. Along the same vein, initial concentrations of arsenic in the sediment should be verified and updated when more data becomes available. In addition, the sediment component should be further evaluated, possibly conducting some localized simulations for a very small area, using new data when available. Additional data required to evaluate the sediment component might include particulate and dissolved concentrations of the different species of arsenic at different depths and corresponding concentrations in the overlying water column. Finally, as with the copper and cadmium models, since solid–liquid partitioning is an influential process in the fate of arsenic in estuaries, improvement of the sediment transport model and the prediction of suspended solids concentrations could improve results for arsenic.

7.6 Sorption Kinetics of Pharmaceuticals

Moving from eutrophication modeling, to metals modeling to pharmaceuticals modeling, a vital mechanism stands out: sorption by particles in the water column and sediment systems. As shown in the eutrophication modeling (Chap. 4), a vital mechanism influencing the fate and transport of phosphorus in the Patuxent Estuary is solid–liquid partitioning between dissolved and particulate forms of phosphorus. Distribution coefficients are utilized to determine the dissolved and particulate fractions of phosphorus. Distribution coefficients, k_d, for the model is estimated by determining the slope of the linear portion of a Langmuir isotherm, using data for dissolved and particulate concentrations of phosphorus in the Patuxent Estuary. In metals modeling of the Patuxent Estuary (Sects. 7.4 and 7.5), partitioning solid and dissolved fractions of copper, cadmium, and arsenic is also based on the distribution coefficient, K_d. The use of a linear isotherm to approximate partitioning is considered adequate for low concentrations of contaminants (Thomann and Mueller 1987; Chapra 1997). The underlying assumption associated with the splitting between the dissolved and particulate portions is that sorption occurs instantly. It implies instantaneous equilibrium is reached between the dissolved and particulate (sorbed) species. The validity of an instantaneous equilibrium has been verified with many contaminants for the majority of the conventional pollutants until recently with emerging chemicals such as pharmaceuticals. While most current water quality models are based on the assumption of instantaneous sorption equilibrium, many emerging chemicals (e.g. some pharmaceutical compounds) exhibit slow sorption kinetics. For example, the sorption of some emerging contaminants onto suspended sediments may take long times (in days) to reach equilibrium (Stein et al. 2008; Yu et al. 2004). Table 7.4 lists the equilibrium times for the sorption of some emerging chemicals onto different types of sorbents summarized by Liu (2012). In addition, the time to reach sorption equilibrium depends on sorbate concentrations. Yu et al. (2004) demonstrated that estrogenic chemicals exhibited slower sorption rate at concentrations below 1 μg/L. Therefore, kinetic models are necessary to quantify the phase distribution of these chemicals with slow sorption behavior.

Water quality models, which are based on the assumption of instantaneous sorption equilibrium, do not consider sorption kinetics, and would thereby leading to inaccurate predictions. Thus, the instantaneous-equilibrium based water quality

Table 7.4 Sorption equilibrium time of emerging chemicals

Chemicals	Sorbent	Sorption to equilibrium	References
β blockers	Sediment	6 h	Ramil et al. (2010)
Estradiol and testosterone	Soils	48–72 h	Sangsupan et al. (2006)
Atorvastatin	Soils/sediment	72 h	Ottmar et al. (2010)
Tetracycline	Humic acid	72 h	Gu et al. (2007)
Estrogens	Soils/sediment	10–14 days	Yu et al. (2004)

models cannot accurately capture the distribution between pharmaceuticals in the dissolved phase and the sorbed phase. Since chemicals in each phase (dissolved or sorbed) exhibit different environmental behaviors (e.g. settling and decomposition), incorporating sorption kinetics into water quality models could lead to significant improvement in predicting distributions of pharmaceuticals in natural water/sediment systems. Another key ingredient in quantifying solid-dissolved partitioning is the TSS levels in the water column and sediment. TSS modeling analyses presented in Sects. 4.5, 7.5, and 7.6 for phosphorus, copper, cadmium, and arsenic are crucial to the modeling of emerging chemicals.

7.6.1 Time to Sorption Equilibrium

Eutrophication and metals modeling of the Patuxent Estuary is easily extended to sorption kinetics modeling of pharmaceuticals since the estuary has been a proving ground for water quality management practices (Weller et al. 2003; Breitburg et al. 2003; Lung and Nice 2007; Nice et al. 2008). A sorption kinetics module and a sediment transport module were developed and linked to the existing hydrodynamic model (see Sects. 7.4 and 7.5) to simulate the pharmaceutical concentrations in the Patuxent Estuary (Liu et al. 2013). The sorption kinetics model developed for the Patuxent Estuary by Liu et al. (2013) is based on the theory of intraparticle pore diffusion (Wu and Gschwend 1986; Rugner et al. 1999; Fan et al. 2006). It has been reported that sorption onto sediment particles initially exhibits high sorption rates, followed by drastically reduced rates (Yu et al. 2004), suggesting multiple-domain features (Liu 2012) associated with varying sorption capacity and diffusion rates. The sorption kinetics data from Weber and Huang (1996) and Huang and Weber (1998) are used to calibrate the sorption kinetics module. The sediment transport module is based largely on Lung and Nice (2007) and Nice et al. (2008) for the Patuxent Estuary. A two-layer sediment compartment is configured: a 10-cm top active layer and an inactive bottom layer. The probability of deposition and resuspension depends on the ambient velocity and sediment particle size. Two critical velocities (0.2 and 0.07 m/sec) are assigned to determine when sediment deposition, transportation, and resuspension occur (Nice et al. 2008).

Figure 7.11 displays the temporal profiles of dissolved concentrations for two pharmaceuticals exhibiting "high" sorption coefficient ($K_d = 10,000$ L/kg) but with different times to reach sorption equilibrium ($t_e = 10$ days and 12 h, respectively). Results from four locations along the Patuxent Estuary are presented: one upstream (Point 1), two mid-estuary (Points 2 and 3), and one at lower estuary (point 4). The left hand column plots of Fig. 7.11 show the comparison of the calculated dissolved pharmaceutical concentrations between the traditional instantaneous equilibrium model (solid lines) and the sorption kinetics model taking ten days to reach sorption equilibrium (dotted lines) at these four locations. Comparisons between the instantaneous equilibrium model (solid lines) and the sorption kinetics model taking twelve hours to reach sorption equilibrium are shown in the right hand column of Fig. 7.11. From

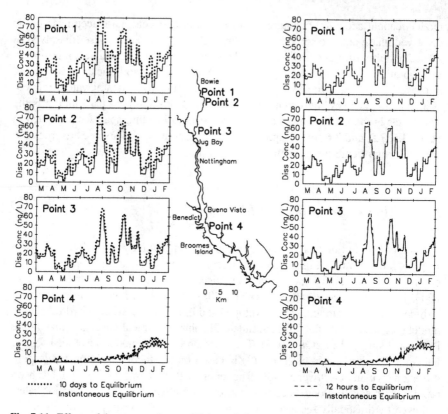

Fig. 7.11 Effects of time to sorption equilibrium of pharmaceuticals in the Patuxent Estuary

these data, the sorption kinetics scenarios predict higher dissolved-phase concentrations than the instantaneous equilibrium scenario, especially for the hypothetical chemical with slow sorption, and for both chemicals at upstream locations. For the chemical with fast sorption (i.e., $t_e = 12$ h), the two modeling approaches yield appreciably different results, although the differences are not as significant as for the chemicals with slow kinetics.

To further explore the model results, differences between the two scenarios are quantified using relative difference (RD) as follows:

$$RD(\%) = \frac{C_{sk} - C_{ie}}{C_{ie}} 100\%$$

where

C_{sk} calculated dissolved concentration from the sorption kinetics model.
C_{ie} calculated dissolved concentration from the instantaneous equilibrium model.

Positive *RD* values indicate that the instantaneous equilibrium model under-predicts the dissolved concentrations, and negative values indicate over-predictions by the

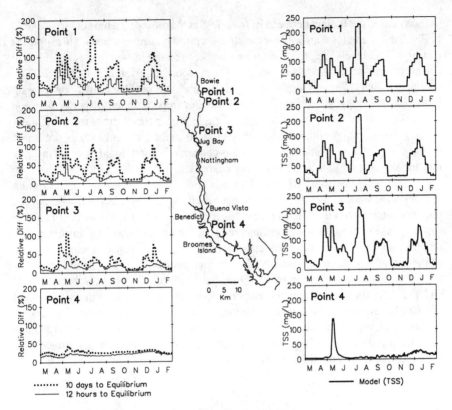

Fig. 7.12 Relative difference in dissolved pharmaceutical concentrations with sorption kinetics and TSS concentrations in the Patuxent Estuary

instantaneous model. Plots in the left hand column of Fig. 7.12 show the comparison of RD values between the two t_e values. At an upstream location 2.4 km below the upper boundary, RDs reach up to 150% for the chemicals showing slow sorption ($t_e = 10$ days) and 80% for the chemicals showing fast sorption ($t_e = 12$ h). In the lower estuary, RDs between the two scenarios decrease because the pharmaceuticals have had longer times to interact with suspended solids and sorption processes are approaching equilibrium. Yet, slight differences still exist. At the most downstream location (66.9 km below the upper boundary), RDs are roughly 25% and 20% for the slowly and quickly sorbed pharmaceuticals, respectively. Thus, the assumption of instantaneous sorption equilibrium, although it is widely used in current water quality models, results in significant underestimation of dissolved-phase concentrations, even for chemicals that reach sorption equilibrium quickly.

The right hand column plots of Fig. 7.12 present the model calculated temporal TSS concentrations at Points 1–4 in the Patuxent Estuary. Comparing plots of the two columns of Fig. 7.12 reveals the similarity in RD and TSS profiles, especially at Points 1 and 2 for pharmaceuticals with slow sorption ($t_e = 10$ days). Quantitatively, the correlation coefficient between RD and TSS at the most upstream station (Point 1) in Fig. 7.12 is greater than 0.98 for the slow sorption pharmaceuticals. This suggests that high TSS levels magnify the difference between the instantaneous equilibrium and sorption kinetics models for locations corresponding to short travel times from the upper estuary. There is less similarity in RD and TSS profiles further downstream in the estuary, suggesting that TSS has a smaller impact on the disparity between these two models for locations corresponding to longer travel times from the upper estuary. The spatial profiles of TSS concentrations in the Patuxent Estuary presented in Fig. 4.23 (in Chap. 4) and Fig. 7.10 of this chapter offer additional physical insights into the key role of TSS in the modeling analysis.

The similarity between RD and TSS profiles in the lower estuary is becoming less apparent since higher TSS at the upper estuary will drive a larger fraction of the pharmaceuticals into the sorbed phase. Since the transition from dissolved phase to sorbed phase does not proceed instantaneously, the inaccuracy of the instantaneous equilibrium assumption becomes clear, especially at the most upstream locations of the estuary. In the lower estuary, the pharmaceuticals have longer time to equilibrate with the TSS, thereby resulting in the decrease of correlation between RD and TSS. The above modeling analysis demonstrates that chemicals in dissolved and sorbed phases exhibit different environmental behaviors (e.g., settling and decomposition), incorporating sorption kinetics into water quality models could lead to significant improvement in predicted distributions of pharmaceuticals in receiving water/sediment systems (Liu 2012).

7.6.2   Sorption/Desorption Kinetics and Isotherms of TCS and ENR

Liu (2012) examined the sorption kinetics of Triclosan (TCS) and Enrofloxacin (ENR) in receiving waters. TCS has been used as an antimicrobial substance in many medical, consumer care, and everyday household products. It is added as a preservative or as an antiseptic in a wide range of products such as hand soaps, medical skin creams, toothpastes, mouthwash, household cleaners, and even textiles (e.g. bed linens and shoes) (Singer et al. 2002). TCS eventually reaches receiving water through WWTP effluents (Richart et al. 2010; Singer et al. 2002). Chalew and Halden (2009) reported that the concentration of TCS in WWTP effluents ranged from 0.027 to 2.7 μg/L, with a maximum concentration of 2.3 being detected in natural waters. Individual personal contributions to receiving waters can be significant. Singer et al. (2002) estimated a TCS load to a lake in Switzerland was 100 mg/day per 1000 persons. In the US, the frequency of TCS detection in untreated drinking water source is 8.1% (Focazio et al. 2008).

ENR is a fluoroquinolone antibiotic used in veterinary medicine (Lizondo et al. 1997; Boxall et al. 2003) most notably in swine and cattle farming (Sturini et al. 2009). In Europe, it was detected in domestic WWTP influents at 122–447 ng/L and effluents at 53.7–212 ng/L with significant seasonal variation (Saifrtova et al. 2008). In the US measured effluent ENR concentrations have been below 34 ng/L (Nakata et al. 2005) to 270 ng/L (Karthikeyan and Meyer 2006). ENR has also been detected in untreated and treated drinking waters: 6.8% of untreated drinking water sources in the US (Focazio et al. 2008); and 35 of treated drinking water sources in Ontario, Canada (Kleywegt et al. 2011). It affinity for sediments could make it significantly accumulated in bottom sediments. It is toxic to certain aquatic organisms.

The sorption kinetics of TCS and ENR were examined at three pH levels. Both chemicals initially show high sorption rates followed by significantly slow sorption kinetics (Fig. 7.13) with initial concentration of $C = C_T$. At pH = 6, ENR has a much lower equilibrium dissolved concentration than TCS. At ambient pH of 7.2 (close to the levels in natural waters), times to reach equilibrium are around 9 and 6 days for TCS and ENR, respectively. This result strongly suggests that instantaneous sorption equilibrium is not a valid assumption for these two pharmaceuticals. Low partition rates of these two chemicals may result from slow mass transfer rates of the chemicals into sediment particles. The mass transfer of sorbates in sediment particles is controlled by diffusion of sorbates in the porous particles, and was retarded

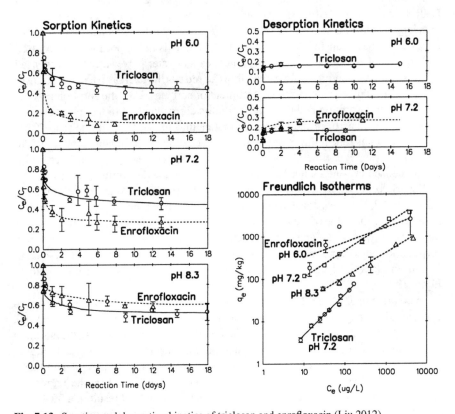

Fig. 7.13 Sorption and desorption kinetics of triclosan and enrofloxacin (Liu 2012)

by the properties of sediment particles and local sorption of sorbates on pore walls, thereby leading to less sorbate molecules available for diffusion (Liu 2012). Sorbate molecules cannot reach sorption sites of sorbents immediately because they need time to travel into sediment particles. The long equilibrium times derived from the experiment are reasonable since initial concentrations of both chemicals are much lower than their water solubility and are at sub-µg/L levels. Yu et al. (2004) pointed out that lower concentration of sorbates lead to longer time to reach sorption equilibrium.

Even though TCS is more hydrophobic than ENR, experimental results indicate that ENR exhibits stronger sorption than TCS, suggesting that the sorption mechanisms of these two chemicals are different from one another. The sorption kinetics models for TCS and ENR are best fitted with data in the left hand column of Fig. 7.13. The sorption of ENR (dotted line in Fig. 7.13) increases sharply at lower pH (pH 6), as the ENR becomes increasingly cationic (Liu 2012). At slightly alkaline conditions (pH 8.3), the sorption of ENR decreases significantly. Upon reaching equilibrium, only 40% of the ENR exists in the sorbed phase.

Liu (2012) conducted desorption kinetics experiments following the sorption processes have reached equilibrium. For TCS, desorption experiments were conducted for all three pH (6.0, 7.2, and 8.3) values while the desorption kinetics of ENR were only explored at pH 7.2 (Liu 2012). The experimental results are presented in the right hand column of Fig. 7.13, showing slow desorption kinetics for both chemicals due to mass transfer. As the amount of chemicals in the pore water decreased, sorbed chemicals detached from sorption sites and enter pore water to compensate for the decrease. Therefore, slow mass transfer retards the decrease of concentrations in pore water thus preventing the detachment of sorbates. Like the sorption process, relative high desorption rates are seen initially, corresponding to the relatively high mass transfer rates from labile organic matter into the aqueous phase (Liu 2012). These high rates reflect direct interaction between the labile organic matter and the aqueous phase. Over time, desorption rates begin to decrease, as molecules within the condensed organic matter began to participate in the desorption process.

For triclosan, the fractions existing in the dissolved-phase are quite similar at pH 6.0 and 7.2. This is consistent with the findings from the sorption experiments. Nevertheless, the value of the sorption coefficient at each pH derived from desorption kinetics experiment is higher than the value of the sorption coefficient derived from sorption kinetics experiment. This is indicative of sorption–desorption hysteresis. Desorption kinetics data and model fits are presented in the upper right panels of Fig. 7.13.

In contrast to TCS, ENR exhibited strongly nonlinear sorption. Within the tested concentration range (50–5000 µg/L), the nonlinearity ($1/n$) was 0.3811 at pH 6.0, 0.6163 at pH 7.2, and 0.56 for pH 8.3. For the nonlinear sorption, the sorption coefficient could be a function of sorbate-sorbent ratio ($C_{Sorbate}/C_{Sorbent}$). Higher sorbate-sorbent ratio leads to smaller fractions of sorbates being sorbed, and therefore results in lower sorption coefficients. Quantitatively, based on the regression equations, every 100% increase in sorbate-sorbent ratio would result in a 39% drop in sorption coefficient at pH 7.2. As mentioned previously and summarized in Table 7.5, a 117% increase was observed for the ENR sorption coefficient computed using

Table 7.5 Kinetics model results of triclosan and enrofloxacin (Liu 2012)

	pH	Diffusion coefficient (cm^2/s)	RMSE
Triclosan	6.0, 7.2, 8.3	3.0×10^{-8}	0.0458
Enrofloxacin	6.0	7.5×10^{-7}	0.0297
	7.2	9.0×10^{-7}	0.0331
	8.3	9.7×10^{-8}	0.0436
	7.2 (0.1 M buffer)	9.0×10^{-8}	0.0121

desorption versus sorption kinetics experiments. This may be largely due to a much lower sorbate-sorbent ratio used in the desorption experiments. The lower concentrations of sorbates used in the desorption experiments relative to the sorption experiments are likely responsible for the large majority (76.5%) of the observed increase in sorption coefficient. The remainder of the increase in K_d (22.9%) is likely due to sorption–desorption hysteresis (Table 7.6).

Isotherms of ENR fitted by Freundlich Equation at three pH values (6.0, 7.2, and 8.3) are shown in the lower right panel of Fig. 7.13. For TCS, a sorption isotherm was derived only at pH 7.2, because sorption kinetics experiments indicated that TCS sorption is less affected by pH over the range 6.0–8.3. For ENR, sorption isotherms were derived for pH of 6.0, 7.2, and 8.3. TCS exhibits generally linear sorption behavior within the tested range, from 20 to 400 μg/L. A Freundlich exponent (1/n) of 0.9484 indicates slightly nonlinear sorption. Thus, the sorption of TCS is not significantly dependent on sorbate concentrations. TCS exhibits relatively high hydrophobicity, based on its low water solubility (10 mg/L), its somewhat high molecular weight (300 g/mol), and its somewhat high octanol–water partitioning coefficient (log K_{oc} = 4.76). Therefore, the measurements are consistent with previous reports that "hydrophobic" compounds have shown linear sorption isotherms for equilibrium dissolved-phase concentrations less than 10^{-5} mol/L or less than one-half of their water solubility (de Maagd et al. 1998; Walters et al. 1989). Moreover, the TCS isotherm results also exhibit sorption coefficients that are consistent with those that were computed using kinetics experiments (Liu 2012).

Table 7.6 Sorption coefficient derived from kinetics experiments (Liu 2012)

	Triclosan			Enrofloxacin
pH	6.0	7.2	8.3	7.2 (0.2 M buffer)
K_d from sorption experiments (L/kg)	418 ± 80 (n = 33)	405 ± 76 (n = 20)	310 ± 119 (n = 9)	559 ± 42 (n = 5)
K_d from desorption experiments (L/kg)	534 ± 79 (n = 12)	503 ± 100 (n = 15)	–	1213 ± 208 (n = 6)

The sorption of pharmaceuticals affects their mobility, reactivity, and bioavailability in natural environments such as rivers, streams, lakes, and sediments, thus significantly influencing their environmental behaviors and exposure levels. Nonlinear sorption and sorption–desorption hysteresis would prohibit their release from the sorbed phase, likely causing them to be more persistent in the natural environment. The sorption and desorption experiments conducted under conditions close to natural environments (e.g. pH, low chemical concentrations, and natural sorbents) have generated results for use in fate and transport modeling of natural waters.

References

Andaluri G et al (2012) Occurrence of estrogen hormones in biosolids, animal manure and mushroom compost. Environ Monit Assess 184(2):1197–1205. https://doi.org/10.1007/s10661-011-2032-8

Anderson PD et al (2004) Screening analysis of human pharmaceutical compounds in US surface waters. Environ Sci Technol 38(3):838–849

Andreae MO (1978) Distribution and speciation of arsenic in natural waters and some marine algae. Deep Sea Res 25(4):291–402

Andreae MO (1979) Arsenic speciation in seawater and interstitial waters: the influence of biological-chemical interactions on the chemistry of a trace element. Limnol Oceanog 24(3):440–452

Bowie GL et al (1985) Rates, constants, and kinetics formulations in surface water quality modeling 2nd ed, EPA/600/3-85/040, Athens, GA

Boxall ABA et al (2003) Peer review: are veterinary medicines causing risk? Environ Sci Technol 37(15):286A-294A

Bradley PM et al (2009) Biodegradation of 17β-estradiol, estrone and testosterone in stream sediments. Environ Sci Technol 43(6):1902–1910

Breitburg DL et al (2003) The pattern and influence of low dissolved oxygen in the Patuxent River, a seasonally hypoxic estuary. Estuaries 26(2A):280–297

Butcher JB (2003) Buildup, washoff, and event mean concentrations. J AWRA 39(6):1521–1528. https://doi.org/10.1111/j.1752-1688.2003.tb04436.x

Caldwell DJ et al (2012) Predicted-no-effect concentrations for the steroid estrogens estrone, 17β-estradiol, estriol, and 17α-ethinylestradiol. Environ Toxicol Chem 31(6):1396–1406. https://doi.org/10.1002/etc.1825

Chalew TEA, Halden RU (2009) Environmental exposure of aquatic and terrestrial biota to triclosan and triclocarbon. J AWRA 45(1):4–13

Chapra SC (1997) Surface water quality modeling. McGraw-Hill, New York

Chen CW et al (1996) Modeling the fate and copper discharged to San Francisco Bay. J Environ Eng 122(10):924–934

Ciffroy P et al (2003) Kinetic partitioning of Co, Mn, Cs, Fe, Ag, Zn and Cd in fresh waters (Loire) mixed and brackish waters (Loire Estuary): experimental and modeling approaches. Marine Pollut Bull 46:626–641

Cutter GA (1992) Kinetic control on metalloid speciation in seawater. Marine Chem 40(1–2):65–80

de Maagd PGJ et al (1998) Sorption coefficients of polycyclic aromatic hydrocarbons for two lake sediments: influence of the bactericide sodium azide. Environ Toxicol Chem 17(10):1899–1907

Di Toro DM (2001) Sediment flux modeling. Wiley-Interscience, New York

Fan Z et al (2006) Fate and transport of1278-TCDD, 1378-TCDD, and 1478-TCDD in soil-water systems. Sci Total Environ 372:323–333

Focazio MJ et al (2008) A national reconnaissance for pharmaceuticals and other organic wastewater contaminants in the United States. II untreated drinking water sources. Sci Total Environ 402:201–216

Francesconi KA, Edmonds JS (1994) Biotransformation of arsenic in the marine environment. In: Nriagu JO (ed) Arsenic in the environ, part I: cycling and characterization. Wiley, New York

Gu C et al (2007) Complexation of the antibiotic tetracycline with humic acid. Environ Toxicol Chem 66:1494–1501

Gurr CJ, Reinhard M (2006) Harnessing natural attenuation of pharmaceuticals and hormones in rivers. Environ Sci Tech 40(9):2872–2876

Hagg WR, Mill T (1989) Hydrolysis kinetics of bis(2-chloroethoxy) methane. SRI Project 6877-1, Menlo Park, CA

Hedlund RT, Youngson CR (1972) The rate of photodecomposition of Picloram in aqueous systems. In: Faust SD (ed) Fate of organic pesticides in the aquatic environment: analyses in chemistry series III. American Chemical Society, Washington, DC, pp 159–172

Hellweger FL, Lall U (2004) Modeling the effect of algal dynamics on arsenic speciation in Lake Biwa. Environ Sci Technol 38(24):6716–6723

Hellweger FL et al (2003) Greedy algae reduce arsenate. Limnol Oceanogr 48(6):2275–2288

Hossain I et al (2010) Development of a catchment water quality model for continuous simulations of pollutants build-up and wash-off. Internat J Environ Chem Ecol Geologic and Geophy Eng 4(1):11–18. scholar.waset.org/1999.6/14989

Howard AG et al (1982) Seasonal variability of biological arsenic methylation in the estuary of the River Beaulieu. Marine Chem 11(5):493–498

Huang W, Weber WJ (1998) A distributed reactivity model for sorption by soils and sediments. 11. slow concentration-dependent sorption rates. Environ Sci Technol 32(22):3549–3555

Jenkins MB et al (2006) Fecal bacteria and sex hormones in soil and runoff from cropped watersheds amended with poultry litter. Sci Total Environ 358(1):164–177. https://doi.org/10.1016/j.scitotenv.2005.04.015

Ji ZG et al (2002) Sediment and metals modeling in shallow rivers. J Environ Eng 128(2):105–119

Juergens MD et al (1999) Fate and behavior of steroid estrogens in rivers: a scoping study. R&D Technical Report No.P161, Environment Agency, UK

Karthikeyan KG, Meyer MT (2006) Occurrence of antibiotics in wastewater treatment facilities in Wisconsin, USA. Sci Total Environ 361:196–207

Kleywegt S et al (2011) Pharmaceuticals, hormones and biphenyl A in untreated source and finished drinking water in Ontario, Canada—occurrence and treatment efficiency. Sci Total Environ 409:1481–1488

Kolpin et al (2002) Pharmaceuticals, hormones, and other organic wastewater contaminants in US streams, 1999–2000: a national reconnaissance. Environ Sci Technol 36(6):1202–1211. https://doi.org/10.1021/es011055j

Lin AYC, Reinhard M (2005) Photodegradation of common environmental pharmaceuticals and estrogens in river water. Environ Toxicol Chem 24(6):1303–1309

Lin AYC et al (2006) Natural attenuation of pharmaceuticals and alkyl phenol polyethoxylate metabolites during river transport: photochemical and biological transformation. Environ Toxicol Chem 25(6):1458–1464

Liu D (2012) Sorption of pharmaceuticals in natural waters. PhD dissertation, University of Virginia

Liu D et al (2013) Effects of sorption kinetics on the fate and transport of pharmaceuticals in estuaries. Chemosphere 92(8):1001–1009

Lizondo M et al (1997) physicochemical properties of Enrofloxacin. J Pharm Biomed Anal 15:1845–1849

Löffler D et al (2005) Environmental fate of pharmaceuticals in water/sediment systems. Environ Sci Technol 39(14):5209–5218

Lung WS, Nice AJ (2007) A eutrophication model for the Patuxent Estuary: advances in predictive capabilities. J Environ Eng 133(9):917–930

Mancini JL (1978) Analysis framework for photodecomposition in water. Environ Sci Technol 12(12):1274–1276

Nakata H et al (2005) Determination of fluoroquinolone antibiotics in wastewater effluents by liquid chromatography-mass spectrometry and fluorescence detection. Chemosphere 58:759–766

Nice AJ et al (2008) Modeling arsenic in the Patuxent Estuary. Environ Sci Technol 42(13):4804–4810

Nice AJ (2006) Developing a fate and transport model for arsenic in estuaries. PhD dissertation, University of Virginia

Nikolaou A et al (2007) Occurrence patterns of pharmaceuticals in water and wastewater environments. Anal Bioanal Chem 387(4):1225–1234

O'Connor DJ (1988) Models of sorptive toxic substances in freshwater systems: I, II,&III. J Environ Eng 114(3):517–574

O'Connor DJ, Lung WS (1981) Suspended solids analysis of estuarine systems. J Environ Eng 107(1):101–119

Ottmar KJ et al (2010) Sorption of statin pharmaceuticals to wastewater treatment biosolids, terrestrial soils, and freshwater sediment. J Environ Eng 136(3):25–264

Pan G et al (1999) Particle concentration effect and adsorption reversibility. Colloids Surf 151:127–133

Perez R et al (2002) Indicators of biases in reporting cloud cover at major airports. Sixth symposium on Integrated observing systems

Ramil M et al (2010) Fate of beta blockers in aquatic-sediment systems: sorption and biotransformation. Environ Sci Technol 44(3):962–970

Riedel GF (1993) The annual cycle of arsenic in temperate estuary. Estuaries 16(3A):533–540

Riedel GF, Sanders JG (2003) The interrelationships among trace element cycling, nutrient loading, and system complexity in estuaries: a mesocosm study. Estuaries 26(2A):339–351

Riedel GF et al (2000) Temporal and spatial patterns of trace elements in the Patuxent River: a whole watershed approach. Estuaries 23:521–535

Riedel GF et al (2003) Seasonal variability in response of estuarine phytoplankton communities to stress: linkages between toxic trace elements and nutrient enrichment. Estuaries 26(2A):323–338

Richart M et al (2010) Triclosan persistence through wastewater treatment plants and its potential toxic effects on river biofilms. Aquat Toxicol 100:346–353

Robinson JA (2017) Degradation and transformation of 17α-estradiol in water-sediment systems under controlled aerobic and anaerobic conditions. Environ Toxicol and Chem 36(3):621–629. https://doi.org/10.1002/etc.3383

Rügner H et al (1999) Long term sorption kinetics of phenanthrene in aquifer materials. Environ Sci Technol 33(4):1645–1651

Saifrtova M et al (2008) Determination of fluoroquinolone antibiotics in hospital and municipal wastewaters in Coimbra by liquid chromatography with a monolithic column and fluorescence detection. Anal Bioanal Chem 391(3):799–805

Sanders JG, Cibik SJ (1985) Adaptive behavior of euryhaline phytoplankton communities to arsenic stress. Marine Ecol Progr Ser 22(2):199–205

Sanders JG et al (1994) Arsenic cycling and its impact in estuarine and coastal marine ecosystems. In: Nriagu JO (ed) Arsenic in the environ, part I: cycling and characterization. Wiley, New York, pp 289–308

Sanders JG, Riedel GF (1993) Trace element transformation during the development of an estuarine algal bloom. Estuaries 16(3A)

Sangsupan HA et al (2006) Sorption and transport of 17 b-estradiol and testosterone in undisturbed soil columns. J Environ Qual 35(6):2261–2272

Sarmah AK et al (2007) Retention of estrogenic steroid hormones by selected New Zealand soils. Environ Int 34(6):749–755. https://doi.org/10.1016/j.envint.2007.12.017

Schoenborn A et al (2015) Estrogenic activity in drainage water: a field study on a Swiss cattle pasture. Environ Sci Europe 27(1):17. https://doi.org/10.1186/s12302-015-0047-4

Servos MR et al (2005) Distribution of estrogens, 17b-estradiol, and estrone in Canadian municipal wastewater treatment plants. Sci Total Environ 336(1–3):155–170

Shaw SB et al (2009) Evaluating urban pollutant buildup/wash-off models using a Madison, Wisconsin catchment. J Environ Eng 136(2):194–203

Singer H et al (2002) Triclosan: occurrence and fate of a widely used biocide in the aquatic environment: field measurements in wastewater treatment plants, surface waters, and lake sediments. Environ Sci Technol 36(23):4998–5004

Smith JD, Butler ECV (1990) Chemical properties of a low-oxygen water column in Port Hacking (Australia): arsenic, iodine and nutrients. Marine Chem 28:353–364

Snyder SA et al (2007) Biological and physical attenuation of endocrine disruptors and pharmaceuticals: implications for water reuse. Ground Water Monit Remediat 24:108–118

Soto AM (2004) Androgenic and estrogenic activity in water bodies receiving cattle feedlot effluent in Eastern Nebraska, USA. Environ Health Perspect 112(3):346. https://doi.org/10.1289/ehp.6590

Stein K et al (2008) Analysis and sorption of psychoactive drugs onto sediment. Environ Sci Technol 42(17):6415–6423

Sturini M et al (2009) Solid phase extraction and HPLC determination of fluoroquinolones in surface waters. J Sep Sci 32:3020–3028

Sumpter JP et al (2006) Modeling effects of mixtures of endocrine-disrupting chemicals at the river catchment scale. Environ Sci Technol 40:5478–5489

Sung W (1995) Some observations on surface partitioning of Cd, Cu, and Zn in estuaries. Environ Sci Technol 29(6):1303–1312

Thomann RV (1987) System analysis in water quality management—a 25 year retrospect. In Beck MB (ed) System Analysis in Water Quality Management, Pergamon Press, Tarrytown, NY, 1–14

Thomann RV et al (1993) Modeling cadmium fate at superfund site: impact of bioturbation. J Environ Eng 119(3):424–441

Thomann RV, Meuller JA (1987) Principles of surface water quality modeling and control. Harper & Row, New York

Thouvenin B et al (1997) Modeling of pollutant behavior in estuaries: application to cadmium in the Loire estuary. Marine Chem 58:147–161

Tixier C et al (2003) Occurrence and fate of carbamazepine, clofibric acid, dislofenac, ibuprofen, ketoprofen, and naproxen in surface waters. Environ Sci Technol 37(6):1061–1068

Turner A (1996) Trace-metal partitioning in estuaries: importance of salinity and particle concentration. Marine Chem 54:27–39

VDEQ (2009) Bacteria and benthic total maximum daily load development for the South River. Rqichmond, VA

Waslenchuk DG (1978) The budget and geochemistry of arsenic in a continental shelf environment. Marine Chem Anal 39–52

Waters RW et al (1989) Sorption of 2,3,7,8-tetrachlorodibenzo-p-dioxin from water by surface soils. Environ Sci Technol 23(4):480–484

Weber WJ, Huang WA (1996) A distributed reactivity model for sorption by soils and sediment. 4. Intraparticle heterogeneity and phase-distribution relationships under nonequilibrium conditions. Environ Sci Technol 30(4):881–888

Weller DE et al (2003) Effects of land-use change on nutrient discharges from the Patuxent River watershed. Estuaries 26(A):244–266

Wu SC, Gschwend PM (1986) Sorption kinetics of hydrophobic organic compounds to natural sediments and soils. Environ Sci Technol 20(2):717–725

Yang YY et al (2012) Steroid hormone runoff from agricultural test plots applied with municipal biosolids. Environ Sci Technol 46(5):2746–2754

Yu Z et al (2004) Sorption of steroid estrogens to soils and sediments. Environ Toxicol Chem 23(3):531–539

Zhao X, Lung WS (2017) Modeling the fate and transport of 17β-estradiol in the South River watershed in Virginia. Chemosphere 186:780–789

Zhao X, Lung WS (2018) Tracking the fate and transport of estrogens following rainfall events. Water Sci Technol 77(10):2474–2481. https://doi.org/10.2166/wst.2018.204

Zhao X, Lung WS (2021) Modeling temporal variation of estrogen levels due to interconversion. Water Air Soil Pollut 232:279. https://doi.org/10.1007/s11270-021-05201-4

Zhao X et al (2019) Attenuation, transport, and management of estrogens: a review. Chemosphere 230:462–478

Zhao X (2018) Developing a quantitative framework to track the fate and transport of estrogens on a watershed scale. PhD dissertation, University of Virginia

Zheng W et al (2012) Anaerobic transformation kinetics and mechanism of steroid estrogenic hormones in dairy lagoon water. Environ Sci Technol 46(10):5471–5478. https://doi.org/10.1021/es301551h

Chapter 8
Mixing Zone Modeling

Historically, water quality-based effluent limits are developed when technology-based effluent limits are not sufficient to meet water quality standards (see the Metro Plant WWTP upgrade with nitrification to meet the DO standard in the upper Mississippi River case in Chap. 2). One of the primary benefits from modeling for water quality management is translating water quality standards into effluent limitations for regulatory use. Work presented prior to this chapter falls into this category. Water quality-based effluent limits are developed from modeling of impaired water systems. Complete mixing of the wastewater flow with the full ambient water flow is assumed in these analyses to meet water quality standards in the receiving water body. The other extreme of regulatory action is meeting water quality standards at the end of the discharge pipe, i.e. not allowing any dilution with the ambient water. Except in an effluent-dominated water body, "complete mixing" does not, in practice, occur, i.e., there will be an area in the water body (i.e. less than 100% of the ambient flow is available for mixing) where water quality criteria are exceeded. In the U.S., the CWA allows mixing zones by delegating such a determination to state regulatory agencies.

Incomplete mixing assumes that there will be a discernible area within the receiving water where mixing occurs. State regulatory agencies would specify conditions on dilution allowances as part of their water quality standards although such a responsibility falls on the discharger (i.e. stakeholder) to demonstrate the dilution allowances, thereby calling for a mixing zone study in the permitting process. Mixing zone determination has become a routine task in implementing water quality standards in the past three decades, thereby requiring a modeling analysis to complete this process. This chapter presents the mixing zone modeling analysis supported by field data with applications to toxics, metals, color, heat, and endocrine disrupting chemicals (EDCs) as part of the regulatory process.

© The Author(s), under exclusive license to Springer Nature Switzerland AG 2022   265
W.-S. Lung, *Water Quality Modeling That Works*,
https://doi.org/10.1007/978-3-030-90483-8_8

8.1 Mixing Small Wastewater Flow with Receiving Water

In general, there are two stages of mixing of wastewater in the receiving water: discharge-induced and ambient-induced. The first stage of mixing is achieved by discharge jet momentum and buoyancy of the effluent. The best example is thermal discharge of cooling water from power generation plants, which have significant flow rates reaching the receiving water at very high velocities (significant momentum). It is particularly important in lakes, impoundments, and slow moving water bodies since ambient mixing in those systems is minimal. This stage generally covers most of the mixing zone allowed by state water quality standards. When the discharge flow encounters a boundary such as surface, bottom, or internal ambient density stratification layer, the near-field region ends and the transition to the far-field begins. In simple terms, the near-field region is typically the region that is controlled by the characteristics of the discharge itself (discharge flow rate, port diameter, etc.). Once the momentum flux from the effluent is exhausted in the receiving water, continuing dilution would come from ambient-induced mixing. Under this circumstance, the advective and dispersive mass transport in the receiving water would play a major role in providing additional dilution for the effluent. The discharge in the far field loses its "memory" of its initial conditions and mixing is now mainly a function of the ambient condition, characterized by the longitudinal advection of the mixed effluent by the ambient velocity. For the past three decades, the U.S. EPA and many state agencies have focused on mostly regulating mixing zones of small wastewater flows (as small as 1 mgd). A unique characteristics of small wastewater flows is lacking momentum-induced mixing. The following calculations demonstrate that the dilution provided by a small wastewater flow is insignificant.

In 1992, the Falling Creek WWTP in Virginia relocated its discharge point, resulting switching the receiving water from the small Falling Creek to the large James River. The jet momentum-dominated regime (or momentum-dominated near field, MDNF), where the jet is not yet bent over by the ambient current, can be defined as: $l_m = \frac{M_0^{0.5}}{u_a}$ (Lee and Chu 2003), where l_m is the distance from the discharge point to the edge of the momentum-induced dilution in ambient water, M_0 is momentum flux $= Q_0 u_0$ (discharge in open water) or $M_0 = 2 Q_0 u_0$ for river bank discharge, Q_0 = wastewater discharge flow, u_0 = wastewater discharge velocity, and u_a = ambient velocity. In the Falling Creek case, l_m is found to be 27 ft. The dilution ratio, s at the edge of MDNF (Lung 2001) is:

$$s = 0.17 \frac{M_0^{0.5} l_m}{2 Q_0} \tag{8.1}$$

The Falling Creek WWTP outfall has a single 4-ft diameter pipe carrying a flow rate, $Q_0 = 10$ mgd and $M_0 = 2 Q_0 u_0$ (river bank discharge). Equation 8.1 yields a dilution ratio, s to be $0.916 < 1$ (i.e. practically no dilution offered by momentum-induced mixing!). Near-field models based on momentum induced mixing are not suitable for small wastewater flow mixing zone calculations as the only available mixing mechanism is turbulence-induced mixing in the far-field.

8.2 Turbulence-Induced Mixing

Hamrick and Neilson (1989) presented a two-dimensional (2-D) steady-state advection-dispersion-reaction model using the following governing equation:

$$uh\frac{\partial C}{\partial x} = \frac{\partial}{\partial x}\left(hD_x\frac{\partial C}{\partial x}\right) + \frac{\partial}{\partial y}\left(hD_y\frac{\partial C}{\partial y}\right) - hk_d C \tag{8.2}$$

where: C is depth-averaged contaminant concentration; u is depth-averaged longitudinal stream velocity in the x-direction; h is average stream depth; x is longitudinal distance from discharge; y is transverse (across stream) distance; D_x is longitudinal dispersion coefficient; D_y is lateral dispersion coefficient, and k_d is a lumped first-order attenuation rate constant. Equation 8.2 has closed form analytical solutions only if the coefficients h, u, D_x, D_y and k_d are constant. Although this is seldom true for field conditions, it is possible to choose representative values, which will give reasonable results in the immediate vicinity of source sites as the sizes of the mixing zones are relatively small. The general solution for Eq. 8.2 presented by Hamrick and Neilson (1989) is:

$$C(x, y) = \frac{M}{\pi h\sqrt{D_x D_y}} \exp\left[\frac{ux}{2D_x}\right] \sum_{i=-\infty}^{i=\infty} K_o\left[\sqrt{1 + \frac{u^2}{4k_d D_x}}\sqrt{\frac{k_d x^2}{D_x} + \frac{k_d(y + 2iB)^2}{D_y}}\right] \tag{8.3}$$

where $K_o[\cdots]$ is the modified Bessel function of the second kind of order zero (Abramowitz and Stegun 1964), M is the contaminant load discharged from a bank, and B is the average width of the stream. The series in Eq. 8.3 may be truncated to the three leading terms ($i = -1, 0, 1$) for narrow to medium width streams (Hamrick and Neilson 1989). This formulation can also been configured for use in modeling mixing zones within streams and rivers exhibiting constant hydraulic geometry and mass transport coefficients (Hamrick and Neilson 1989; Lung 1995, 2001; Pagsuyoin et al. 2012).

For narrow to medium width, B, use $i = -1, 0, 1$. The solution then becomes

$$C(x, y) = \frac{M}{\pi h\sqrt{D_x D_y}} \exp\left[\frac{ux}{2D_x}\right]$$

$$\left\{K_o\left[\sqrt{1 + \frac{u^2}{4k_d D_x}}\sqrt{\frac{k_d x^2}{D_x} + \frac{k_d(y - 2B)^2}{D_y}}\right]\right.$$

$$+ K_o \left[\sqrt{1 + \frac{u^2}{4k_d D_x}} \sqrt{\frac{k_d x^2}{D_x} + \frac{k_d y^2}{D_y}} \right]$$

$$+ K_o \left[\sqrt{1 + \frac{u^2}{4k_d D_x}} \sqrt{\frac{k_d x^2}{D_x} + \frac{k_d (y + 2B)^2}{D_y}} \right] \Bigg\}$$ (8.4)

There are a number of simplifications of the above solution. If the receiving water channel is sufficiently wide, satisfying $B^2 \gg (D_y/k_d)$ or $B^2 > 300(D_y/k_d)$, the solution may truncated at the $i = 0$ term:

$$C(x, y) = \frac{M}{\pi h \sqrt{D_x D_y}} \exp\left[\frac{ux}{2D_x} \right] K_o \left[\sqrt{1 + \frac{u^2}{4k_d D_x}} \sqrt{\frac{k_d x^2}{D_x} + \frac{k_d y^2}{D_y}} \right]$$ (8.5)

Further, for conservative substances, $k_d = 0$, the solution truncated at $i = -1, 0, 1$ becomes:

$$C(x, y) = \frac{M}{\pi h \sqrt{D_x D_y}} \exp\left[\frac{ux}{2D_x} \right]$$

$$\left\{ K_o \left[\sqrt{\frac{u^2}{4D_x}} \sqrt{\frac{x^2}{D_x} + \frac{(y - 2B)^2}{D_y}} \right] + K_o \left[\sqrt{\frac{u^2}{4D_x}} \sqrt{\frac{x^2}{D_x} + \frac{y^2}{D_y}} \right] \right.$$

$$\left. + K_o \left[\sqrt{\frac{u^2}{4D_x}} \sqrt{\frac{x^2}{D_x} + \frac{(y + 2B)^2}{D_y}} \right] \right\}$$ (8.6)

Further simplification for a wide channel (i.e. large B), the solution is:

$$C(x, y) = \frac{M}{\pi h \sqrt{D_x D_y}} \exp\left[\frac{ux}{2D_x} \right] K_o \left[\sqrt{\frac{u^2}{4D_x}} \sqrt{\frac{x^2}{D_x} + \frac{y^2}{D_y}} \right]$$ (8.7)

The last equation is particularly suitable for relatively wide estuarine and tidal river systems.

For wide riverine systems with insignificant longitudinal dispersion (i.e. $D_x \approx 0$), Eq. 8.2 is reduced to:

$$uh \frac{\partial C}{\partial x} = \frac{\partial}{\partial y}\left(h D_y \frac{\partial C}{\partial y} \right) - h k_d C$$

The solution for large B is:

$$C(x, y) = \frac{M}{2ud\left(\pi D_y x/u\right)^{1/2}} \exp\left(\frac{-y^2 u}{4D_y x}\right) \exp^{-k_d \frac{x}{u}} \qquad (8.8)$$

Further simplification for a conservative substance (i.e. $k_d = 0$) yields:

$$C(x, y) = \frac{M}{2ud\left(\pi D_y x/u\right)^{1/2}} \exp\left(\frac{-y^2 u}{4D_y x}\right) \qquad (8.9)$$

which is the equation used by Fischer et al. (1979), Neely (1982), and more recently by Wu et al (2017) for turbulence induced mixing.

The following application demonstrates this simplified analysis in regulatory use. The VPDES permit for the Blacksburg/VPI WWTP in Virginia contained ammonia limitations based on a wasteload allocation (WLA) that was calculated by multiplying the acute water quality standard (1.35 mg/L) for ammonia by a factor of 2, a rather arbitrary determination. In short, the Virginia Department of Environmental Quality (VDEQ) assumed that the initial dilution immediately following discharge is 50%. This was their routine (i.e. somewhat arbitrary) practice where site-specific mixing characteristics were unknown during the early 1990s, resulting in an ammonia limit of 2.7 mg/L (= 1.35 mg/L × 2) at the end of the outfall pipe. To appeal and to demonstrate that additional initial dilution is available, the discharger submitted a mixing zone modeling analysis based on Eq. 8.3.

Key information on the wastewater and receiving water is listed as follows:

(a) Wastewater flow rate, $Q_o = 6$ mgd
(b) Cross-sectional area of discharge pipe, $A(D = 4.5 \text{ ft}) = 15.9 \text{ ft}^2$
(c) Discharge velocity at end of pipe, $u_o = Q_o/A = 0.584$ ft/s
(d) Momentum flux, $M_o = 2Q_o u_o = 10.83 \text{ ft}^4/\text{s}^2$
(e) Dilution ratio at the edge of allocated impact zone ($y = 28$ ft) using Eq. 8.1 = 0.85.

The dilution ratio is below 1.0 (i.e. no dilution), suggesting insignificant momentum in the vicinity near the outfall. Therefore, models based on momentum-induced mixing do not apply. This example represents the majority of the NPDES permit related mixing zone analyses in many states as regulatory agencies are focusing on small wastewater flow dischargers at the present time. Results from Eq. 8.3 are plotted in Fig. 8.1, showing a number of ammonia concentration contours ranging from 1.17 to 3.50 mg/L associated with an effluent ammonia concentration of 7 mg/L. The important point is that the 1.35 mg/L concentration contour is located well within the allocated impact zone (AIZ) of a semicircle with radius of 28 ft allowed by VDEQ. The maximum monthly average of ammonia levels may not exceed 7.0 mg/L, as that is the limit driven by the chronic ammonia standard. The mixing zone modeling calculation showed that the effluent ammonia level of 7 mg/L would yield ammonia levels in the AIZ below 1.35 mg/L (Fig. 8.1), thereby demonstrating significant dilution in the New River. Based on the modeling results, the Blacksburg/VPI received an effluent ammonia limit of **7** mg/L instead of 2.7 mg/L from VDEQ. This is a success

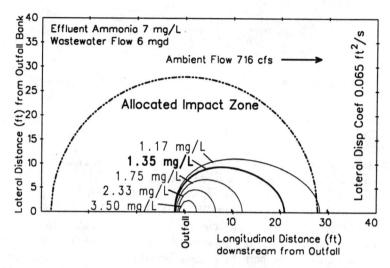

Fig. 8.1 Calculated ammonia contours in the mixing zone of Blacksburg/VPI WWTP effluent in New River

story that the ambient water quality standard is adopted for the end of the pipe effluent limit. Note that the ammonia concentration contours start slightly upstream of the outfall location ($x = 0$), indicating that a small longitudinal dispersion coefficient, D_x is factored in the calculation. In this calculation, ammonia is treated as a conservative substance (i.e. $k_d = 0$), an assumption justified for small spatial areas as in this case, yielding conservative results favored by the regulatory agency.

Applying Eq. 8.3 to a color mixing zone from a textile plant effluent on the LaGrange River in Georgia is presented in Fig. 8.2, showing the color levels of 64,

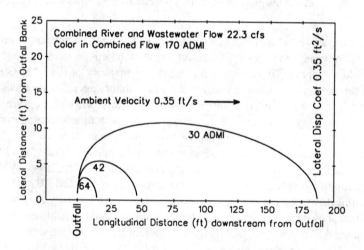

Fig. 8.2 Calculated color contours in the mixing zone in the LaGrange River

42, and 30 in ADMI (American Dye Manufacturers Institute) unit. Again, color is treated as a conservative substance in this analysis. Note that the discharge flow is quite small compared with the ambient river flow in this case.

8.3   Regulatory Terms: CMC, CCC, and WET

To control toxic discharges, U.S. EPA recommends an integrated approach to implementing water quality standards and developing WQBELs (water quality-based effluent limits). This approach includes three elements: a chemical-specific approach, a whole effluent toxicity (WET) approach, and a biological criteria or bioassessment approach. The chemical-specific approach uses the chemical-specific criteria for protection of aquatic life, human health, and wildlife adopted into a state's water quality standards. Chemical-specific WQBELs in permits involve a site-specific evaluation of the discharge and its effect upon the receiving water. The WET approach protects the receiving water quality from the aggregate toxic effect of a mixture of pollutants in the effluent. The WET approach is useful for complex effluents where it may be infeasible to identify and regulate all toxic pollutants in the discharge where chemical-specific pollutant limits are set, but synergistic effects are suspected to be problematic. There are two types of WET tests: acute and chronic. An acute toxicity test is usually conducted over a short time period (48–96 h) and the endpoint measured is mortality. A chronic toxicity test is usually conducted over a longer period (often greater than 28 days) and the endpoints measured are mortality and sub lethal effects, such as changes in reproduction and growth.

For toxic discharges, U.S. EPA (1991) recommends careful evaluation of mixing to prevent zones of chronic toxicity that extend for excessive distances because of poor mixing. Two water quality criteria for the allowable magnitude of toxic substances are used: a criterion maximum concentration (CMC) to protect against acute or lethal effects, and a criterion continuous concentration (CCC) to protect against chronic effects. In rivers or tidal rivers with a persistent through flow in the downstream direction that do not exhibit significant natural density stratification, hydrologically based flows 1Q10 (1-day, 10-year low flow) and 7Q10 (7-day, 10-year low flow) for the CMC and CCC, respectively, have been used traditionally in steady-state mixing zone modeling analysis (U.S EPA 1991). The acute criteria should be met at the edge of the zone of initial dilution (ZID) and the chronic criteria should be met at the edge of the overall (regulatory) mixing zone (Fig. 8.3).

For WET modeling, the percent effluent measurements must be converted to toxicity units that can be directly related to mass. When comparing toxicity among chemicals, the relationship between toxicity and concentration is inverse—chemicals that have toxic effects at low concentrations have a greater toxicity than chemicals that have toxic effects at higher concentrations. Modeling toxic effluents is based on mass balance principles; therefore, toxicity needs to be in units that increase when the percent of the effluent of the receiving water increases. Thus, a toxicity unit is the reciprocal of the dilution that produces the test endpoint, i.e., acute toxicity

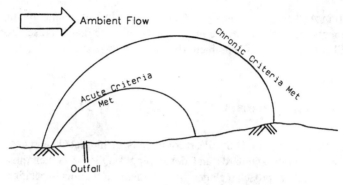

Fig. 8.3 Acute (CMC) and chronic (CCC) criteria for mixing zones

endpoint (ATE) or chronic toxicity endpoint (CTE). An acute toxicity unit (TU_a) is
the reciprocal of an ATE. A chronic toxic unit (TU_c) is the reciprocal of a CTE. The
acute toxicity unit is therefore defined as:

$$TU_a = \frac{100}{LC_{50}}$$

where LC_{50} is the percent effluent that causes 50% of the organisms to die by the end
of the acute exposure period. For example, an effluent that is found to have an LC_{50}
of 5% is an effluent containing 20 TU_a. U.S. EPA (1991) recommends 0.3 TU_a be
met at the edge of the ZID. Similarly, the chronic toxicity unit is defined as:

$$TU_c = \frac{100}{NOEC}$$

where $NOEC$ (no observable effect concentration) is the highest tested concentration of the effluent at which no adverse effects are observed and is expressed in
terms of percent effluent. The CCC for toxicity measured with chronic tests is recommended as the following: CCC = 1.0 TU_c (US EPA 1991) at the edge of the overall
regulatory mixing zone (Fig. 8.3).

8.4 Mixing Zones in Tidal Freshwater

Figure 8.4 shows the Falling Creek WWTP discharging into the James River outside
of Richmond, VA (continuing from the example of initial dilution calculation in
Sect. 8.1) in a tidal fresh portion of the river. A mixing zone modeling was conducted
for the wastewater flow of 10 mgd in 1992 (Lung 1995, 2001). When the discharger
requested an increase of flow to 12 mgd in their new permit application in 2014,
VDEQ requested an update on the mixing zone modeling analysis to address the

Fig. 8.4 Falling creek WWTP outfall and James River

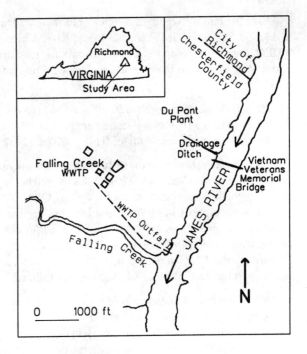

flow increase. The 0.916: 1 dilution (calculated in Sect. 8.1) due to the impact from the momentum of the discharge is so small and can be neglected. Therefore, mixing would rely solely on the turbulence in the ambient water (in this case, the James River) to provide dilution and to meet the CMC at the edge of the allocated impact zone, thereby offering a conservative assumption from the regulatory standpoint. Another assumption is $k_d = 0$ (no decay), treating WET in this case as a conservative substance. Equation 8.8 is therefore used in this analysis for the Falling Creek discharge into the James River Estuary. These assumptions form the basis of a simplified mixing zone modeling approach yielding conservative results for regulatory purpose. Combining this modeling analysis with federal and state regulations leads to the development of technical information to be submitted to regulatory agencies for permit action.

The General Criteria specified in Virginia Mixing Zone Regulations, 9 VAC 25-260-Virginia Water Quality Standards in January 2011, provides that mixing zones evaluated or established by the board in fresh water shall not:

(a) Prevent movement of or cause lethality to passing and drifting aquatic organisms through the water body in question;

(b) Constitute more than one half of the width of the receiving watercourse nor constitute more than one third of the area of any cross section of the receiving watercourse;

(c) Extend downstream at any time a distance more than five times the width of the receiving watercourse at the point of discharge.

Using the James River hydrographic information obtained from Virginia Institute of Marine Science (VIMS) for the James River Chlorophyll a modeling study by VDEQ, one can develop the size of the mixing zone downstream from the Falling Creek WWTP (called the overall regulatory mixing zone) based on the following data:

- One half of the river width (130 m) = 65 m (213 ft);
- One third of the river cross-section area (1032 m^2) = 344 m^2 (3700 ft^2);
- Five times of the river width (130 m) = 650 m (2132 ft).

Note that the near shore shallow (3-ft depth) area is not included in the above calculations, as the Falling Creek WWTP outfall is located below mean sea level (MSL). The overall regulatory mixing zone for the Falling Creek WWTP discharge is therefore sized as a rectangle of 142 ft (from the 1/3 cross-section area) by 2130 ft. According to the Federal guidelines set forth by the U.S. EPA (1991) for toxics, the CCC must be met within the *overall* regulatory mixing zone.

Further, the CMC must be met within the *allocated impact zone*, which is determined for the Falling Creek discharge as follows:

- 10% of the lateral distance from the edge of the outfall structure to the edge of the mixing zone = 14.2 ft (10% of 142 ft);
- a distance of 50 times the discharge length scale in any spatial direction = 177 ft (50 times 3.55 ft, which is the discharge pipe length scale for a 4-ft pipe);
- a distance of five times the local depth of the receiving water = 139 ft (5 times 27.9 ft).

Since the Falling Creek WWTP outfall is located in a tidal fresh portion of the James River, the total longitudinal length of the allocated impact zone can be quantified as twice of 139 ft, i.e. 278 ft. Based on the above calculation, a rectangle of 14.2 ft by 278 ft, centered at the discharge location, is determined as the allocated impact zone for the Falling Creek WWTP discharge.

The mixing zone modeling is conducted for low flows per U.S. EPA regulations. Historically, the 1Q10 flow is used to evaluate whether the CMC is met at the edge of the allocated impact zone and 7Q10 flow for CCC compliance at the edge of the overall regulatory mixing zone (US EPA 1991). Low flow frequency analyses of Virginia streams can be found in the literature (Virginia State Water Control Board 1979). Another low flow frequency study for Virginia streams by Austin et al. (2011) provides an update for the low flows, which are used for this study. The low flows for the mixing zone modeling analysis are based on the data collected at the USGS station of Cartersville, VA (02035000), where the river flow is not regulated (see Fig. 2.40 for the low flow frequency curves for this station). The unregulated flow at Cartersville is used to derive the flow at Richmond by a ratio (1.08) of drainage areas in the earlier study (Lung 1995). In addition, flows withdrawal from the James in the Richmond area for both municipal and industrial uses are returned to the river. This procedure was confirmed with VDEQ for this study. The low flows at the Cartersville gage reported in Austin et al. (2011) are listed in Table 8.1 along with the flow adjusted for the Richmond area for use in this mixing zone study. The 1Q10

Table 8.1 Low flows in the James River

Low flow	At Cartersville, VA[a]	For this study
1Q10 (cfs)	581	627
7Q10 (cfs)	665	707
30Q10 (cfs)	826	892

[a]From Austin et al. (2011)

low flow compares well with the 1Q10 low flow of 605 cfs used in the previous study (Lung 1995).

Next, the longitudinal and lateral dispersion coefficients for the study area need to be assigned. Results from a hydrodynamic model of the James Estuary by Hamrick (1992) yield the longitudinal and lateral dispersion coefficients for the study area: $D_x = 10$ ft^2/s and $D_y = 1.0$ ft^2/s, respectively. Another consideration in the longitudinal dispersion is the compensation from spatial averaging, particularly in the vertical direction, which is relatively small in the study area near the fall line of the James River. The ambient condition in the study area is therefore quite close to a riverine environment. As such, the second component contributing to longitudinal dispersion is also small in the study area. The lateral dispersion coefficient of 1.0 ft^2/s $(= 0.093$ m^2/s) selected for this analysis is consistent with some other values in rivers reported in the literature (Lung 2001).

Since an earlier study by Lung (1995) demonstrated that the CMC of 0.3 TU_a would be met at the edge of the allocated impact zone with an effluent concentration of 2.0 TU_a (i.e. LC_{50} of 50% effluent) and a plant flow of 10.1 mgd under the 1Q10 low flow in the James River, the first task was to check if this result is still valid for a plant flow of 12 mgd under the 1Q10 low flow of 627 cfs. The model calculated concentrations are presented in the top left panel in Fig. 8.5, showing a number of concentration contours of 0.1 TU_a, 0.15 TU_a, 0.2 TU_a, 0.3 TU_a, and 0.4 TU_a. It is clear that the 0.3 TU_a contour is located within the allocated impact zone (shown in the dashed line rectangle in Fig. 8.5), thereby meeting the CMC. The bottom left panel of Fig. 8.5 presents the CCC results, indicating that the 1 TU_c level can be met within the overall regulatory mixing zone under a 7Q10 low flow in the James River. It is clear that a 31 TU_c concentration in the effluent of 12 mgd would generate a 1 TU_c contour at the edge of the red overall regulatory mixing zone under the 7Q10 low flow of 707 cfs in the James River, thereby meeting the CCC. [The upper and lower boundaries of the regulatory mixing zone are located at -2132 ft and 2132 ft, respectively, per Virginia regulations.]

Effort was made to compute the dilution ratios under the 1Q10 and 7Q10 low flow conditions. The model results for 1Q10 are shown in the top right panel of Fig. 8.5. The maximum dilution ratio achieved within the allocated impact zone is 6.67 (under the 1Q10 low flow of 627 cfs it is approximately 34:1. Other dilution ratios such as 30:1, 25:1, 20:1, and 15:1 are also shown in Fig. 8.5). The dilution ratio contour inside the allocated impact is 6.67, corresponding to the 0.3 TU_a resulting from the effluent concentration of 2 TU_a (see the top left panel of Fig. 8.5). The calculated dilution ratios under the 7Q10 low flow condition are presented in the bottom right

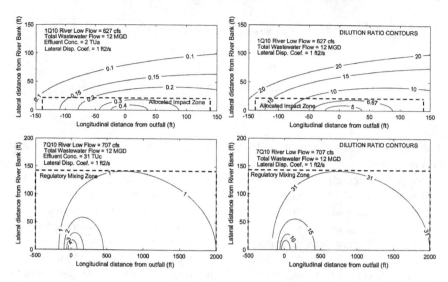

Fig. 8.5 Meeting CMC in allocated impact zone and CCC in regulatory mixing zone and dilution ratios of falling creek WWTP discharge in the James River

panel of Fig. 8.5, showing dilution ratios of 31:1, 15:1, and 10:1. Note that the 31:1 ratio is at the edge of the regulatory mixing zone, matching the results in the bottom left panel of Fig. 8.5. The dilution contours are useful to calculate concentrations of contaminants such as metals, residue chlorine, ammonia, etc. in the allocated impact zone and the overall regulatory mixing zone.

A comparison of the modeling results with those from the earlier study is summarized in Table 8.2, showing that the 2.0 TU_a WET level calculated in this study is substantially higher than the effluent WET limit of 0.6 TU_a issued by VDEQ in the current NPDES permit.

A mixing zone modeling study was performed for an industrial facility near the Falling Creek WWTP in Fig. 8.4 to support their NPDES permit renewal. This facility discharges its effluents into a drainage ditch about 75-ft long, which in turn enters the

Table 8.2 Summary of effluent WET limits and dilution ratios

	Wastewater flow (mgd)	Effluent WET (TU_a)	Dilution ratio	Effluent WET (TU_c)	Dilution ratio
The 1992 study	10.1	>2	>6.67:1		
2008 VDEQ worksheet	10.1	2	6.67:1	30	30:1
2014 VDEQ permit	12	0.6	2:1	30	30:1
Permit renewal study	12	2.0	6.67:1	31	31:1

tidal fresh portion of the James River Estuary (see map in Fig. 8.4). The discharge point from the standpoint of VDEQ permitting is where this drainage ditch meets the James River Estuary, the receiving waterbody. Again, mixing zone models developed for point source discharges are not applicable due to lack of momentum provided by the low ditch velocity when entering the James River. In addition, tidal influence due to the James River Estuary reaches the effluent drainage ditch, resulting in a complicated mass transport pattern. Following a consultation with the regulatory staff of VDEQ, a conservative approach of analysis was adopted. That is, initial dilution due to momentum mixing, which is insignificant, would be ignored. No credit would be given to the discharger for the momentum-induced mixing. Further, the effluent being modeled is considered as a conservative substance, given the fine spatial and temporal scales associated with the mixing zone. A mixing zone modeling study was conducted to quantify the dilution ratio in the regulatory mixing zone (RMZ). The dilution ratios were then used to determine whether the criterion maximum concentration (CMC) and criterion continuous concentration (CCC) could be met at the edges of the allocated impact zone (AIZ) and of the regulatory mixing zone (RMZ), respectively. Equation 8.3 was utilized for this analysis with the following data:

(a)   Effluent flow rate = 40 mgd
(b)   Effluent acute toxicity = 1.0 TU_a (LC_{50} = 100%)
(c)   Effluent chronic toxicity = 1.0 TU_c, meeting the limit at the end of discharge,
(d)   Ambient velocity in the river = 0.046 ft/s (1Q10 low flow condition) and 0.053 ft/s (7Q10 low flow condition)
(e)   Longitudinal dispersion coefficient, D_x = 10 ft^2/s
(f)   Lateral dispersion coefficient D_y = 1.0 ft^2/s
(g)   River depth = 25 ft derived from hydrographic charts.

In an earlier mixing zone modeling study of the Falling Creek treatment plant, located very close to this study area along the same receiving water, a longitudinal dispersion coefficient of 10 ft^2/s was used by Lung (1995). It should be pointed out that the VDEQ used a longitudinal dispersion coefficient of 325 ft^2/s (30.2 m^2/s) in their permit, a reasonable estimate for a one-dimensional tidal river without vertical and lateral concentration gradients in the open water when the effluent and ambient river is completely mixed. For a near shore mixing zone in a sizable river like the James, the smaller value of 10 ft^2/s is justified to yield conservative results. As indicated in the Falling Creek WWTP mixing zone study in the early part of this section, dilution ratios are useful in quantifying concentrations of water quality constituents in the mixing zones. Figure 8.6 presents the dilution ratio contours calculated in the RMZ for this discharge of 40 mgd and an effluent concentration of 1 mg/L. A dilution ratio of 10:1 is achieved within the RMZ. Table 8.3 presents the calculated concentrations of other contaminants at the edge of the allocated impact zone (AIZ) and regulatory mixing zone (RMZ) for comparison with the permit requirements for a plant effluent at 40 mgd.

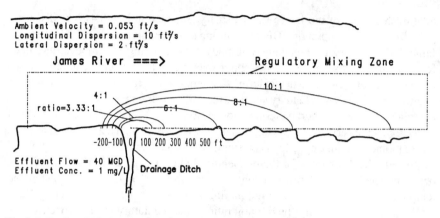

Fig. 8.6 Predicted dilution ratio contours in RMZ

8.5 Thermal Mixing Zones

With construction completion and commercial operation of Unit 6, a 550 MW combined cycle gas turbine unit in 2003, several new wastewater streams entered the exiting treatment pond system at the Possum Point power station of Virginia Dominion Power in Dumfries, VA. A low volume waste stream from this facility, regulated as outfall 004 under the VPDES permit, enters Quantico Creek near its confluence with the Potomac River (Fig. 8.7). One of these new wastewater streams, cooling-tower blow-down quench water, is thermally enriched. Because of this new thermal source, VDEQ expressed concern about the impact of the thermal loading from outfall 004 to the existing thermal mixing zone in Quantico Creek. In response to the VDEQ's concern, a thermal mixing zone modeling of the discharge was conducted to determine if the thermal loading posed a regulatory compliance risk.

The existing VPDES permit stipulates that temperature rise over the ambient should not be greater than 3°C within the regulatory mixing zone. Temperature monitoring to record the thermal impact of outfall 004 has been conducted since 1985 on an annual basis and it has become a semi-annual event since 2001. Historical data from the monitoring indicate that temperature rises over the ambient has been consistently below 1°C. Thus, the key question to be addressed in this study is whether the new thermally enriched water from outfall 004 would significantly increase this temperature rise beyond that level. That is, would the combined temperature rise still be below 3°C? Historical temperature measurements data at monitoring transects were analyzed to indicate that the highest temperature increase over the ambient through the two decades prior to this study is about 1.2°C. Therefore, a maximum of 1.8°C (= 3.0–1.2°C) would be the allowable temperature increase from the outfall 004 effluent.

Equation 8.3 was used to simulate the temperature rises in the mixing zone. Two key assumptions for the modeling analysis are:

Table 8.3 Summary of mixing zone modeling of plant effluent at 40 mgd

Water quality constituent	Effluent concentration	Calculated concentration at edge of AIZ[a]	Acute standard	Calculated concentration at edge of RMZ[b]	Chronic standard
Acute toxicity (TU$_a$)	1.0	0.16	0.30		
Chronic toxicity (TU$_c$)	1.0			0.10	1.0
Temperature (°C)	37.61	$\Delta T = 6.27$		$\Delta T = 3.76$	
Arsenic (μg/L)	3.0	0.5	360	0.30	190
Cadmium (μg/L)	0.48	0.080	2.50	0.048	0.83
Copper (μg/L)	5.0	0.833	12.154	0.50	8.397
Nickel (μg/L)	16.7	2.783	1010.671	1.67	112.356
Selenium (μg/L)	4.0	0.667	20	0.40	5
Silver (μg/L)	0.09	0.015	2.038	0.009	
Zinc (μg/L)	49	8.167	83.349	4.90	75.493
Chromium (μg/L)	1.2	0.20		0.12	
Triv. chromium (μg/L)	5.7	0.95	1251	0.57	149.1
Hex. chromium (μg/L)	6.0	1.00	16.0	0.60	11.0
Lead (μg/L)	1.0	0.167	49.037	0.100	1.911
Mercury (μg/L)	0.23	0.038	2.4	0.023	0.012
Ammonia (mg/L)	2.1	0.350	7.45	0.210	1.04
Residue chlorine (mg/L)	0.33	0.055		0.033	

[a] Allocated Impact Zone (AIZ) where acute standard must be met
[b] Regulatory Mixing Zone (RMZ) where chronic standard must be met.

1. Initial dilution of the thermal discharge due to momentum impact is neglected. The flow rate of outfall 004 is very small and delivers little momentum on impact to the receiving water.
2. Surface dissipation of heat in the ambient water is not included.

These assumptions form the basis of a conservative approach from a regulatory standpoint. Values of the key model coefficients such as D_x and D_y were obtained from the EPA's fully calibrated and verified Chesapeake Bay (including the Potomac River) hydrodynamic model in lieu of a field study. The advective and dispersive

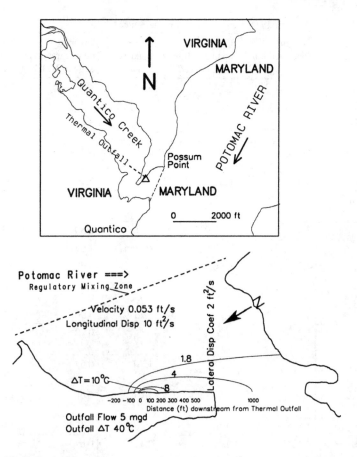

Fig. 8.7 Thermal mixing zone from power plant at Quantico, VA

patterns derived from that model was readily accepted by VDEQ [Note that the width and depth on the Potomac River in Fig. 8.7 is much bigger than the discharge site of the Falling Creek WWTP on the James River in Sect. 8.4]. Two model simulation scenarios were developed for the modeling analysis:

1. The worse-case thermal load from the effluent at outfall 004 with a flow rate of 3.5 mgd and temperature of 20°C above the ambient water (i.e., $\Delta T = 20°C$).
2. The extreme worse case thermal load scenario with a flow rate of 5.0 mgd and a ΔT of 40°C.

The station operation records indicate that the ΔT at the outfall has been rarely over 10°C. The flow rate of 3.5 mgd is also an extreme event. As such, both thermal load scenarios are on the conservative side. In this study, values of key model coefficients in Eq. 8.3 such as u, D_x, and D_y were obtained from the U.S. EPA's Chesapeake Bay hydrodynamic model. Another key parameter, the average water depth, was derived from depth profile of the hydrographic measurements. For this study, an average depth of 4 ft was used.

Results from the model run for the extreme worst-case scenario (effluent flow rate of 5.0 mgd and temperature over the ambient, ΔT of 40°C) are shown in Fig. 8.7. The 1.8°C contour of ΔT is well within the regulatory mixing zone. Model results indicate that the thermal impact of outfall 004 is insignificant due to the small temperature rises generated by the effluent. The modeling study demonstrated that there is no reasonable potential for the discharge from outfall 004 to cause or contribute to a violation of the water quality standards for temperature in the Potomac River. As a result, an effluent temperature limit was not assigned to the NPDES permit of outfall 004 by VDEQ.

8.6   Derivation of Dispersion Coefficients

The first step of the modeling analysis using Eqs. 8.7 and 8.9 is to quantify D_x and D_y. This can be accomplished with a variety of approaches for each parameter. Rutherford (1994) provided robust quantitative relationships between D_x and river flow and relationships between D_y and river flow, respectively. Examples of mixing zone modeling presented in earlier sections of this chapter show that the size of most regulatory mixing zones has the lateral dimension as the limiting factor, i.e. the longitudinal dimension of the mixing zones are sufficiently long that the longitudinal dispersion coefficient, D_y, is not a key factor as far as regulatory concerns.

One of commonly used equation to estimate D_y is by Fischer et al. (1979):

$$D_y = 0.25hu^* \qquad (8.10)$$

where h is average water depth and u^* is shear velocity $= \sqrt{ghs}$; g is gravitational acceleration; and s is slope of the river channel. The empirical coefficient, 0.25 could range from 0.15 to 0.6 (Socolofsky and Jirka 2005) depending on the site conditions. For example, Fischer et al. (1979) recommended 0.15 for a uniform straight channel and 0.6 for natural streams.

D_y can also be derived from field data of a conservative substance such as electrical conductivity by using Eq. 8.9. This is a convenient approach because most wastewaters have conductivity levels that are 4–5 times higher than the ambient levels (i.e., 400–500 µmho/cm in effluent versus 80–130 µmho/cm in ambient waters). Such a significant difference creates an appreciable, easily measurable spatial gradient of conductivity. Iterative model predictions can then be used to determine what value of D_y best reproduces the field measured conductivity. Figure 8.8 shows the conductivity measurements in the Banister River in Virginia following the discharge of an industrial wastewater. The top left plot in Fig. 8.8 presents the measured conductivity in the discharge mixing zone where the effluent conductivity level from a single outfall pipe is 730 µmho/cm and the background conductivity is 80 µmho/cm, thereby offering a good setting to calibrate D_y. The total wastewater flow is 0.044 mgd

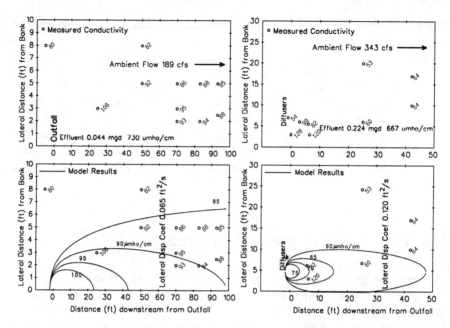

Fig. 8.8 Measured conductivity (μmho/cm) and model results of mixing zones in Banister River

and the ambient river flow is 189 cfs (5.38 cms). The ambient velocity near the bank is 0.30 ft/s (0.092 m/s). The river depth is 2 ft (0.61 m). Following a series of model runs using Eq. 8.9, the model results versus data are presented in the bottom left plot of Fig. 8.8 in which four isopleth conductivity contours (labelled at 85, 90, 95, and 100 μmho/cm) are displayed. The measured conductivity levels are shown with slanted letters. The model calculated conductivity contours resulted from a lateral dispersion coefficient of 0.065 ft^2/s (60.4 cm^2/s). Based on a river slope of 0.0003 measured in the field, Eq. 8.10 yields a D_y value of 0.069 ft^2/s, thereby substantiating the calibrated value.

The single outfall pipe was later replaced with a diffuser system of four ports. The top right plot of Fig. 8.8 displays the measured conductivity associated with one of the four ports. Other field measurements include the following:

Total wastewater flow = 0.224 mgd
Effluent conductivity = 667 μmho/cm
Ambient conductivity = 53 μmho/cm
River velocity near the bank = 0.50 ft/s (associated with a river flow of 343 cfs)
River depth = 3 ft.

While the total wastewater flow of 0.224 mgd is 5 times the single pipe discharger earlier, each diffuser port has a flow of 0.056 mgd (one quarter of 0.224 mgd). The river flow of 343 cfs is almost twice that of the earlier river flow of 189 cfs. These factors make a difference in the mixing zone calculations. Model results from

Eq. 8.9 are shown in isopleth contours of conductivity: 60, 65, 70, and 75 μmho/cm. Comparing these two data sets and model results shows that a dilution ratio about 8 to 1 (730 μmho/cm to 90 μmho/cm) is achieved within 100 ft downstream of the single pipe outfall while the diffuser system could reach a dilution ratio of 11:1 within 50 ft downstream from the discharge location. The diffuser system outperforms the single pipe outfall.

8.7 Limitation and Enhancement

Domestic WWTPs are known as an important source of estrogens releasing significant loads of endocrine disrupting chemicals (EDCs) to the receiving water. A field monitoring program of Moores Creek in Charlottesville, VA was launched to investigate the mixing zone of the effluent from this WWTP. Figure 8.9 shows conductivity and 17β-estradiol concentrations measured at 15 transects along the Moores Creek, following the WWTP discharge. The WWTP effluent conductivity is 435 μmho/cm and the background level is 150 μmho/cm while the 17β-estradiol levels in the effluent and background are 253 ng/L and 15 ng/L, respectively. Note that the WWTP flow of 0.355 cms is significantly higher than the upstream ambient river flow of 0.195 cms during the field survey. The conductivity data show small lateral concentration gradients across the stream due to the strong WWTP flow. That is, the mass is spreading from the outfall bank toward the opposite bank very slowly because the ambient river flow offers limited mixing. The lateral concentration gradients for 17β-estradiol are somewhat greater than the conductivity gradients due to attenuation while conductivity is considered a conservative substance without attenuation.

A number of assumptions in Eq. 8.3 are violated when applying the simplified 2-D steady-state mixing zone model to this site. First, the ambient river flow, which has a direct impact on the mixing zone size, is only indirectly incorporated into the calculations using a constant velocity. Second, during the survey the discharge flow dominates the receiving water flow. With the WWTP flow almost twice the upstream ambient flow, the impact of the wastewater flow is significant, thereby making the simplified analytical solution invalid for this mixing zone calculation. This is not a small wastewater flow problem. The strong WWTP flow generates a flow pattern, which is not accounted for in Eq. 8.3. A different modeling approach is needed for this situation (Lung 2001). The U.S. EPA WASP model was configured with a 2-D segmentation to form a mass transport model for the study area. Field measurements of velocities in Moores Creek were used to derive the advective flows between segments. The mass transport model would be first calibrated to back calculate the lateral dispersion coefficients using the conductivity measured in the field.

Results of the conductivity mass transport model are shown in the top half of Fig. 8.10 in dotted lines, matching the field data well at all transects in Moores Creek. As suggested earlier, the dispersion from the outfall bank to the other bank is slow, resulting in very mild lateral concentration gradients as shown by the data and the model results. Also presented in Fig. 8.10 are the model results (in solid lines)

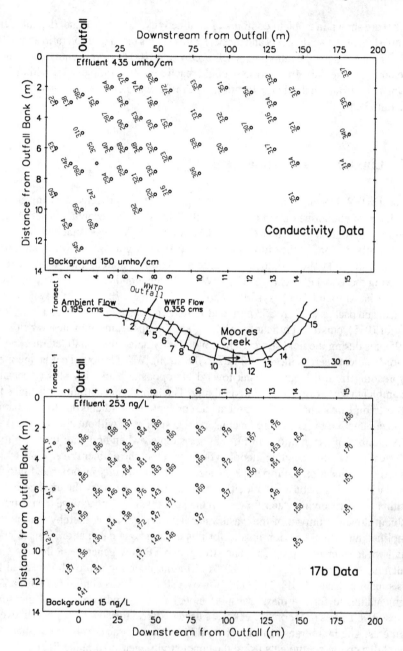

Fig. 8.9 Measured conductivity (μmho/cm) and 17β Estradiol (ng/L) in Moores creek following the WWTP discharge

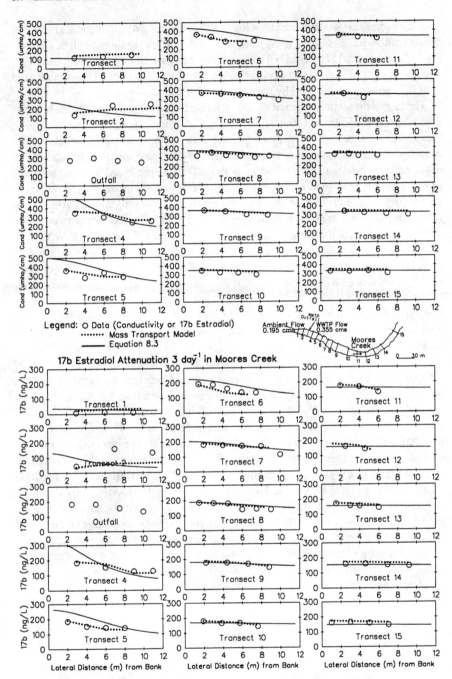

Fig. 8.10 Modeling conductivity and 17β Estradiol mixing zones in Moores creek

from the simplified analysis based on Eq. 8.3 for comparison with the mass transport model results. Results of the mass transport model match the data much better than Eq. 8.3 at transects 1, 2, 4, 5, and 6 (the area close to the outfall), clearly highlighting the deficiency of Eq. 8.3 in resolving the spatial pattern of conductivity in that area with flow patterns dominated by the strong WWTP flow.

The lateral dispersion coefficient, D_y calibrated from the conductivity mass transport model is 0.056 m^2/s (0.6 ft^2/s), which was also used in the simplified analysis, Eq. 8.3. This calibrated D_y was incorporated into the mass transport model of 17β-estradiol to back calculate the attenuation rate by matching the data. Results of the mass transport model and Eq. 8.3 along with the data are presented in the bottom half of Fig. 8.10. Like the conductivity results, the mass transport model results match the spatial pattern of the 17β-estradiol data very well in the vicinity of the outfall while results from Eq. 8.3 miss the data at transects 4, 5, and 6. Further downstream from the outfall both models match the data well as the influence of the strong WWTP flow gradually minimizes. The calibrated attenuation rate, k_d of 17β-estradiol is 3 day^{-1}, comparable to the value found in the Redwood River in Minnesota by Writer et al. (2012). The above physical insights into the modeling results reveal the deficiency of the simplified analysis of using Eq. 8.3, which is more appropriate for wastewater flows significantly smaller than the ambient river flows. The mass transport modeling approach works well regardless of the relative magnitudes of outfall versus ambient flows.

The lateral dispersion coefficient, D_y in Eq. 8.3 characterizing the spread rate of water quality constituents across the river is assumed constant in space, which may not be totally valid, particularly along a sizable river. This allows Eq. 8.3 to be solved analytically and greatly facilitates problem solving for a straight, uniform channel in which the lateral dispersion coefficient is unlikely to vary significantly along the channel (Rutherford 1994). Recently, Wu et al. (2017) presented an analysis of quantifying the increase of the lateral dispersion coefficient in the longitudinal direction. The study yielded the following equation: $D_y = 0.0114x^{1.67}$ using field data from the Yangtze River near Huangshaxi, China. The exponent 1.67 is calling for an increase of D_y in the downstream direction from the source. It should be pointed out that the Yangtze River is a sizable river, significantly larger than any of the receiving water systems presented in this Chapter. On the other hand, using a spatially constant lateral dispersion for the modeling analysis of small wastewater flows would retard spreading, thereby offering conservative results acceptable by regulatory agencies for mixing zone permitting exercise.

References

Abramowitz M, Stegun I (1964) Handbook of mathematical functions with Formulas, graphs, and mathematical tables. U.S. National Bureau of Standards, Washington, DC

Austin SH et al (2011) Low-flow characteristics of Virginia streams. USGS scientific investigation report 2011-5143. http://pubs.usgs.gov/sir/2011/5143/

Fischer HB et al (1979) Mixing in inland and coastal waters. Academic Press, London

Hamrick JM (1992) A three-dimensional environmental fluid dynamics computer code: theoretical and computational aspects. Virginia Institute of Marine Science, Gloucester Point, VA

Hamrick JM, Neilson BJ (1989) Determination of marina buffer zones using simple mixing and transport models. Virginia Institute of Marine Science, Gloucester Point, VA

Lee JHW, Chu V (2003) Turbulent jets and plumes-a Lagrangian approach. Springer

Lung WS (1995) Mixing zone modeling for toxic waste load allocations. J Environ Eng 121(11):839–842

Lung WS (2001) Water quality modeling for wasteload allocations and TMDLs. Wiley, New York

Neely WB (1982) The definition and use of mixing zones. Environ Sci Technol 16:518–521

Pagsuyoin SA et al (2012) Predicting EDC concentrations in a river mixing zone. Chemosphere 87:1111–1118

Rutherford JC (1994) River mixing. Wiley, Chichester, England

Socolofsky SA, Jirka GH (2005) Special topics in mixing and transport processes in the environment, 5th edn. Texas A & M University, College Station, TX

US EPA (1991) Technical support document for water quality-based toxics control. EPA/505/2-90-001:77-78

Virginia State Water Control Board (1979) Flow characteristics of Virginia streams, South Atlantic Slope Basin. Basic Data Bulletin 34A:160–161

Writer JH et al (2012) Fate of 4-nonylphenol and 17β-estradiol in the Redwood River of Minnesota. Environ Sci Technol 46:860–868

Wu W et al (2017) Calculation method for steady-state pollutant concentration in mixing zones considering variable lateral diffusion coefficient. Water Sci Technol 76(1):201–209. https://doi.org/10.2166/wst.2017.206

Chapter 9
Making It Work

The model is well calibrated and verified with field data. The modeler proceeds to produce model predictions under future conditions for water quality management. Can the modeler claims a victory and go home? No, not by a long shot! A key question the modeler can expect to face from decision makers is: How do you know your model prediction results are correct? Strictly speaking, no one knows if the model results are 100% correct based upon assumptions resulting from future condition uncertainties. A model verified under existing conditions cannot fully operate as a framework free of uncertainty. The 1983 Potomac Estuary algal bloom reported in Chap. 1 is a good example, where only after a model post-audit was a missing model mechanism revealed. How about the deoxygenation rates in the receiving river under future conditions? Similarly, open boundary conditions are difficult to assign for tidal systems (i.e. an estuary) under future conditions when interior load reductions are being evaluated. Is the sediment system in equilibrium with the external loads? Reality check applications are presented in this chapter to illustrate the difficulties a modeler can potentially face these days. Perhaps in the future machine learning can be utilized to tackle this situation!

9.1 Model Calibration and Verification

The normal process of calibrating and verifying a steady-state water quality model is having the model results matching two different sets of data with the same model parameter values. Ideally, the two sets of data would be obtained at different hydrologic and environmental conditions to substantiate the validity of the process. A classic example of steady-state calibration and verification is presented in Fig. 9.1 for the James Estuary using two sets of field data (July and September) under different freshwater flow and water temperature conditions. All the model coefficients are identical for both model runs. Model results match the spatial trend of key water quality constituents CBOD, organic nitrogen, ammonia, nitrite/nitrate, organic phosphorus,

© The Author(s), under exclusive license to Springer Nature Switzerland AG 2022
W.-S. Lung, *Water Quality Modeling That Works*,
https://doi.org/10.1007/978-3-030-90483-8_9

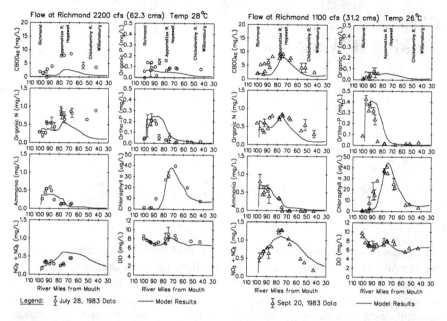

Fig. 9.1 Calibration and verification of the James River model under different freshwater flows

orthophosphate, chlorophyll *a*, and dissolved oxygen very well. The July flow is twice as that of the September flow, further enhancing the credibility of the modeling analysis. As noted in Chap. 4, the eutrophication endpoint in the James River Estuary is phytoplankton growth with substantial levels of chlorophyll *a* in the upper estuary near Hopewell, VA. The peak chlorophyll *a* level is higher in September than July due to lower freshwater flow entering the estuary. The lower flow is also reflected in the higher orthophosphate levels, partially responsible for the higher peak chlorophyll *a* concentrations. The decline of chlorophyll *a* levels downstream from Hopewell is caused by the depletion of orthophosphate in the estuary.

Selecting data sets for steady-state model calibration and verification is based on a number of factors. Figure 9.2 shows the flow, temperature, conductivity, and dissolved oxygen data at Moorefield, WV to be used for a TMDL modeling study of the South Fork South Branch of the Potomac River. While daily flow data are available at this USGS station, key water quality parameters are only measured on a monthly basis. The river flows are plotted in a log scale to accommodate the wide range of flows. The two low flow periods in the summer and fall of 1994 were picked for the model calibration and verification analysis. Slow time of travel or long residence time would make the kinetics coefficients (also time constants) stand out, thereby leading to optimum model calibration settings. The summer low flow was associated with high water temperature at 30°C while the fall low flow period had low water temperature at 10°C. Such a large temperature difference is ideal for the calibration and verification analysis. The fall low flow yielded an annual high conductivity of

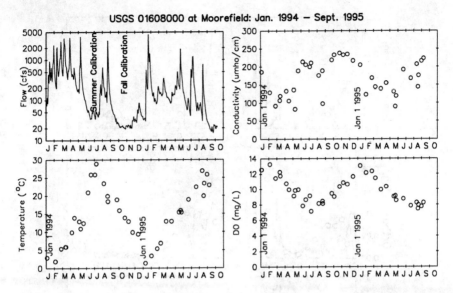

Fig. 9.2 Selection of steady-state calibration periods

250 μmho/cm. The low DO level during the summer months of 1994 indicates low water quality for the river, providing a good test of the modeling analysis.

The approach to calibrating and verifying time-variable models with data are much different from the steady-state models. Annual simulations need not be run on a discrete, year-by-year basis. Instead, continuous, multiple-year runs should be made consecutively without restarting each year. Once the initial conditions are specified, the simulation is launched with no modifications of model parameters except renewing the hydrologic and environmental conditions for each year. Figure 9.3 shows the model results of ammonia, orthophosphate, and DO versus data of a two-year run from August 1997 to July 1999 for the Patuxent Estuary. Note that the initial conditions for the second year run are from the results of the last day of the first year. The initial conditions of subsequent years should not be reset. The Chesapeake Bay model was calibrated and verified with a ten-year run to match the data, quite a successful effort (Cerco 1995; Cerco and Noel 2005). Once the initial conditions for the simulation are set, the model is run for 10 years continuously without changing any model parameters except loading rates and environmental conditions for individual years. This process is fundamentally different from the steady-state model analysis.

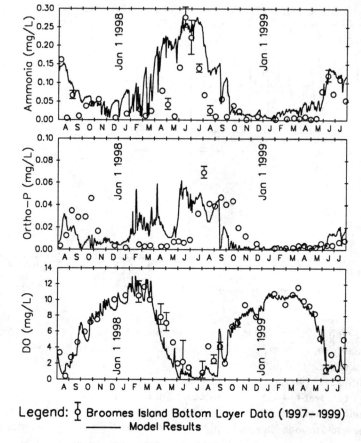

Fig. 9.3 Two-year continuous model run for time-variable simulations

9.2   Model Post-audit

The initial operating rule of the post-audit analysis is simple: "Without any change in the model parameters (e.g., BOD deoxygenation rate, phytoplankton growth, nutrient recycle, and settling parameters), but with updated meteorology, hydrology, and loads, run the model and compare to the observed data." If the comparison is satis-factory in some sense, then the model is considered to have been successful in the post-audit condition. This is theoretical and wishful thinking. In reality, a model post-audit is much more sophisticated and sometimes can be quite involved. In the upper Mississippi River model post-audit case (in Chap. 3), the dissolved oxygen levels in the receiving water were expected to improve with the installation of the nitrification process at the Metro plant. Yet, the model results with the reduced ammonia load alone failed to reproduce the data of improved DO downstream of the Metro plant. The key is the reduction of the in-stream CBOD deoxygenation coefficient to match the DO data. Did changing this coefficient value violate the rule of model post-audit? Not if an independent justification can be found. It turned out that the side effect of the nitrification process is the change of the organic carbon from a more labile to more refractory state, i.e. lowering the CBOD deoxygenation rate. An independent derivation of this key coefficient from the receiving water data, filtered CBOD in the receiving water, gave a reduced rate of deoxygenation, thereby consuming less DO (see Fig. 3.3 in Sect. 3.2). The outcome of this model post-audit analysis demonstrates that the path to achieve model verification may be different from one case to another. If the comparison is unsatisfactory in some sense, then plausible hypotheses are sought that might account for any differences that have been noted between the first prediction and the observed data.

In the PEM audit case reported in Chap. 1, initially the model failed to reproduce the 1983 blue-green algal bloom. Then the second operating rule, in the form of a question is invoked: "What are the minimum number of changes necessary for the model to provide a satisfactory post-audit comparison?" There is not a fixed answer to this question. One of the lessons learned from the 1983 Potomac Estuary algal bloom is that the sediment is a standby phosphorus source all the time even under unusual conditions. Let us look at the Potomac Estuary again. For more than a decade prior to 1980, studies, investigations, and hearings were carried out relative to eutrophication. This effort was initiated as a result of the occurrence of significant phytoplankton blooms and accompanying nuisance of mats of floating blue-green algae in the mid to late 1960s. Subsequent studies, analyses, and hearings centered primarily on the discharge from the Washington, DC Blue Plains WWTP, the major input into the estuary. That work resulted in the issuance of a discharge permit to that plant in May 1974, calling for reductions in both phosphorus and nitrogen. Subsequent discussions and further analyses dropped the nitrogen removal requirement, primarily over the cost as well as the uncertainty of the water quality benefit. The relatively low capital and operating cost of phosphorus removal coupled with the promising benefit of controlling phytoplankton blooms by lowering phosphorus levels in the Potomac Estuary resulted in an initial regulatory action: phosphorus removal at WWTPs. Effluent phosphorus concentrations of the Blue Plains WWTP dropped in stages from 1.6 to 0.23 mg/L by June 1981. As a result, the phosphorus load from Blue Plans WWTP had decreased by a factor of more than three during the period 1971–1979.

Figure 9.4 shows the longitudinal plots of chlorophyll a and phosphorus in the Potomac Estuary during a previous bloom period of August 22–29, 1977. Note that the peak of the bloom is located near mile 30, approximately 20 miles downstream from the Blue Plains WWTP outfall. Total phosphorus concentrations rise to about 0.25 mg/L around the bloom and the orthophosphate increases in the downstream direction to slightly above 0.1 mg/L, thereby suggesting additional distributed input of phosphorus. Could it be the release of phosphorus from the sediment and/or the recycling of phosphorus from the sharp decline of phytoplankton in the vicinity of mile 40–50? Let us look at the orthophosphate levels in the area upstream of the bloom. At mile 10 where the Blue Plains WWTP comes in, the orthophosphate levels of 0.04–0.06 mg/L could support only a level of 60–90 μg/L of chlorophyll a (based on a chlorophyll a to phosphorus ratio of 1.5 in the algal biomass). This indicates

Fig. 9.4 Chlorophyll *a* and phosphorus in the Potomac estuary (data from HydroQual 1980)

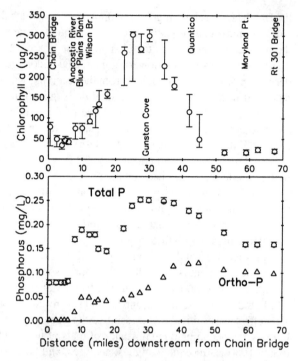

Legend: Chl a & Total P (Aug 22–29 1977)
         △ Ortho–P (Aug 22 1977)

that a secondary source (sediment release) is supplying phosphorus to the overlying water, resulting in an increase of about 0.2–0.3 mg/L orthophosphate, which in turn produces an algal bloom of 200–300 μg/L chlorophyll *a*. A data analysis offering intriguing physical insights into the algal bloom is what a modeler should conduct and is what this book is striving to convey, not just pushing buttons of running models. The alarm did not go off after examining the above data at that time as the water quality conditions improved significantly in the Potomac Estuary during the late 1970s and early 1980s. Nevertheless, a lesson has been learned that the above data analysis already suggests a second source of phosphorus was present to support the August 1977 bloom.

The analysis of the 1977 and 1983 blue-green algal blooms in the Potomac Estuary clearly points out two important aspects of water quality modeling: science and data. As indicated in Chap. 1, aerobic phosphorus release from the sediment under high pH conditions provided the needed extra phosphorus supply to fuel the blooms, a new finding leading to existing model enhancement. The model kinetics needed to be modified to account for the aerobic sediment phosphorus release under high pH conditions in the water column. Perhaps more intriguing is that the algal blooms have not happened in the Potomac Estuary since that 1983 event!

9.3 Open Boundary Conditions

The difficulty of assigning values at an open-mouth boundary of an estuarine water quality model when considering hypothetical loading scenarios is frequently encountered in modeling studies. The developers of the well-documented Chesapeake Bay Estuarine model were required to address this very dilemma. In earlier versions of the Chesapeake Bay model, modelers developed a mass balance algorithm to couple the reduction of loads upstream to the inflow of nutrients at the mouth of the estuary, effectively reducing the open-mouth boundary conditions with load reductions (Cerco 1995) using a grid as shown in the left panel of Fig. 9.5. Later revisions extended the model grid into the continental shelf area, well beyond the influence of nutrient loads from the Bay (Cerco and Noel 2005). While solving the problem of assigning boundary conditions at the mouth of the Chesapeake Bay, the modelers were faced with a new problem of having to assign open boundary conditions in the continental shelf where a paucity of observed data was available.

Eutrophication modeling of the Potomac, Patuxent, and James Estuaries (in Fig. 9.5), tributaries to the Chesapeake Bay, are presented in Sects. 1.1, 4.4, and 4.3, respectively. Assigning open boundary conditions at the mouth of each estuary is still a problem in model prediction analyses for individual estuaries. In the Patuxent

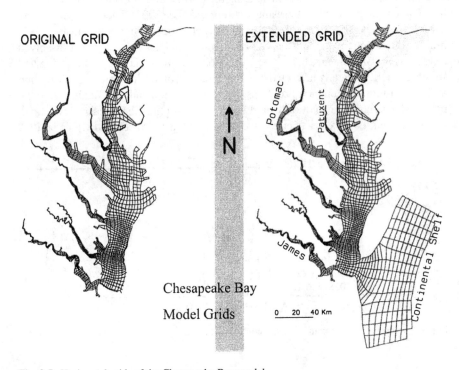

Fig. 9.5 Horizontal grids of the Chesapeake Bay model

Estuary, the lack of sensitivity of dissolved oxygen levels in the lower estuary to tributary loads prompted a sensitivity analysis of boundary conditions at the mouth of
the estuary. During the execution of the Patuxent Estuary eutrophication model, the
open boundary conditions at the mouth for all state variables, while time variant, were
identical between scenarios considered. Model sensitivity runs were performed by
scaling the downstream boundary conditions for nutrients and CBOD using three
different sets of boundary conditions at mouth; (1) the current conditions scenario
with downstream boundary conditions used during calibration, (2) the current conditions scenario with downstream boundary conditions for nutrients and CBOD scaled
by a factor of 2, and (3) the current conditions scenario with downstream boundary
conditions for nutrients and CBOD scaled by a factor of 0.25 (Lung and Nice 2007).
The results are summarized in Fig. 9.6, which displays the percent of total hypoxic
volume (dissolved oxygen below 2 mg/l) in the estuary from August 1997 to August
1999. It is apparent that the downstream boundary conditions have a significant
impact on levels of dissolved oxygen in the lower estuary. Also notable is the trend
of lower dissolved oxygen levels following higher river flows (and nutrient loading),
as discussed by Hagy et al. (2004). Flows during the spring of the first year were
substantially higher than those in the second year. Likewise, hypoxic volume was
larger in the summer following the first year than in the summer following the second
year. In some cases, the effects of changing the downstream boundary conditions are
felt 60 km from the mouth of the estuary (Lung and Nice 2007). Their analysis clearly
demonstrated the sensitivity of dissolved oxygen to downstream boundary conditions at this location. In addition, their modeling analysis predicted that dissolved
oxygen at Broomes Island, which has been noted for hypoxia in the late spring and
summer, might be greatly impacted by concentrations of nutrients and organic carbon
at the open-mouth boundary. Problems like this present a daunting challenge to the
modeler! The ultimate solution would be to run the entire Chesapeake Bay model,
as the Patuxent Estuary is a tributary to the Bay. Is this a realistic and viable option?

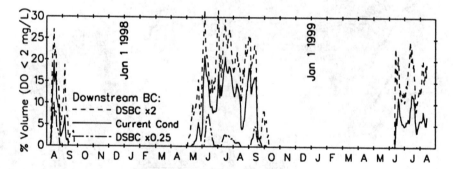

Fig. 9.6 Impact of downstream boundary conditions on anoxic conditions in the Patuxent Estuary

9.4 Mass Transport Modeling—a Necessary Step

The role of hydrodynamic models in water quality modeling is to provide information on the movement of modeled pollutants. Since hydrodynamic models are not one hundred percent physically based (still having the turbulence closure issue), they must be fully calibrated and verified with data prior to model projection analyses. The task of calibrating hydrodynamic models is comprehensive, when compared with the subsequent water quality modeling analysis. For one thing, the time scales of hydrodynamic models are much smaller than the water quality models, thereby requiring a much more significant computation effort. In light of the fact that the ultimate goal of the analysis is to perform water quality simulations, a mass transport modeling analysis supported by field data offers an advantage in reaching the calibration of key kinetics coefficients at a much reduced effort prior to getting the hydrodynamic model fully calibrated and verified—the short cut.

Recalling the turbidity maximum modeling in Sect. 4.5, a two-layer estuarine mass transport can be developed relatively easily given available two-dimensional (longitudinal/vertical) salinity distributions. Figure 9.7 shows the advective flows of the steady-state two-layer mass transport model in a 38-segment configuration developed by Lung (1990, 1992) for the Patuxent Estuary (see the map of the study area). The freshwater flow at the head of the estuary at Bowie, Maryland is 100 cfs, splitting

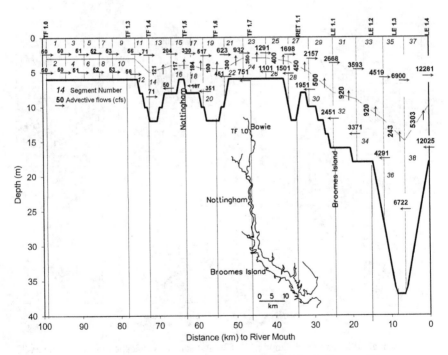

Fig. 9.7 Steady-state two-layer mass transport (advective flows) in the Patuxent Estuary

into 50 cfs in each of the two layers. The two-layer flows are in the downstream direction and are evenly split from the upstream to segment 14, meeting the bottom layer flow traveling upstream. Flow balance therefore generates vertical upward flows from this point on to the mouth of the estuary. Vertical mixing and vertical flows eventually lead to very significant flows at the mouth, reaching 12,281 cfs in the top layer going out the Chesapeake Bay and 12,025 cfs in the bottom layer from the Bay to the estuary. The difference of these two flows is 256 cfs, the net freshwater flow into the Bay. Note that the upstream freshwater flow of 100 cfs has increased to 256 cfs, i.e. a net gain of 156 cfs from the watershed following a journey of 100 km. The vertical flows are developed to maintain the flow balance throughout the estuarine system. Derivation of the advective flows is based on the approach developed by Lung and O'Connor (1984) using salinity data. To complete the mass transport, vertical mixing (i.e. dispersion) between the two layers is also quantified. The essence of Lung and O'Connor's analysis of converting from a one-dimensional (1-D) to two-dimensional (2-D) configuration is substituting one parameter (the 1-D longitudinal dispersion) with three parameters (2-layer adventive flows, vertical flows, and vertical dispersion).

This two-layer mass transport can be expanded into time-variable simulations. Figure 9.8 shows the model verification by reproducing the time-variable salinity in the Patuxent Estuary in 1983. The freshwater flows at the USGS gaging station 01594440 in Bowie, Maryland are plotted, showing seasonal variation with high

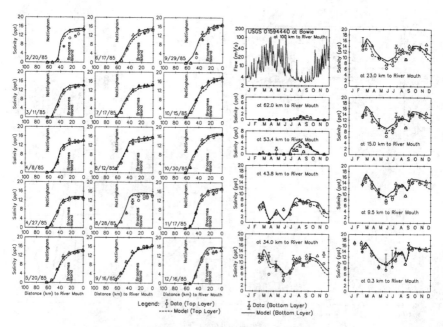

Fig. 9.8 Spatial (1985 data) and temporal (1983 data) salinity verification of the two-layer, 38-segment mass transport model of the Patuxent Estuary

flows in spring and low flows in the summer months. The flows pick up again in the fall and remain higher until the end of the year. Also plotted are time-variable salinity distributions at 8 monitoring stations, responding to the freshwater flows. In the meantime, the influence of tides is progressing from the mouth to the upstream. High freshwater flows tend to minimize the exchange (i.e. mixing) between the two layers, thereby pulling the salinity levels apart between the two layers while low flows tend to mix salinity well. Additional demonstration of the success of the two-layer Patuxent Estuary mass transport model is presented in Fig. 9.8, showing the salinity distributions on 15 dates in 1985. Again, this methodology works well. The verified two-layer mass transport can now be used to calibrate kinetics coefficients in the water quality model, sidestepping the lengthy process of hydrodynamic model calibration. Calibrated water quality coefficients were then used per se in subsequent modeling efforts which require more sophisticated spatial configurations to meet various requirements.

The upgrade of two-layer, 38-segment configuration mass transport to a multi-layer, 225-segment setup using the CE-QUAL-W2 modeling framework for the Patuxent Estuary was made by Lung (1992) for eutrophication modeling. This fine vertical resolution is warranted to address the summer anoxic conditions in the deep portion of the estuary, particularly at Broomes Island. Figure 9.9 presents the salinity and temperature verification of this 225-sgement configuration using 1994 data. With the vertical DO profiles verified by the model results in Fig. 4.16 (in Sect. 4.4.2), this fine-grid model is able to generate the results in Fig. 9.6 showing the temporal pattern of percentage volume of the DO below 2 mg/L from 1997 to 1999 for regulatory purposes. By now, the simplified two-layer mass transport analysis is replaced by the multi-layer hydrodynamic model in CE-QUAL-W2.

Further upgrade of the Patuxent Estuary model to 1993-segment configuration by Lung and Bai (2003), Bai and Lung (2005); Lung and Nice (2007), and Nice et al (2008) is necessary for the next level of water quality modeling including sediment–water interactions and the fate and transport of trace metals to address additional management questions. The hydrodynamic model in CE-QUAL-W2 is linked up with the WASP/EUTRO kinetics to become W2EUTRO (Nice 2006). Figure 9.10 shows the segmentation and salinity verification for a three-year period from 1997 to 1999 using W2EUTRO. Although the simplified two-layer mass transport analysis is not involved in this exercise, its early role of providing input for a speedy calibration of water quality kinetics cannot be denied. The calibrated kinetics coefficients from the two-layer configuration serve as the initial estimates for the 225-segment and 1993-segment model recalibrations of the Patuxent Estuary water quality model (Lung and Nice 2007).

9.5   Computational Frameworks and Model Accuracy

The Water Quality Analysis Simulation Program (WASP) is an enhancement of the original WASP (Di Toro et al. 1983; Ambrose et al. 1993). This model

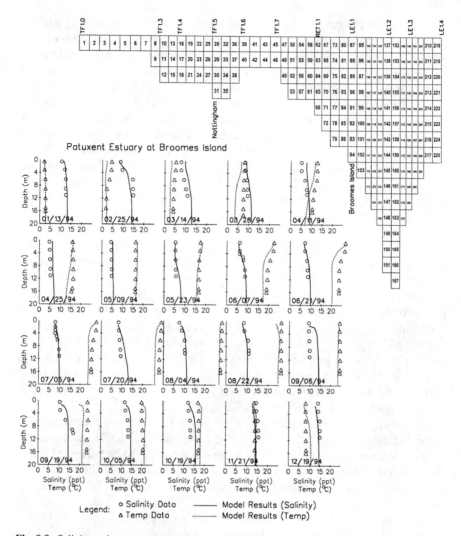

Fig. 9.9 Salinity and temperature verification of 2-D 225-seg model of the Patuxent Estuary

helps users interpret and predict water quality responses to natural phenomena and manmade pollution for various pollution management decisions. WASP is a dynamic compartment-modeling program for aquatic systems, including both the water column and the underlying benthos. WASP8.32 was released on April 2, 2019.

WASP allows the user to investigate 1, 2, and 3 dimensional systems and a variety of pollutant types. The time varying processes of advection, dispersion, point and diffuse mass loading, and boundary exchange are represented in the model. WASP also can be linked with hydrodynamic and sediment transport models that can provide mass transport, temperature, salinity, and sediment fluxes. This release of WASP

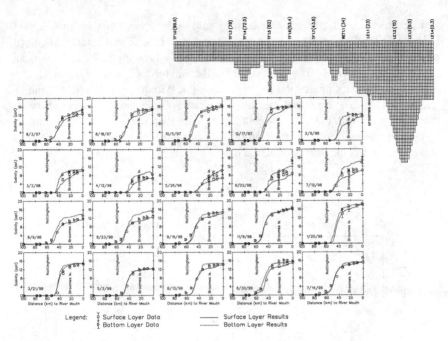

Fig. 9.10 1993 segments configuration and salinity verification of 2-D Patuxent Estuary model

contains the inclusion of the sediment diagenesis model linked to the Advanced Eutrophication sub model, which predicted sediment oxygen demand and nutrient fluxes from the underlying sediments. WASP is one of the most widely used water quality models in the United States and throughout the world. Because of the model's capabilities of handling multiple pollutant types it has been widely applied in the development of TMDL. WASP has capabilities of linking with hydrodynamic and watershed models which allows for multi-year analysis under varying meteorological and environmental conditions.

Modelers constantly face a question from stakeholders and regulatory water quality managers: "how accurate are your model predictions?" While plotting data and model results can be a most graphical measure of model credibility—easily understood and clearly visual, such a qualitative measure is simply not sufficient (Thomann 1980). This is particularly so for time variable models of several state variables and multi-dimensional systems. Some quantitative (i.e. statistical) comparisons are needed. Thomann (1980) listed a number of statistical analyses between observed and computed values:

(a) Regression analyses
(b) Relative error
(c) Comparison of means
(d) Root mean square error.

Recalling the eutrophication modeling study of the upper Mississippi River and Lake Pepin (see Figs. 4.2, , 4.3 and 4.4), regression of the model calculated chlorophyll *a* and DO concentrations versus measured values for 1992 and 1996 are presented in Fig. 9.11. The regression coefficient for DO is significantly higher than chlorophyll *a*, showing the DO regression line with a slope close to 1:1 and the intercept close to 0. Note that the chlorophyll *a* levels are much different between these two years with the 1992 concentrations more than twice the levels in 1996.

Further analysis of the regression results is to test the significance on the slope and intercept. Let the testing equation be

$$C_o = \alpha + \beta C_m + \varepsilon \qquad (9.1)$$

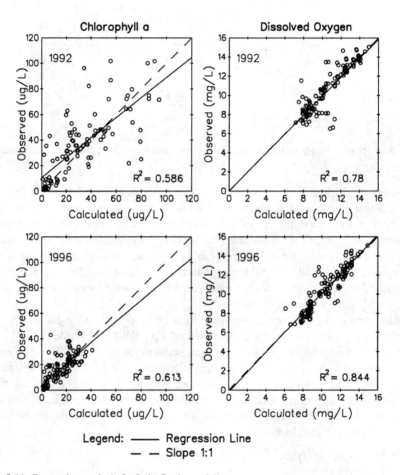

Fig. 9.11 Regression analysis for Lake Pepin modeling

where α and β are the true intercept and slope, respectively, between the calculated concentrations, C_m and the observed values, C_o, and ε is the error of C_m. With Eq. 9.1, standard linear regression statistics can be computed, including:

- The square of the correlation coefficient, R^2,
- Standard error of estimate, representing the residual error between model and data,
- Slope estimate, b of β and intercept estimate, a of α, and
- Test of significance on the slope and intercept.

The null hypothesis on the slope and intercept is given by $\beta = 1.0$ and $\alpha = 0.0$. A two-tailed "t" test is conducted on b and a, separately, with a 5% probability in each tail, i.e., a critical value of t of about 2 provides the rejection limit of the null hypothesis (Thomann 1980). The results of the two-tailed t test for the upper Mississippi River and Lake Pepin (in Fig. 9.11) indicate that the intercept is not significantly different from 0 and the slope is not significantly different from 1 for DO (Lung 2001). For chlorophyll a, the intercept and slope are significantly different from 0 and 1, respectively.

Excellent regression for DO has been reported for many rivers and other water systems. Thomann (1980) reviewed dissolved oxygen models of nineteen water systems following robust calibration/verification procedures without curve fitting the model to the data. For each water body, the error represents the median relative error where 50% of the stations (or times) had errors less than the values calculated. Across the board, one-half of the models had median relative errors greater than 10% and one-half of the models had median relative errors less than 10%, suggesting an overall median relative error of 10% for DO models. This level of median relative error has not been improved over the past few decades. Why? The amount of collected data has not improved over the past two decades. One of the key factors is the in-stream CBOD deoxygenation rate, k_d which is difficult to determine without data. Another key parameter is nitrification rate, k_n. The Danshui River (in Taiwan) modeling study reported in Sect. 4.4 points out the need for gathering field data to support the derivation of these two rates.

Other quantitative measures of model calibration and verification are median percent relative error and relative percent root-mean-squared error. Median percent relative error is calculated (Thomann 1980) as:

$$Med - Re = \left(\frac{C_o - C_m}{C_o} \right)_{median} \times 100 \tag{9.2}$$

While a simple and easily understood measure of error, this method tends to behave poorly when the observed data is insufficient to offer data variability. Relative percent root mean square error is calculated (Ji et al. 2002) as:

$$R - RMSE = \frac{\left(\frac{1}{n} \sum_{i=1}^{n} [C_o - C_m]^2 \right)^{0.5}}{C_{o\,max} - C_{o\,min}} \times 100 \tag{9.3}$$

Table 9.1 Performance of the Patuxent estuary eutrophication model

	No. of data points	$Med - Re$ (%)	$R - RMSE$ (%)
Top layer chlorophyll a (Fig. 4.14)	187	33.8	14.0
Bottom layer DO (Fig. 4.15)	115	22.0	14.2
Vertical profiles of DO (Fig. 4.16)	207	11.6	9.3

Limitations of this metric are that (1) root mean square error can exaggerate a small number of large errors and (2) relevance to the variance in observed data can yield values which are inflated when variance is very small and deflated when variance is very large (Nice 2006). Regardless of limitations, the use of these metrics by other modelers provided a basis for their use in the Patuxent Estuary eutrophication modeling study (see Sect. 4.4). For the modeling results presented in Figs. 4.14, 4.15, and 4.16 for the top layer chlorophyll a, bottom layer DO, and vertical DO profiles in the Patuxent Estuary, respectively, a summary is presented in Table 9.1. Again, DO outperforms chlorophyll a in both $Med - Re$ and $R - RMSE$. The number of data points matters also with more data points improving the model accuracy—not surprising at all. Although the $R - RMSE$ values of the chlorophyll a in the top layer of the Patuxent Estuary are low, they are still above 10%. The amount of the water quality data in the Patuxent Estuary is overwhelmingly rich with excellent quality, yet it can at best yield a 14% relative root mean square error, probably the current limit for eutrophication model for the time being.

9.6 Monitoring to Support Modeling

In using modeling to address a specific water quality problem for management purpose, it is rare that existing data is sufficient to support model configuration, calibration, and verification. For example, the eutrophication model developed for the upper Mississippi River and Lake Pepin (see map in Sect. 4.2) was designed to quantify a cause-and-effect relationship between the Metro Plant phosphorus input and water quality responses in Lake Pepin. While the model was calibrated with data collected in 1988, 1990, and 1991, the only data from a low flow year in 1988 were found deficient in many aspects, particularly during the summer months in Lake Pepin. Modeling analyses have provided data needs insights for further calibration of the model. In the BOD/DO modeling of the Yamuna River (Sect. 3.9), a full-scale monitoring of the Najafgarh Drain was recommended to allow the expansion of the model to cover the Drain. A comprehensive monitoring program was carried out in 2017, producing a significant amount of water quality data (FICCI 2017) for the model expansion effort. The data gathering effort to support the fecal coliform modeling of the Washington Ship Channel and Tidal Basin assembled a significant amount of data but still had several data deficient areas (Sect. 2.12). The daily E $coli$ monitoring program for a select number of beaches out of 44 marine beaches in Hong Kong in 2007 and 2008 was a huge effort to support the development of a real time beach water quality forecast system (Thoe 2010). All of these monitoring programs are big undertakings, time consuming, and costly.

Only until recently, remote sensing has entered the water quality monitoring arena (Smith and Bernard 2020; Gernez et al. 2021; Moses and Miller 2019). For example, Huang et al. (2017) demonstrated the measurements of particulate organic carbon in Taihu Lake, China. Wolny et al. (2020) demonstrated the use of remote sensing to study harmful algal blooms in the Chesapeake Bay to support the shellfish industry. The quality of the data (reliability and accuracy) collected using schemes other than traditional field work is another concern. While remote sensing is promising, the bulk of water quality data collection still relies on field sampling at the present time.

9.7 Closing the Loop

The spatially variable CBOD deoxygenation rates, k_d in the Danshui River (Taiwan) were field measured to improve the model calibration (see Chap. 3). While this is an accurate and reliable effort to obtain this model parameter, it presents the modeler another challenge. Acknowledging that this key variable would change with the treatment upgrades of the domestic WWTPs along the river as well as with the introduction of new plants, the modeler would not be able to assign *a prior* this variable for the future conditions, particularly with multiple point source discharges. The current design of the CBOD-DO kinetics in most water quality models prevents the modeler from doing so.

The dilemma can be seen in the model prediction to determine the allowable BOD loads following a treatment upgrade for the Alta Vista WWTP on the Roanoke River in Virginia. Not knowing the exact CBOD deoxygenation, k_d rate under the future conditions, the modeler ran the calculations for a range of k_d rates (Fig. 9.12)

Fig. 9.12 BOD load projection over a range of deoxygenation rates

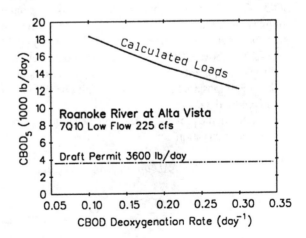

to determine the allowable loads under the 7Q10 low flow condition, which are substantially higher than the draft permit loads issued by the permiter (Lung 2001). In this case, the high k_d rates can be traced back to loads of less treated effluents, thereby resulting in lower allowable $CBOD_5$ loads. Eventually, the somewhat arbitrarily assigned $CBOD_5$ load of 3600 lb/day by the regulatory agency was lifted based on this analysis. An even better example is the upper Mississippi River BOD/DO modeling study (see Fig. 3.4 in Sect. 3.2), in which the in stream k_d value decreases from 0.35 day^{-1} when the Metro Plant had primary treatment only, to 0.25 day^{-1} with secondary treatment, and eventually to 0.07 day^{-1} at secondary with nitrification. Progressive reduction of k_d could not be predicted a priori.

This is because the water quality models are not one hundred percent physically based. Polus et al. (2011) argued that the Streeter-Phelps model and many other water quality models such as WASP, QUAL2, etc. are physically based models as they are based on physical, chemical, and biochemical processes and some parameter values can be obtained through in situ or lab experiments. However, measuring the k_d rate under future conditions cannot be made in the Roanoke River. That is, does a modeling framework free of model parameter adjustment exist? To explore this question, we need to compare water quality models with hydrodynamic models. Simply stated, a hydrodynamic model has five fundamental equations: three momentum equations, one continuity equation, and a mass balance equation. These five equations have been around for several centuries and together they come very close to a physically based modeling framework. Their difficulty still lies in the turbulence closure issue. There have been many turbulence schemes developed over the past century to resolve this problem. The closest the hydrodynamicists have come is to develop a new turbulence closure scheme to eliminate the last parameter (i.e. a knob), yet they ended up adding two more undermined parameters; the search for a completely physically based model would continue.

A water quality model fairs worse. It has only one equation—mass conservation, a very rock solid and unshakable principle. The rest of a water quality model, however, has many empirical relationships, e.g. the 1st order deoxygenation k_d rate in the BOD/DO model of the Roanoke River, the phytoplankton growth kinetics, and the nonlinear nutrient limitation scheme, to name a few. All these formulations have many parameters and coefficients, making a physically based model impossible. Just like the above CBOD deoxygenation coefficient for the future conditions, there is always one parameter in the hydrodynamic models which cannot be assigned a priori. It is impossible to have a physically based water quality model because they are driven by data.

As stated in the previous paragraph, water quality models have only one equation, i.e., mass conservation:

$$Accumulation = Input - Output$$

$$V\frac{dC}{dt} = J + \sum R + \sum T + \sum W \tag{9.4}$$

where

$V \frac{dC}{dt}$ represents the mass rate change with respect to time

J mass change rate due to mass transport

$\sum R$ mass change rate due to kinetics, e.g. deoxygenation

$\sum T$ mass change rate through phase shift, e.g. oxygen reaeration

$\sum W$ loading rates from point and nonpoint sources.

Formulating each term of the right-hand side of Eq. 9.4 is based on science (i.e. research) and variable degrees of parameterization. Phosphorus release from the sediment under aerobic conditions in the Potomac Estuary blue-green algal blooms (Chap. 1) was uncovered in a scientific investigation. BOD/DO kinetics and algal growth kinetics require a significant amount of field data to quantify these processes. It is clear that water quality modeling requires the strong support of science (i.e. research) and data (monitoring), the critical pillars. Using sophisticated water quality models with insufficient data support does not compensate for lack of data (Loucks and van Beek 2017).

As pointed out by Thomann (1998), the growth in model size over the history of modeling has been significant and parallels the increase in computing power in the past two decades, leading to the "Golden Age" of water quality modeling. However, computing power does not mean over spatial discretization and expansion of state variables in model configuration without the concurrent increase in data support. It is true that the Chesapeake Bay data base (see Sect. 2.12) has not expanded in spatial and temporal coverage in the past twenty years. Therefore, increasing the model discretization without concurrent expansion of data support would not be worthwhile in many cases. Thomann (1998) cautioned us that the predictive water quality and ecosystem models will not be judged by how big the model is or its degree of complexity. These models will have to survive relentless confirmation with field and laboratory data and, most importantly, critical review by the scientific and engineering community to determine suitability for decision-making, leading to the lessening of future conflicts between environmental interests, regulatory managers, and stake holders. At the end of the day, we still rely on field data to support water quality modeling for management use. It is a general consensus that eighty percent of the total cost of a water quality modeling study is expected to spend on field sampling work and data collection. The model execution effort is not expected to exceed twenty percent of the budget.

References

Ambrose RB et al (1993) The water quality analysis simulation program, WASP5. U.S. EPA, Athens, GA

Bai S, Lung WS (2005) Numerical modeling of algae in the Patuxent Estuary. Int J River Basin Manag 3(4):273–281

Cerco CF (1995) Response of Chesapeake Bay to nutrient load reductions. J Environ Eng 121(8):549–557

Cerco CF, Noel MR (2005) Incremental improvements in Chesapeake Bay environmental model package. J Environ Eng 131(5):745–754

Di Toro DM et al (1983) Water quality analysis simulation program (WASP) and model verification program (MVP) documentation. Report submitted by Hydroscience, Inc. to US EPA Environmental Research Lab, Duluth, MN

FICCI (2017) Najafgarh drain project. FICCI Research & Analysis Centre. Delhi, India

Gernez P et al (2021) Editorial: remote sensing for aquaculture. Front Mar Sci 7:638156. https://doi.org/10.3389/fmars.2020.638156

Hagy JD et al (2004) Hypoxia in Chesapeake Bay, 1950–2001: long-term change in relation to nutrient loading and river flow. Estuaries 27(4):634–658

Huang C et al (2017) Spatiotemporal variation in particulate organic carbon based on long-term MODIS observations in Taihu Lake, China. Remote Sens 9 624. https://doi.org/10.3390/rs9060624

HydroQual (1980) Overview of Potomac Estuary modeling, tasks I and II—dissolved oxygen and eutrophication. Report prepared for Potomac Studies Technical Committee, Washington, DC

Ji ZG et al (2002) Sediment and metals modeling in shallow river. J Environ Eng 128(2):105–119

Loucks DP, van Beek E (2017) Water resources systems planning and management. Springer, Cham, Switzerland

Lung WS (1990) Development of a water quality model for the Patuxent Estuary in coastal and estuarine studies. In: Michaelis W (ed) Estuarine water quality management, vol 36. Springer-Verlag, New York, pp 49–54

Lung WS (1992) A water quality model for the Patuxent Estuary. Report submitted to Maryland Office of Environment. Department of Civil Engineering, University of Virginia

Lung WS (2001) Water quality modeling for wasteload allocations and TMDLs. Wiley, New York

Lung WS, Bai S (2003) A water quality model for the Patuxent Estuary: current conditions and predictions under changing land-use scenarios. Estuaries 26(2A):267–279

Lung WS, Nice AJ (2007) Eutrophication model for the Patuxent Estuary: advances in predictive capabilities. J Environ Eng 133(9):917–930

Lung WS, O'Connor DJ (1984) Two-dimensional mass transport in estuaries. J Hyd Eng 110(10):1340–1357

Moses WJ, Miller WD (2019) Editorial for the special issue "remote sensing of water quality". Remote Sens 11:2178. https://doi.org/10.3390/rs11182178

Nice AJ (2006) Developing a fate and transport model for arsenic in estuaries. PhD dissertation, University of Virginia

Nice AJ et al (2008) Modeling arsenic in the Patuxent Estuary. Environ Sci Technol 42(13):4804–4810

Polus E et al (2011) Geostatistics for assessing the efficiency of a distributed physically-based water quality model: application to nitrate in the Seine River. Hydrol Process 25:217–233. https://doi.org/10.1002/hyp.7838

Smith ME, Bernard S (2020) Satellite ocean color based harmful algal bloom indicators for aquaculture decision support in the southern Benguela. Front Mar Sci 7:61. https://doi.org/10.3389/fmars.2020.00061

Thoe W (2010) A daily forecasting system of marine beach water quality in Hong Kong. PhD dissertation, University of Hong Kong

Thomann RV (1980) Measure of verification. Workshop on verification of water quality models. EPA-600/9-80-016: 37–61

Thomann RV (1998) The future "Golden Age" of predictive models for surface water quality and ecosystem management. J Environ Eng 124(2):94-103

Wolny JL et al (2020) Current and future remote sensing of harmful algal blooms in the Chesapeake Bay to support the shellfish industry. Front Mar Sci 7:337. https://doi.org/10.3389/fmars.2020.00337

CPSIA information can be obtained
at www.ICGtesting.com
Printed in the USA
LVHW080538240322
714271LV00004B/54

9 783030 904821